Hematopoietic Stem Cell Protocols

METHODS IN MOLECULAR BIOLOGY™

John M. Walker, SERIES EDITOR

METHODS IN MOLECULAR BIOLOGY™

Hematopoietic Stem Cell Protocols

Second Edition

Edited by

Kevin D. Bunting

*Department of Medicine, Division of Hematology-Oncology,
Center for Stem Cell and Regenerative Medicine,
Case Western Reserve University, Cleveland, OH*

 Humana Press

Editor
Kevin D. Bunting
Department of Medicine
Division of Hematology & Oncology
Case Western Reserve University
Cleveland, OH, USA
kdb10@case.edu

Series Editor
John M. Walker
University of Hertfordshire
Hatfield, Herts
United Kingdom

ISBN: 978-1-58829-868-3   e-ISBN: 978-1-59745-182-6

Library of Congress Control Number: 2007935117

Cover Illustration: Neonatal chimeras generated with GFP transgenic gynogenetic embryonic stem cells. Fetal liver hematopoietic stem cell derivatives of these gynogenetic ES cells were transplanted into irradiated adult hosts and eventually replaced the entire hematopoietic system.
Photo: K. John McLaughlin and Sigrid Eckardt

Printed on acid-free paper

9 8 7 6 5 4 3 2 1

springer.com

Preface

The hematopoietic stem cell (HSC) field has rapidly grown in the past several years as new technologies have been developed and the older tried and true methods have been used in new ways. Major advances in the isolation of HSC and progenitor subsets have been sparked by the gene expression array technologies, leading to the identification of new markers for prospective isolation. The first edition of *Hematopoietic Stem Cell Protocols* is a very thorough resource covering historically the basic techniques that have been published extensively over many years. To put together, a follow-up to this excellent book has been a hard act to follow. The intent of this book is to be distinct and not duplicate but to build on the strong basic assays and include updated approaches for the combination of phenotypic and functional analyses. This second edition of *Hematopoietic Stem Cell Protocols* thus aims to provide timely protocols developed in the HSC field since the last comprehensive review in this series.

The approaches described in this book should be readily useful for the basic science researcher, especially those familiar with the use of mouse genetic models. The chapters in this book are geared toward the laboratory-based researcher and the development of pre-clinical studies to apply HSCs toward treating blood diseases more effectively. An overview chapter links the other chapters in a larger view including areas that may not yet be ready for this edition but which may rapidly yield the new techniques of the future. The methods chapters include (1) descriptions of newer stem cell purification techniques, (2) methods of in vitro stem cell culture and expansion, (3) murine and human HSC transplantation assays, and (4) application of genomic and imaging technologies of stem cells. The author list is made up of leading researchers who have made major contributions toward these technical advances and who have provided a needed technical resource for new stem cell investigators.

I thank all the contributors for their time and efforts in this project. I appreciate the support from my laboratory members in preparation of this book. To provide updates in such a rapidly developing field has been a challenge, but I believe that we have distilled a set of useful and timely techniques that will facilitate rigorous and quantitative HSC research. That is the goal, and I am pleased to have had the opportunity to contribute my part in organizing this information into the user-friendly *Methods in Molecular Biology* format.

Kevin D. Bunting

Contents

Contributors

GÉNÈVE AWONG • *Department of Immunology, University of Toronto, Sunnybrook Research Institute, Toronto, Ontario, Canada*

CHRISTOPHER BAUM • *Department of Experimental Hematology, Hannover Medical School, Hannover, Germany, Division of Experimental Hematology, Cincinnati Children's Research Foundation, Cincinnati, OH*

MATTHEW BOYER • *Department of Radiation Oncology, University of Pittsburgh, Pittsburgh, PA*

KEVIN D. BUNTING • *Department of Medicine, Division of Hematology-Oncology, Center for Stem Cell and Regenerative Medicine, Case Western Reserve University, Cleveland, OH*

FERNANDO D. CAMARGO • *Whitehead Institute for Biomedical Research, Cambridge, MA*

BEN CAPOCCIA • *Department of Internal Medicine, Division of Oncology, Hematopoietic Development and Malignancy Group, Washington University School of Medicine, St. Louis, MO*

LINZHAO CHENG • *Institute for Cell Engineering, Johns Hopkins University School of Medicine, Baltimore, MD*

TAO CHENG • *Department of Radiation Oncology, University of Pittsburgh, Pittsburgh, PA*

GERALD DE HAAN • *Department of Stem Cell Biology, University Medical Center Groningen, the Netherlands*

BERTIEN DETHMERS-AUSEMA • *Department of Stem Cell Biology, University Medical Center Groningen, the Netherlands*

SIGRID ECKARDT • *Center for Animal Transgenesis and Germ Cell Research, New Bolton Center, University of Pennsylvania, Kennett Square, PA*

BORIS FEHSE • *Clinic for Stem Cell Transplantation, University Medical Center Hamburg-Eppendorf, Hamburg, Germany*

SEIJI FUKUDA• *Department of Microbiology and Immunology and the Walther Oncology Center, Indiana University School of Medicine, and the Walther Cancer Institute, Indianapolis, IN*

STANTON L. GERSON • *Department of Medicine, Division of Hematology-Oncology, Center for Stem Cell and Regenerative Medicine, Case Western Reserve University, Cleveland, OH*

PETER HAVIERNIK • *Department of Medicine, Division of Hematology-Oncology, Center for Stem Cell and Regenerative Medicine, Case Western Reserve University, Cleveland, OH*

DAVID HESS • *Department of Internal Medicine, Division of Oncology, Hematopoietic Development and Malignancy Group, Washington University School of Medicine, St. Louis, MO*

SARAH HOHM • *Department of Internal Medicine, Division of Oncology, Hematopoietic Development and Malignancy Group, Washington University School of Medicine, St. Louis, MO*

JONATHAN B. JOHNNIDIS • *Whitehead Institute for Biomedical Research, Cambridge, MA*

TARJA A. JUOPPERI • *Department of Pathology, Johns Hopkins University School of Medicine, Baltimore, MD*

DAN S. KAUFMAN • *Stem Cell Institute and Department of Medicine, University of Minnesota, Minneapolis, MN*

WILLIAM G. KERR • *Immunology Program, Moffitt Cancer Center and Research Institute, Tampa, FL*

OLGA S. KUSTIKOVA • *Department of Experimental Hematology, Hannover Medical School, Hannover, Germany, Engelhardt Institute of Molecular Biology, Russian Academy of Sciences, Moscow, Russia, Division of Experimental Hematology, Cincinnati Children's Research Foundation, Cincinnati, OH*

ROSS N. LA MOTTE-MOHS • *Department of Immunology, University of Toronto, Sunnybrook Research Institute, Toronto, Ontario, Canada*

ZHENGHONG LEE • *Department of Radiology, Division of Nuclear Medicine, Case Western Reserve University, Cleveland, OH*

YUAN LIN • *Department of Medicine, Division of Hematology-Oncology, Case Western Reserve University, Cleveland, OH*

DAN LINK • *Department of Internal Medicine, Division of Oncology, Hematopoietic Development and Malignancy Group, Washington University School of Medicine, St. Louis, MO*

K. JOHN MCLAUGHLIN • *Center for Animal Transgenesis and Germ Cell Research, New Bolton Center, University of Pennsylvania, Kennett Square, PA*

JOE MOLTER • *Department of Radiology, University Hospitals of Cleveland, Cleveland, OH*

RICHARD MORIGGL • *Ludwig Boltzmann Institute for Cancer Research, Vienna, Austria*

JAN NOLTA • *Department of Internal Medicine, Division of Oncology, Hematopoietic Development and Malignancy Group, Washington University School of Medicine, St. Louis, MO*

Louis M. Pelus • *Department of Microbiology and Immunology and the Walther Oncology Center, Indiana University School of Medicine, and the Walther Cancer Institute, Indianapolis, IN*

Cheng-Kui Qu • *Department of Medicine, Division of Hematology-Oncology, Center for Stem Cell and Regenerative Medicine, Case Western Reserve University, Cleveland, OH*

Simon N. Robinson • *University of Texas, MD Anderson Cancer Center, Houston, TX*

Saul J. Sharkis • *Department of Pathology, Johns Hopkins University School of Medicine, Baltimore, MD*

Hongmei Shen • *Department of Radiation Oncology, University of Pittsburgh, Pittsburgh, PA*

Jinhua Shen • *Department of Medicine, Division of Hematology-Oncology, Center for Stem Cell and Regenerative Medicine, Case Western Reserve University, Cleveland, OH*

Xinghui Tian • *Stem Cell Institute and Department of Medicine, University of Minnesota, Minneapolis, MN*

William Tse • *Department of Medicine, Division of Hematology-Oncology, Center for Stem Cell and Regenerative Medicine, Case Western Reserve University, Cleveland, OH*

Andriy Tsyrulnyk • *Ludwig Boltzmann Institute for Cancer Research, Vienna, Austria*

Ronald P. van Os • *Department of Stem Cell Biology, University Medical Center Groningen, the Netherlands*

Zhengqi Wang • *Department of Medicine, Division of Hematology-Oncology, Center for Stem Cell and Regenerative Medicine, Case Western Reserve University, Cleveland, OH*

Louisa Wirthlin • *Department of Internal Medicine, Division of Oncology, Hematopoietic Development and Malignancy Group, Washington University School of Medicine, St. Louis, MO*

Zhaohui Ye • *Institute for Cell Engineering, Johns Hopkins University School of Medicine, Baltimore, MD*

Xiaobing Yu • *Institute for Cell Engineering, Johns Hopkins University School of Medicine, Baltimore, MD*

Yi Zhao • *Gene Therapy Laboratories, Norris Cancer Center, University of Southern California School of Medicine, Los Angeles, CA*

Yuxia Zhan • *Gene Therapy Laboratories, Norris Cancer Center, University of Southern California School of Medicine, Los Angeles, CA*

YI ZHANG • *Department of Medicine, Division of Hematology-Oncology, Center for Stem Cell and Regenerative Medicine, Case Western Reserve University, Cleveland, OH*

PING ZHOU • *Department of Internal Medicine, Division of Oncology, Hematopoietic Development and Malignancy Group, Washington University School of Medicine, St. Louis, MO*

JUAN CARLOS ZÚÑIGA-PFLÜCKER • *Department of Immunology, University of Toronto, Sunnybrook Research Institute, Toronto, Ontario, Canada*

I ————————————————————

OVERVIEW

1

The Expanding Tool Kit for Hematopoietic Stem Cell Research

William Tse and Kevin D. Bunting

Summary

Hematopoietic stem cells (HSC) play critical roles in maintaining blood cell production for the lifetime of the organism. Considerable progress has been made in their isolation from mouse bone marrow to high levels of purity based on a combination of cell-surface phenotype and functional characteristics. In addition, in vitro assays have been established that provide important tools for study of hematopoietic differentiation from HSC and for differentiation to generate HSC from embryonic stem cells. Although these in vitro studies provide a window on the temporal function and differentiation of HSC progeny, the transplantation assay still serves as the gold standard for quantitative and qualitative analysis of murine HSC biology. There are now many flavors of syngeneic and xenogeneic HSC transplant, all focused on quantitative assessment of repopulating function. As a vehicle for genetic modification of HSC, retroviral-mediated gene transfer followed by transplantation has had a major impact upon our understanding of genetic disorders, gene therapy, and leukemogenesis. This overview chapter summarizes the growing number of tools available for HSC research and specifically ties together the methods in chapters of the second edition of *Hematopoietic Stem Cell Protocols*.

Key Words: Hematopoietic stem cell; flow cytometry; survival; proliferation; transplantation; embryonic stem cells; retroviral vector.

1. Stem Cell Purification and Analysis

The basic assays for hematopoietic stem cell (HSC)-repopulating ability were borne in the 1950s because of the efforts to develop radioprotective strategies. It was clear that bone marrow transplant provided a survival advantage to mice given otherwise lethal doses of irradiation. We now know that the radioprotective fraction of mouse bone marrow is derived from the myeloerythroid

From: *Methods in Molecular Biology, vol. 430: Hematopoietic Stem Cell Protocols*
Edited by: K. D. Bunting © Humana Press, Totowa, NJ

progenitors *(1)* with most of the day 8 CFU-S activity *(2)*. The simple assay of bone marrow transplant in the mouse model provided a boom for HSC studies by providing a quantitative and robust platform for assessing the blood cell regenerative ability of HSC. The application of congenic mouse strains, especially the CD45.1 polymorphism, provided the key tool to track donor leukocyte contribution to all hematopoietic lineages. The *ptprc* gene encodes a protein tyrosine phosphatase expressed on all leukocytes *(3)*, and an allelic variant was serially backcrossed from SJL/J onto C57BL/6 with selection at each generation by serotyping *(4)*.

Eventual elimination of the non-HSC components of the bone marrow graft led to the definition of the HSC phenotypes widely used today. This systematic dissection of the cell-surface phenotype was pioneered by Irving Weissman and colleagues beginning in the late 1980s and was followed up by many groups using labor intensive, costly, and time-consuming flow cytometry methods to sort fractions based on cell-surface phenotype *(1–10)* followed by transplantation into recipient mice and analysis of long-term reconstitution potential over multiple hematopoietic lineages. More recent studies have relied on the combined use of the cell-surface markers and microarray analysis *(11,12)* to identify additional unique HSC-specific markers. **Table 1** lists various stem cell markers that have been reported. This list demonstrates the diversity of molecules associated with important HSC functions as defined using rigorous methodologies. **Table 2** lists markers that can be excluded to facilitate HSC isolation because of their tightly controlled down-regulation associated with hematopoietic differentiation. Note that CD34 is included in both the tables. CD34 is a variable marker of HSCs depending on the age of development and activation status *(13–16)*. In addition to the cell-surface markers, additional functional methods of isolation based on the quiescence of HSC *(17)* have utilized Rhodamine *(18)* or Hoechst 33342 *(19,20)* dye efflux principally by ATP-binding cassette transporters such as MDR1 and ABCG2 *(21)*. 5-Fluororucail (5-FU) treatment is perhaps the cheapest and simplest method based on relatively quiescent cell-cycle status of the HSC *(22)*. Remarkably, quiescent HSC home to the mouse bone marrow 24–48 h after transplant *(23)*, and this attribute can be used to improve isolation of HSC when combined with elutriation methods *(24,25)*.

With knowledge about HSC isolation, characterization of the phenotypically identified HSC populations has also been facilitated by advanced flow cytometry techniques. Because HSC are very rare and cannot be easily isolated in large numbers, flow cytometry permits analysis of single cells within a phenotypically defined HSC fraction. Assays for cell cycle *(26,27)* and apoptosis *(28)* endpoints have been used to study knockout mice to derive information regarding gene function in HSC fractions. Developing technologies that capture

Table 1
Hematopoietic Stem Cell Markers Positively Identifying Mouse or Human Hematopoietic Stem Cells (HSC)

Marker	Description	References
Sca-1	Stem cell antigen-1	*(9,10,70)*
c-Kit	Receptor tyrosine kinase, product of the White spotting (W) locus	*(71–74)*
Thy1.1	Low expression is found on HSC fractions that are also Sca-1$^+$c-Kit$^+$Lin$^-$	*(75)*
CD34	Glycoprotein expressed on human HSC and expressed on murine fetal liver stem cells, expressed on adult murine HSC upon activation	*(13–15,76–78)*
CD150	Glycoprotein expressed on T, B, NK and dendritic cells as a co-activation marker	*(31)*
CD41, Endomucin	An endothelial sialomucin closely related to CD34 that is highly expressed in Sca-1$^+$c-Kit$^+$Lin$^-$CD34$^-$ cells	*(79)*
ABCG2, side population, SP	Breast cancer resistance gene (BCRP) responsible for side-population	*(80,81)*
Endoglin	Marker of activated endothelium and a modulator of TGF-beta signaling	*(12)*
Prion protein	GPI-anchored prion protein product responsible for chronic Creutzfeldt–Jakob disease	*(82)*
Tie-2	Vascular endothelial marker expressed on common hematopoietic and vascular intermediates	*(83,84)*
c-MPL	Myeloproliferative leukemia gene, not unique to HSC but highly expressed	*(85,86)*
Aldehyde dehydro-genase	Xenobiotic detoxifying enzyme that can be detected by catalytic activity, shown in human cells	*(87–92)*
C1qRp	Expressed on CD34$^+$ or CD34$^-$ but CD38$^-$ human SRC	*(93)*

phosphorylation of intracellular signaling molecules *(29,30)* promise to one day provide even greater understanding of signal transduction in the most highly refined phenotypic populations of HSCs. The advantage of all of these flow cytometry-based approaches is the ability to define unique subpopulations of cells with particular differences in their response to cytokines or growth factors.

The process of HSC engraftment involves multiple steps (**Fig. 1**). Study of HSC at all phases of this process can now be done. A potential advantage of the

Table 2
Hematopoietic Cell Markers with no Expression on Adult Mouse or Human
Hematopoietic Stem Cells (HSC)

Marker	Description	References
Lineage cocktail – Gr1, Mac1, B220, CD4, CD8, NK1.1	Well-characterized hematopoietic differentiation antigens representing lymphoid, myeloid, and erythroid cell lineages	*(94,95)*
CD34	Glycoprotein not expressed on adult murine Sca-1$^+$c-Kit$^+$Lin$^-$Thy1.1^{low} cells	*(13–15,76–78)*
CD38	The SRC is enriched in the CD34$^+$CD38$^-$ fraction	*(96,97)*
CD48, CD244	SLAM family markers that are not expressed on the HSC but become expressed at the MPP (CD244) or the LRP (CD48) stage of differentiation	*(31)*
Flt3/Flk2*	Type IV cytokine receptor containing an immunoglobulin-like repeat in the extracellular domain and a split tyrosine kinase domain	*(6,98)*

*Flt3/Flk2 is expressed on mouse short-term repopulating HSC and human NOD/SCID repopulating cells.

SLAM family markers is that simple immunostaining to determine localization of HSC can be done under various experimental conditions *(31)*. The mouse also provides a very useful tool for development of new mobilization strategies for increasing the circulating mass of HSC. Common approaches using cytokine or chemokine treatment are available to disengage HSC from the niche interactions and promote mobilization to the circulation at greater numbers than the typical percentage present during steady state hematopoiesis *(17)*. New methods based on antagonism of CXCR4 are developing rapidly *(32,33)*. A CXCR4 inhibitor, AMD3100, is currently under active clinical evaluation for augmentation of peripheral blood stem cell mobilization in combination with granulocyte colony-stimulating factor. Initial results look very promising. As the SDF-1/CXCR4 signaling is critical for hematopoietic cell migration, assay for transmigration *(34)* measures a function that is relevant for the in vivo process of HSC lodgment *(35)*.

2. In Vitro Assays and Differentiation

Although the methods of isolating murine ES cells were developed in the 1980s, the optimization of HSC differentiation has lagged behind. Embryoid bodies could be isolated and differentiated to make hematopoietic cells, but HSC

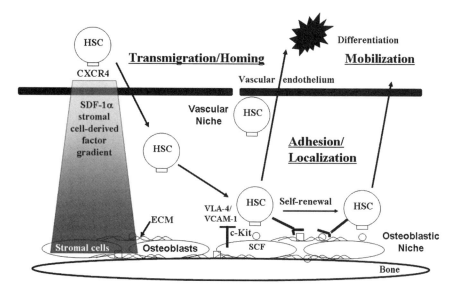

Fig. 1. Complex biological processes associated with functional hematopoietic stem cell activity. A number of methods are available for assay of hematopoietic stem cells (HSC) at various steps in the engraftment process. These include assays of adhesion and migration on endothelial membranes, chemotaxis to SDF-1 gradients, or adhesion to extracellular matrix in vitro. In vivo localization within the bone marrow microenvironment by immunohistological methods has been used to define osteoblastic and vascular niches. HSC can also be isolated from the bone marrow and studied further by flow cytometry at the single cell level to obtain information about survival and proliferation. Likewise, differentiation ability is measured by flow cytometry on peripheral blood lineages and mobilization following treatment with cytokines or chemokines results in increased blood cell counts and HSC-repopulating activity. It is important to keep in mind that the most robust HSC assays incorporate many of these processes and must show evidence for self-renewal and long-term multilineage hematopoietic differentiation.

isolation remained elusive. This was in part due to rejection of embryonic HSC lacking major histocompatibility complex (MHC) class I *(36)* and due to other homing/engraftment defects. Novel approaches to overcome these transplant barriers have been developed that incorporate transplant into immune-deficient recipients and/or overexpression of HoxB4 *(37)*. Additional experiments have reported using newborn mice with busulfan conditioning *(38)*. The *HoxB4* gene has proven to be particularly useful for studies of HSC expansion, and new protein-based delivery methods could provide novel therapeutics for clinical application *(39)*. The isolation of human ES cells in 1998 *(40)* re-invigorated the stem cell field and led to considerably increased interest in developing HSC in vitro for potential therapeutic application as either a corrected blood cell source or a method to introduce tolerance to a particular non-hematopoietic organ replacement or repair approach *(41)*. Additional approaches show that parthenogenic stem cells may be useful for overcoming MHC barriers *(36)* and that androgenetic and gynogenetic stem cells can be used to derive transplantable HSC in vitro and in vivo *(42)*. Importantly, these HSC retained parent-of-origin methylation marks at the Igf2/H19 locus. Recent demonstration that multipotent adult progenitor cells (MAPC) could generate transplantable HSC in vitro *(43)* provides a particularly elegant method for overcoming immunologic barriers to stem cell therapies by the concomitant generation of HSC ready for transplantation. This approach is unique from those mentioned earlier in that the source of the HSC is from adult tissue and not from germ cell or embryonic origin. Whereas these types of approaches for resetting the immune system will take time to reach the clinic, the directed differentiation of ES or ES-like cells into HSC will provide the backbone in which new stem cell therapies will grow. In addition to regenerative medicine, derived HSC may be particularly useful for defining the early events leading to leukemic transformation.

In addition to these methods, it has been critical to uncover methods that differentiate HSC into desired hematopoietic cell types. Much of this work has built upon classical Dexter cultures *(44,45)* that have served as in vitro surrogates for maintenance and assay of HSC. The cobblestone-area-forming-cell (CAFC) *(46)* and the long-term culture-initiating cell (LTC-IC) *(47,48)* assays rely on stromal cells for hematopoietic support. The LTC-IC assay provides myeloid CFU-C at varying times of culture on a stromal layer and can be used in a quantitative manner to estimate the primitive cell pool. Variations of this assay allowing detection of lymphoid/myeloid output have been described *(49)*. These have been particularly useful for identifying primitive human cells. A particularly widely used system has been to culture HSC or ES cells on a feeder cell line expressing Notch to stimulate T-lymphocyte differentiation from various cell sources *(50,51)*. This new methodology has provided substantial new insights into T cell development and provided an easy alternative to the tricky fetal thymic organ culture approach for defining T cell potential.

3. Transplantation Assays for Mouse and Human HSC

Because HSCs are functionally defined, the gold-standard assay is the competitive repopulating unit. This assay is classically done in lethally irradiated hosts using small amounts of compromised competitor cells. When injected at limiting dilution, this assay can reliably quantitate the lympho-myeloid repopulating cell pool by applying Poisson statistics *(52)*. The limiting dilution approach has also been taken to the extreme, and single HSC have been transplanted to assess the clonal progeny and differentiation potential *(24)*. The Hoechst SP facilitates such approaches because of the high purity for HSC; however, even these cells are unable to engraft recipient mice with absolute efficiency *(53)*. The transplant assays for human HSC have benefited from immune-deficient mouse models that are conditioned with low-dose radiation and/or methods to suppress innate immune response originating from natural killer (NK) cells and macrophage. Even more recent techniques such as intrafemoral injection *(54)* have permitted detection of new populations of rapidly repopulating human HSC *(55)* and have been a useful tool to overcome engraftment defects that may be related to homing or migration.

The transplant assay has now been used in a wide variety of mutant mouse models. **Fig. 2** includes many genes that have been studied and which play major roles in HSC function. These include genes acting at different levels throughout

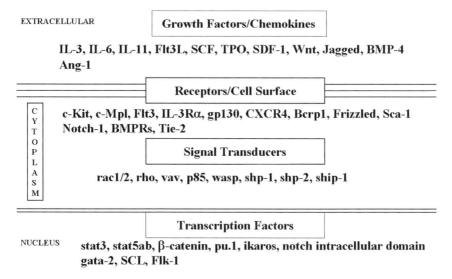

Fig. 2. Genes associated with hematopoietic stem cell phenotypes in knockout mouse models. Through the methodologies described in this book, the role of various molecules has been defined in the hematopoietic stem cells (HSC). Summarized here are many well-characterized molecules that act at various levels inside and outside the cell to promote HSC function. It is clear that many molecules are essential for normal HSC function and many play important cell-intrinsic functions in the signaling response to extracellular signals such as early acting cytokines. This is not an exhaustive list and is meant to illustrate how diverse signals can be important for overall HSC activity. Defining the underlying molecular mechanisms responsible for HSC functional defects requires continued studies and development of new methods such as the ones described in this book.

the cell, and many are connected with signaling transduction pathways. In particular, the Wnt *(56)*, Notch *(57)*, and Bmi-1 *(58,59)* signaling pathways are clearly important in HSC development and self-renewal.

4. Genetic Modification of HSCs and Imaging Engraftment

Retroviral- and lentiviral-mediated gene transfer was borne out of the gene therapy field and the desire to express a transferred gene in HSC to correct a hematologic disorder for the lifetime of the patient. This approach has provided a novel way to complement genetic defects in mutant mice and to study oncogene-mediated leukemogenesis. New methods using lentiviral vectors have overcome some of the obstacles of cell-cycle-dependent retroviral vectors. However, all oncoretroviral vectors have the ability to integrate into the genome randomly, and there is surprisingly strong bias toward integration sites especially those near active sites of chromatin and around start sites of genes *(60)*. This can be problematic and lead to insertional mutagenesis *(61)*. Elegant methods to clone and sequence integration sites have been extremely useful for not only understanding safety of gene therapy but also identifying cooperative mutations in leukemia models *(62,63)*. A database of insertional sites associated with clonal dominance *(64,65)* has recently been developed through the collaboration of several laboratories. This promises to provide a resource for future identification of pathways common to oncogenic events and may have important implications in cancer and stem cell research fields.

In addition to the more random genetic events associated with retroviral marking, the bacterial artificial chromosome (BAC) cloning techniques have revolutionized the already complex transgenic mouse field. Cre-recombinase-mediated deletion of DNA segments can be achieved by parallel loxP sites placed in the same orientation *(66)* or by inversion of DNA segments flanked by loxP sites placed in an inverted orientation *(67)* [review of "Flip-flop" mechanism *(68)*]. The combined use of engineered mouse strains expressing molecules of interest and reporter molecules for tracking gene expression promises to provide a growing set of tools for the future. Engineering of luciferase represents a simple genetic modification with wide application in molecular imaging. Small animal-imaging technologies have facilitated analysis of individual mice serially and in a non-invasive manner. The application of bioluminescence detection for tracking HSC clonal progeny provides a relatively quick and cost-efficient method *(69)*. More sophisticated techniques using microPET/SPECT, MRI, or quantum dots are available for the more experienced researcher. Combined multimodal imaging promises to provide the best advantages from all the technologies in an improved methodology.

Overall, these new molecular and imaging techniques have great potential to re-define HSC migration behavior in different settings and in real time.

5. Summary

This book provides a comprehensive set of chapters that describe many of the state-of-the-art techniques that have emerged in HSC research and which are briefly mentioned in this overview. This overview chapter is designed to set the tone for the methods chapters and to place them in an overall context. Combined approaches using tissue culture assays and animal models for study of HSC biology have the greatest promise for rapid advances in the field. The development of new clinical therapies based on HSC will require continued pre-clinical model development and validation. The tools that we now have at hand promise to usher in the new generation of stem cell therapies.

Acknowledgments

The authors thank the members of the Tse and Bunting laboratories for their helpful discussions. It is inevitable in this format that omissions have been made and the authors apologize for any references that could not be included.

References

1. Na, N. T., Traver, D., Weissman, I. L., and Akashi, K. (2002) Myeloerythroid-restricted progenitors are sufficient to confer radioprotection and provide the majority of day 8 CFU-S. *J. Clin. Invest.* **109**, 1579–1585.
2. Gregory, C. J., McCulloch, E. A., and Till, J. E. (1975) The cellular basis for the defect in haemopoiesis in flexed-tailed mice. III. Restriction of the defect to erythropoietic progenitors capable of transient colony formation in vivo. *Br. J. Haematol.* **30**, 401–10.
3. Charbonneau, H., Tonks, N. K., Walsh, K. A., and Fischer, E. H. (1988) The leukocyte common antigen (CD45): a putative receptor-linked protein tyrosine phosphatase. *Proc. Natl. Acad. Sci. U.S.A.* **85**, 7182–7186.
4. Shen, F. W., Saga, Y., Litman, G., Freeman, G., Tung, J. S., Cantor, H., and Boyse, E. A. (1985) Cloning of Ly-5 cDNA. *Proc. Natl. Acad. Sci. U.S.A.* **82**, 7360–7363.
5. Akashi, K., Traver, D., Miyamoto, T., and Weissman, I. L. (2000) A clonogenic common myeloid progenitor that gives rise to all myeloid lineages. *Nature* **404**, 193–197.
6. Christensen, J. L. and Weissman, I. L. (2001) Flk-2 is a marker in hematopoietic stem cell differentiation: a simple method to isolate long-term stem cells. *Proc. Natl. Acad. Sci. U.S.A.* **98**, 14541–14546.

7. Kondo, M., Weissman, I. L., and Akashi, K. (1997) Identification of clonogenic common lymphoid progenitors in mouse bone marrow. *Cell* **91,** 661–672.

8. Spangrude, G. J., Heimfeld, S., and Weissman, I. L. (1988) Purification and characterization of mouse hematopoietic stem cells. *Science* **241,** 58–62.

9. Uchida, N., Weissman, I. L. (1992) Searching for hematopoietic stem cells: evidence that Thy-1.1lo Lin- Sca-1+ cells are the only stem cells in C57BL/Ka-Thy-1.1 bone marrow. *J. Exp. Med.* **175,** 175–184.

10. van de Rijn, M., Heimfeld, S., Spangrude, G. J., and Weissman, I. L. (1989) Mouse hematopoietic stem-cell antigen Sca-1 is a member of the Ly-6 antigen family. *Proc. Natl. Acad. Sci. U.S.A.* **86,** 4634–4638.

11. Chen, C. Z., Li, L., Li, M., and Lodish, H. F. (2003) The endoglin(positive) sca-1(positive) rhodamine(low) phenotype defines a near-homogeneous population of long-term repopulating hematopoietic stem cells. *Immunity* **19,** 525–533.

12. Chen, C. Z., Li, M., de, G. D., Monti, S., Gottgens, B., Sanchez, M. J., Lander, E. S., Golub, T. R., Green, A. R., and Lodish, H. F. (2002) Identification of endoglin as a functional marker that defines long-term repopulating hematopoietic stem cells. *Proc. Natl. Acad. Sci. U.S.A.* **99,** 15468–15473.

13. Ito, T., Tajima, F., and Ogawa, M. (2000) Developmental changes of CD34 expression by murine hematopoietic stem cells. *Exp. Hematol.* **28,** 1269–1273.

14. Sato, T., Laver, J. H., and Ogawa, M. (1999) Reversible expression of CD34 by murine hematopoietic stem cells. *Blood* **94,** 2548–2554.

15. Zanjani, E. D., Almeida-Porada, G., Livingston, A. G., Zeng, H., and Ogawa, M. (2003) Reversible expression of CD34 by adult human bone marrow long-term engrafting hematopoietic stem cells. *Exp. Hematol.* **31,** 406–412.

16. Krause, D. S., Ito, T., Fackler, M. J., Smith, O. M., Collector, M. I., Sharkis, S. J., and May, W. S. (1994) Characterization of murine CD34, a marker for hematopoietic progenitor and stem cells. *Blood* **84,** 691–701.

17. Bradford, G. B., Williams, B., Rossi, R., and Bertoncello, I. (1997) Quiescence, cycling, and turnover in the primitive hematopoietic stem cell compartment. *Exp. Hematol.* **25,** 445–453.

18. Bertoncello, I., Hodgson, G. S., and Bradley, T. R. (1988) Multiparameter analysis of transplantable hemopoietic stem cells. II. Stem cells of long-term bone marrow-reconstituted recipients. *Exp. Hematol.* **16,** 245–249.

19. Reddy, G. P., Tiarks, C. Y., Pang, L., Wuu, J., Hsieh, C. C., and Quesenberry, P. J. (1997) Cell cycle analysis and synchronization of pluripotent hematopoietic progenitor stem cells. *Blood* **90,** 2293–2299.

20. Goodell, M. A., Brose, K., Paradis, G., Conner, A. S., and Mulligan, R. C. (1996) Isolation and functional properties of murine hematopoietic stem cells that are replicating in vivo. *J. Exp. Med.* **183,** 1797–1806.

21. Bunting, K. D. (2002) ABC transporters as phenotypic markers and functional regulators of stem cells. *Stem Cells* **20,** 11–20.

22. Berardi, A. C., Wang, A., Levine, J. D., Lopez, P., and Scadden, D. T. (1995) Functional isolation and characterization of human hematopoietic stem cells. *Science* **267,** 104–108.

23. Srour, E. F., Jetmore, A., Wolber, F. M., Plett, P. A., Abonour, R., Yoder, M. C., and Orschell-Traycoff, C. M. (2001) Homing, cell cycle kinetics and fate of transplanted hematopoietic stem cells. *Leukemia* **15**, 1681–1684.
24. Krause, D. S., Theise, N. D., Collector, M. I., Henegariu, O., Hwang, S., Gardner, R., Neutzel, S., and Sharkis, S. J. (2001) Multi-organ, multi-lineage engraftment by a single bone marrow-derived stem cell. *Cell* **105**, 369–377.
25. Lanzkron, S. M., Collector, M. I., and Sharkis, S. J. (1999) Hematopoietic stem cell tracking in vivo: a comparison of short-term and long-term repopulating cells. *Blood* **93**, 1916–1921.
26. Cheng, T., Rodrigues, N., Shen, H., Yang, Y., Dombkowski, D., Sykes, M., and Scadden, D. T. (2000) Hematopoietic stem cell quiescence maintained by p21(cip1/waf1). *Science* **287**, 1804–1808.
27. Yuan, Y., Shen, H., Franklin, D. S., Scadden, D. T., and Cheng, T. (2004) In vivo self-renewing divisions of haematopoietic stem cells are increased in the absence of the early G1-phase inhibitor, p18INK4C. *Nat. Cell Biol.* **6**, 436–442.
28. Desponts, C., Hazen, A. L., Paraiso, K. H., and Kerr, W. G. (2006) SHIP deficiency enhances HSC proliferation and survival but compromises homing and repopulation. *Blood* **107**, 4338–4345.
29. Sachs, K., Perez, O., Pe'er, D., Lauffenburger, D. A., and Nolan, G. P. (2005) Causal protein-signaling networks derived from multiparameter single-cell data. *Science* **308**, 523–529.
30. Irish, J. M., Hovland, R., Krutzik, P. O., Perez, O. D., Bruserud, O., Gjertsen, B. T., and Nolan, G. P. (2004) Single cell profiling of potentiated phospho-protein networks in cancer cells. *Cell* **118**, 217–228.
31. Kiel, M. J., Yilmaz, O. H., Iwashita, T., Yilmaz, O. H., Terhorst, C., and Morrison, S. J. (2005) SLAM family receptors distinguish hematopoietic stem and progenitor cells and reveal endothelial niches for stem cells. *Cell* **121**, 1109–1121.
32. Larochelle, A., Krouse, A., Metzger, M., Orlic, D., Donahue, R. E., Fricker, S., Bridger, G., Dunbar, C. E., and Hematti, P. (2006) AMD3100 mobilizes hematopoietic stem cells with long-term repopulating capacity in nonhuman primates. *Blood* **107**, 3772–3778.
33. Chen, J., Larochelle, A., Fricker, S., Bridger, G., Dunbar, C. E., and Abkowitz, J. L. (2006) Mobilization as a preparative regimen for hematopoietic stem cell transplantation. *Blood* **107**, 3764–3771.
34. Fukuda, S., Broxmeyer, H. E., and Pelus, L. M. (2005) Flt3 ligand and the Flt3 receptor regulate hematopoietic cell migration by modulating the SDF-1alpha(CXCL12)/CXCR4 axis. *Blood* **105**, 3117–3126.
35. Cancelas, J. A., Lee, A. W., Prabhakar, R., Stringer, K. F., Zheng, Y., and Williams, D. A. (2005) Rac GTPases differentially integrate signals regulating hematopoietic stem cell localization. *Nat. Med.* **11**, 886–891.
36. Kim, K., Lerou, P., Yabuuchi, A., Lengerke, C., Ng, K., West, J., Kirby, A., Daly, M. J., and Daley, G. Q. (2007) Histocompatible embryonic stem cells by parthenogenesis. *Science* **315**, 482–486.

37. Kyba, M., Perlingeiro, R. C., and Daley, G. Q. (2002) HoxB4 confers definitive lymphoid-myeloid engraftment potential on embryonic stem cell and yolk sac hematopoietic progenitors. *Cell* **109**, 29–37.
38. Yoder, M. C., Cumming, J. G., Hiatt, K., Mukherjee, P., and Williams, D. A. (1996) A novel method of myeloablation to enhance engraftment of adult bone marrow cells in newborn mice. *Biol. Blood Marrow Transplant.* **2**, 59–67.
39. Antonchuk, J., Sauvageau, G., and Humphries, R. K. (2002) HOXB4-induced expansion of adult hematopoietic stem cells ex vivo. *Cell* **109, 39–45.**
40. Thomson, J. A., Itskovitz-Eldor, J., Shapiro, S. S., Waknitz, M. A., Swiergiel, J. J., Marshall, V. S., and Jones, J. M. (1998) Embryonic stem cell lines derived from human blastocysts. *Science* **282**, 1145–1147.
41. Kaufman, D. S., Hanson, E. T., Lewis, R. L., Auerbach, R., and Thomson, J. A. (2001) Hematopoietic colony-forming cells derived from human embryonic stem cells. *Proc. Natl. Acad. Sci. U.S.A.* **98**, 10716–10721.
42. Eckardt, S., Leu, N. A., Bradley, H. L., Kato, H., Bunting, K. D., and McLaughlin, K. J. (2007) Hematopoietic reconstitution with androgenetic and gynogenetic stem cells. *Genes Dev* **21**, 409–419.
43. Serafini, M., Dylla, S. J., Oki, M., Heremans, Y., Tolar, J., Jiang, Y., Buckley, S. M., Pelacho, B., Burns, T. C., Frommer, S., Rossi, D. J., Bryder, D., Panoskaltsis-Mortari, A., O'Shaughnessy, M. J., Nelson-Holte, M., Fine, G. C., Weissman, I. L., Blazar, B. R., and Verfaillie, C. M. (2007) Hematopoietic reconstitution by multipotent adult progenitor cells: precursors to long-term hematopoietic stem cells. *J. Exp. Med.* **204**, 129–139.
44. Dexter, T. M., Allen, T. D., and Lajtha, L. G. (1977) Conditions controlling the proliferation of haemopoietic stem cells in vitro. *J. Cell Physiol.* **91**, 335–344.
45. Dexter, T. M., Moore, M. A., and Sheridan, A. P. (1977) Maintenance of hemopoietic stem cells and production of differentiated progeny in allogeneic and semiallogeneic bone marrow chimeras in vitro. *J. Exp. Med.* **145**, 1612–1616.
46. de Haan, G., Bystrykh, L. V., Weersing, E., Dontje, B., Geiger, H., Ivanova, N., Lemischka, I. R., Vellenga, E., and Van Zant, G. (2002) A genetic and genomic analysis identifies a cluster of genes associated with hematopoietic cell turnover. *Blood* **100**, 2056–2062.
47. Lemieux, M. E., Rebel, V. I., Lansdorp, P. M., and Eaves, C. J. (1995) Characterization and purification of a primitive hematopoietic cell type in adult mouse marrow capable of lymphomyeloid differentiation in long-term marrow "switch" cultures. *Blood* **86**, 1339–1347.
48. Sutherland, H. J., Eaves, C. J., Lansdorp, P. M., Thacker, J. D., and Hogge, D. E. (1991) Differential regulation of primitive human hematopoietic cells in long-term cultures maintained on genetically engineered murine stromal cells. *Blood* **78**, 666–672.
49. Coulombel, L. (2004) Identification of hematopoietic stem/progenitor cells: strength and drawbacks of functional assays. *Oncogene* **23**, 7210–7222.
50. Zuniga-Pflucker, J. C. (2004) T-cell development made simple. *Nat. Rev. Immunol.* **4**, 67–72.

51. Schmitt, T. M., Zuniga-Pflucker, J. C. (2002) Induction of T cell development from hematopoietic progenitor cells by delta-like-1 in vitro. *Immunity* **17**, 749–756.
52. Szilvassy, S. J., Humphries, R. K., Lansdorp, P. M., Eaves, A. C., and Eaves, C. J. (1990) Quantitative assay for totipotent reconstituting hematopoietic stem cells by a competitive repopulation strategy. *Proc. Natl. Acad. Sci. U.S.A.* **87**, 8736–8740.
53. Camargo, F. D., Chambers, S. M., Drew, E., McNagny, K. M., and Goodell, M. A. (2006) Hematopoietic stem cells do not engraft with absolute efficiencies. *Blood* **107**, 501–507.
54. Zhong, J. F., Zhan, Y., Anderson, W. F., and Zhao, Y. (2002) Murine hematopoietic stem cell distribution and proliferation in ablated and nonablated bone marrow transplantation. *Blood* **100**, 3521–3526.
55. Mazurier, F., Doedens, M., Gan, O. I., and Dick, J. E. (2003) Rapid myeloerythroid repopulation after intrafemoral transplantation of NOD-SCID mice reveals a new class of human stem cells. *Nat. Med.* **9**, 959–963.
56. Reya, T., Duncan, A. W., Ailles, L., Domen, J., Scherer, D. C., Willert, K., Hintz, L., Nusse, R., and Weissman, I. L. (2003) A role for Wnt signalling in self-renewal of haematopoietic stem cells. *Nature* **423**, 409–414.
57. Varnum-Finney, B., Purton, L. E., Yu, M., Brashem-Stein, C., Flowers, D., Staats, S., Moore, K. A., Le, R., I, Mann, R., Gray, G., Artavanis-Tsakonas, S., and Bernstein, I. D. (1998) The Notch ligand, Jagged-1, influences the development of primitive hematopoietic precursor cells. *Blood* **91**, 4084–4091.
58. Park, I. K., Qian, D., Kiel, M., Becker, M. W., Pihalja, M., Weissman, I. L., Morrison, S. J., and Clarke, M. F. (2003) Bmi-1 is required for maintenance of adult self-renewing haematopoietic stem cells. *Nature* **423**, 302–305.
59. Iwama, A., Oguro, H., Negishi, M., Kato, Y., Morita, Y., Tsukui, H., Ema, H., Kamijo, T., Katoh-Fukui, Y., Koseki, H., van Lohuizen M., and Nakauchi, H. (2004) Enhanced self-renewal of hematopoietic stem cells mediated by the polycomb gene product Bmi-1. *Immunity* **21**, 843–851.
60. Schroder, A. R., Shinn, P., Chen, H., Berry, C., Ecker, J. R., and Bushman, F. (2002) HIV-1 integration in the human genome favors active genes and local hotspots. *Cell* **110**, 521–529.
61. Baum, C., Dullmann, J., Li, Z., Fehse, B., Meyer, J., Williams, D. A., and von Kalle, C. (2003) Side effects of retroviral gene transfer into hematopoietic stem cells. *Blood* **101**, 2099–2114.
62. Li, Z., Dullmann, J., Schiedlmeier, B., Schmidt, M., von Kalle, C., Meyer, J., Forster, M., Stocking, C., Wahlers, A., Frank, O., Ostertag, W., Kuhlcke, K., Eckert, H. G., Fehse, B., and Baum, C. (2002) Murine leukemia induced by retroviral gene marking. *Science* **296**, 497.
63. Modlich, U., Kustikova, O. S., Schmidt, M., Rudolph, C., Meyer, J., Li, Z., Kamino, K., von Neuhoff, N., Schlegelberger, B., Kuehlcke, K., Bunting, K. D., Schmidt, S., Deichmann, A., von Kalle, C., Fehse, B., and Baum, C. (2005) Leukemias following retroviral transfer of multidrug resistance 1 (MDR1) are driven by combinatorial insertional mutagenesis. *Blood* **105**, 4235–4246.

64. Kustikova, O., Fehse, B., Modlich, U., Yang, M., Dullmann, J., Kamino, K., von Neuhoff, N., Schlegelberger, B., Li, Z., and Baum, C. (2005) Clonal dominance of hematopoietic stem cells triggered by retroviral gene marking. *Science* **308**, 1171–1174.

65. Kustikova, O. S., Geiger, H., Li, Z., Brugman, M. H., Chambers, S. M., Shaw, C. A., Pike-Overzet, K., Ridder, D., Staal, F. J., Keudell, G., Cornils, K., Nattamai, K. J., Modlich, U., Wagemaker, G., Goodell, M. A., Fehse, B., and Baum, C. (2007) Retroviral vector insertion sites associated with dominant hematopoietic clones mark "stemness" pathways. *Blood* **109**, 1897–1907.

66. Jonkers, J. and Berns, A. (2002) Conditional mouse models of sporadic cancer. *Nat. Rev. Cancer* **2**, 251–265.

67. Forster, A., Pannell, R., Drynan, L. F., Codrington, R., Daser, A., Metzler, M., Lobato, M. N., and Rabbitts, T. H. (2005) The invertor knock-in conditional chromosomal translocation mimic. *Nat. Methods* **2**, 27–30.

68. Branda, C. S. and Dymecki, S. M. (2004) Talking about a revolution: the impact of site-specific recombinases on genetic analyses in mice. *Dev. Cell* **6**, 7–28.

69. Cao, Y. A., Wagers, A. J., Beilhack, A., Dusich, J., Bachmann, M. H., Negrin, R. S., Weissman, I. L., and Contag, C. H. (2004) Shifting foci of hematopoiesis during reconstitution from single stem cells. *Proc. Natl. Acad. Sci. U.S.A.* **101**, 221–226.

70. Ito, C. Y., Li, C. Y., Bernstein, A., Dick, J. E., and Stanford, W. L. (2003) Hematopoietic stem cell and progenitor defects in Sca-1/Ly-6A-null mice. *Blood* **101**, 517–523.

71. Chabot, B., Stephenson, D. A., Chapman, V. M., Besmer, P., and Bernstein, A. (1988) The proto-oncogene c-kit encoding a transmembrane tyrosine kinase receptor maps to the mouse W locus. *Nature* **335**, 88–89.

72. Orlic, D., Bodine, D. M. (1992) Pluripotent hematopoietic stem cells of low and high density can repopulate W/Wv mice. *Exp. Hematol.* **20**, 1291–1295.

73. Miller, C. L., Rebel, V. I., Lemieux, M. E., Helgason, C. D., Lansdorp, P. M., and Eaves, C. J. (1996) Studies of W mutant mice provide evidence for alternate mechanisms capable of activating hematopoietic stem cells. *Exp. Hematol.* **24**, 185–194.

74. Ikuta, K., Weissman, I. L. (1992) Evidence that hematopoietic stem cells express mouse c-kit but do not depend on steel factor for their generation. *Proc. Natl. Acad. Sci. U.S.A.* **89**, 1502–1506.

75. Morrison, S. J., Lagasse, E., and Weissman, I. L. (1994) Demonstration that Thy(lo) subsets of mouse bone marrow that express high levels of lineage markers are not significant hematopoietic progenitors. *Blood* **83**, 3480–3490.

76. Kawashima, I., Zanjani, E. D., Almaida-Porada, G., Flake, A. W., Zeng, H., and Ogawa, M. (1996) CD34+ human marrow cells that express low levels of Kit protein are enriched for long-term marrow-engrafting cells. *Blood* **87**, 4136–4142.

77. Zanjani, E. D., Almeida-Porada, G., Livingston, A. G., Flake, A. W., and Ogawa, M. (1998) Human bone marrow CD34- cells engraft in vivo and undergo

multilineage expression that includes giving rise to CD34+ cells. *Exp. Hematol.* **26,** 353–360.

78. Civin, C. I., Trischmann, T., Kadan, N. S., Davis, J., Noga, S., Cohen, K., Duffy, B., Groenewegen, I., Wiley, J., Law, P., Hardwick, A., Oldham, F., and Gee, A. (1996) Highly purified CD34-positive cells reconstitute hematopoiesis. *J. Clin. Oncol.* **14,** 2224–2233.

79. Matsubara, A., Iwama, A., Yamazaki, S., Furuta, C., Hirasawa, R., Morita, Y., Osawa, M., Motohashi, T., Eto, K., Ema, H., Kitamura, T., Vestweber, D., and Nakauchi, H. (2005) Endomucin, a CD34-like sialomucin, marks hematopoietic stem cells throughout development. *J. Exp. Med.* **202,** 1483–1492.

80. Zhou, S., Morris, J. J., Barnes, Y., Lan, L., Schuetz, J. D., and Sorrentino, B. P. (2002) Bcrp1 gene expression is required for normal numbers of side population stem cells in mice, and confers relative protection to mitoxantrone in hematopoietic cells in vivo. *Proc. Natl. Acad. Sci. U.S.A.* **99,** 12339–12344.

81. Zhou, S., Schuetz, J. D., Bunting, K. D., Colapietro, A., Sampath, J., Morris, J. J., Lagutina, I., Grosveld, G. C., Osawa, M., Nakauchi, H., and Sorrentino, B. P. (2001) The ABC transporter Bcrp1/ABCG2 is expressed in a wide variety of stem cells and is a molecular determinant of the side-population phenotype. *Nat. Med.* **7,** 1028–1034.

82. Zhang, C. C., Steele, A. D., Lindquist, S., and Lodish, H. F. (2006) Prion protein is expressed on long-term repopulating hematopoietic stem cells and is important for their self-renewal. *Proc. Natl. Acad. Sci. U.S.A.* **103,** 2184–2189.

83. Hsu, H. C., Ema, H., Osawa, M., Nakamura, Y., Suda, T., and Nakauchi, H. (2000) Hematopoietic stem cells express Tie-2 receptor in the murine fetal liver. *Blood* **96,** 3757–3762.

84. Takakura, N., Huang, X. L., Naruse, T., Hamaguchi, I., Dumont, D. J., Yancopoulos, G. D., and Suda, T. (1998) Critical role of the TIE2 endothelial cell receptor in the development of definitive hematopoiesis. *Immunity* **9,** 677–686.

85. Solar, G. P., Kerr, W. G., Zeigler, F. C., Hess, D., Donahue, C., de Sauvage, F. J., and Eaton, D. L. (1998) Role of c-mpl in early hematopoiesis. *Blood* **92,** 4–10.

86. Ninos, J. M., Jefferies, L. C., Cogle, C. R., and Kerr, W. G. (2006) The thrombopoietin receptor, c-Mpl, is a selective surface marker for human hematopoietic stem cells. *J. Transl. Med.* **4,** 9.

87. Jones, R. J., Collector, M. I., Barber, J. P., Vala, M. S., Fackler, M. J., May, W. S., Griffin, C. A., Hawkins, A. L., Zehnbauer, B. A., Hilton, J., Colvin, O. M., and Sharkis, S. J. (1996) Characterization of mouse lymphohematopoietic stem cells lacking spleen colony-forming activity. *Blood* **88,** 487–491.

88. Storms, R. W., Trujillo, A. P., Springer, J. B., Shah, L., Colvin, O. M., Ludeman, S. M., and Smith, C. (1999) Isolation of primitive human hematopoietic progenitors on the basis of aldehyde dehydrogenase activity. *Proc. Natl. Acad. Sci. U.S.A.* **96,** 9118–9123.

89. Hess, D. A., Wirthlin, L., Craft, T. P., Herrbrich, P. E., Hohm, S. A., Lahey, R., Eades, W. C., Creer, M. H., and Nolta, J. A. (2006) Selection based on CD133

and high aldehyde dehydrogenase activity isolates long-term reconstituting human hematopoietic stem cells. *Blood* **107,** 2162–2169.

90. Storms, R. W., Green, P. D., Safford, K. M., Niedzwiecki, D., Cogle, C. R., Colvin, O. M., Chao, N. J., Rice, H. E., and Smith, C. A. (2005) Distinct hematopoietic progenitor compartments are delineated by the expression of aldehyde dehydrogenase and CD34. *Blood* **106,** 95–102.

91. Armstrong, L., Stojkovic, M., Dimmick, I., Ahmad, S., Stojkovic, P., Hole, N., and Lako, M. (2004) Phenotypic characterization of murine primitive hematopoietic progenitor cells isolated on basis of aldehyde dehydrogenase activity. *Stem Cells* **22,** 1142–1151.

92. Hess, D. A., Meyerrose, T. E., Wirthlin, L., Craft, T. P., Herrbrich, P. E., Creer, M. H., and Nolta, J. A. (2004) Functional characterization of highly purified human hematopoietic repopulating cells isolated according to aldehyde dehydrogenase activity. *Blood* **104,** 1648–1655.

93. Danet, G. H., Luongo, J. L., Butler, G., Lu, M. M., Tenner, A. J., Simon, M. C., and Bonnet, D. A. (2002) C1qRp defines a new human stem cell population with hematopoietic and hepatic potential. *Proc. Natl. Acad. Sci. U.S.A.* **99,** 10441–10445.

94. Fleming, W. H., Alpern, E. J., Uchida, N., Ikuta, K., Spangrude, G. J., and Weissman, I. L. (1993) Functional heterogeneity is associated with the cell cycle status of murine hematopoietic stem cells. *J. Cell Biol.* **122,** 897–902.

95. Fleming, W. H., Alpern, E. J., Uchida, N., Ikuta, K., and Weissman, I. L. (1993) Steel factor influences the distribution and activity of murine hematopoietic stem cells in vivo. *Proc. Natl. Acad. Sci. U.S.A.* **90,** 3760–3764.

96. Larochelle, A., Vormoor, J., Hanenberg, H., Wang, J. C., Bhatia, M., Lapidot, T., Moritz, T., Murdoch, B., Xiao, X. L., Kato, I., Williams, D. A., and Dick, J. E. (1996) Identification of primitive human hematopoietic cells capable of repopulating NOD/SCID mouse bone marrow: implications for gene therapy. *Nat. Med.* **2,** 1329–1337.

97. Terstappen, L. W., Huang, S., Safford, M., Lansdorp, P. M., and Loken, M. R. (1991) Sequential generations of hematopoietic colonies derived from single nonlineage-committed CD34+. *Blood* **77,** 1218–1227.

98. Adolfsson, J., Borge, O. J., Bryder, D., Theilgaard-Monch, K., Astrand-Grundstrom, I., Sitnicka, E., Sasaki, Y., and Jacobsen, S. E. (2001) Upregulation of Flt3 expression within the bone marrow Lin(-)Sca1(+)c-kit(+) stem cell compartment is accompanied by loss of self-renewal capacity. *Immunity* **15,** 659–669.

II

Stem Cell Enrichment and Analysis

2

Isolation of Quiescent Murine Hematopoietic Stem Cells by Homing Properties

Tarja A. Juopperi and Saul J. Sharkis

Summary

A major challenge facing investigators working in the field of hematopoietic stem cell (HSC) biology has been to develop a strategy to purify rare primitive HSCs from bone marrow. Several methods have been available including the commonly used technique of isolating HSCs based on a specific cell-surface phenotype. As surface marker expression is dynamic and may fluctuate depending on the proliferative or activation state of the cell, our laboratory has established a unique functional in vivo assay (the 2-day homing assay) to isolate murine HSCs. This protocol selects for HSCs on the basis of their ability to home to bone marrow and yields a population that can reconstitute the murine hematopoietic system with the transplantation of a single cell. In contrast to other methods that use specific cell-surface antigens to acquire HSCs, our functional assay aids in obtaining a primitive HSC that exhibits both hematopoietic and epithelial engraftment capabilities. The 2-day homing protocol involves harvesting whole bone marrow and performing a physical separation method (elutriation) to acquire a fraction of small-sized cells (fraction 25). Fraction 25 cells are then depleted of later progenitors and differentiated hematopoietic cells, labeled with a fluorescent tracking dye and transplanted into lethally irradiated recipient mice. Two days after transplantation, the bone marrow is harvested from the primary recipient, and HSCs that have homed to the bone marrow are collected by fluorescence-activated cell sorting. In addition to the traditional 2-day homing protocol, we have included in this chapter our recently developed method of using density gradient centrifugation to replace the elutriation step that also selects for a primitive HSC.

Key Words: Elutriation; density gradient separation; homing; hematopoietic stem cells; murine.

From: *Methods in Molecular Biology, vol. 430: Hematopoietic Stem Cell Protocols*
Edited by: K. D. Bunting © Humana Press, Totowa, NJ

1. Introduction

Hematopoiesis is the tightly regulated process by which blood cells are maintained at physiological levels in the body. The hematopoietic stem cell (HSC) having extensive proliferative capacity is ultimately responsible for this task and can produce differentiated cells of all hematopoietic lineages as well as additional HSCs. HSCs are found in relatively low numbers (i.e., 1 in 10,000 murine bone marrow cells), and numerous strategies have been employed to isolate this rare cell type *(1–6)*.

Our laboratory has developed a functional assay to isolate murine HSC from bone marrow *(7)*. The assay relies upon the ability of HSC when transplanted, to home to the marrow. The 2-day homing protocol enriches for unique cells that have the ability not only to reconstitute the hematopoietic system at limiting dilutions but also to engraft non-hematopoietic tissues such as the gastrointestinal tract, lung, and skin *(8)*. By convention, adult stem cells such as HSCs have been considered to have the ability to differentiate only into cells specific to their tissue of origin. However, there is accumulating evidence to support the concept of hematopoietic plasticity *(9,10)*. Research findings from our laboratory have shown that HSCs can convert into cells with a hepatic phenotype *(11)*. Using an injured liver model, we have demonstrated that HSCs could convert into cells expressing liver-specific proteins such as albumin, transferrin, and fibrinogen, as well as expressing epithelial markers such as the cytokeratins and E-cadherin.

The initial step of our isolation protocol involves a physical separation technique known as counter-flow elutriation. This centrifugation step separates whole bone marrow on the basis of size and density and enables the acquisition of a fraction of small, relatively quiescent cells. The fraction collected at a flow rate of 25 ml/min contains cells with long-term hematopoietic repopulating capabilities *(5–8)*.

In addition to the traditional 2-day homing protocol, we also describe the use of density gradient separation as an alternative to elutriation to provide a technology that will not require this specialized equipment. The investigator has the option of fractionating whole bone marrow either by elutriation or by density gradient separation. We have shown that high-density bone marrow cells collected using a four-layer discontinuous gradient, most resemble fraction 25 cells in that they have the capacity for homing and the ability to provide long-term hematopoietic engraftment with the injection of only ten cells *(12)*. Density gradient separation may be used in conjunction with the 2-day homing protocol to enrich for cells that can provide hematopoietic reconstitution; however, the ability of these cells to engraft non-hematopoietic tissues has not yet been determined.

After the physical separation step (elutriation or density gradient centrifugation), the resulting fraction must be depleted of later progenitors and differentiated cells. Our laboratory employs immuno-panning for the depletion step; however, other methods such as magnetic bead separation may also be utilized. The lineage-depleted cells are subsequently labeled with PKH26 dye, a fluorescent marker that stably incorporates into lipid regions of the cell membrane. This fluorescent cell linker has been used for long-term in vivo cell-tracking applications and is the primary means by which the homed HSCs are identified in the bone marrow. After labeling, the cells are injected into lethally irradiated syngeneic mice. Forty-eight hours post-transplantation, the mice are killed, bone marrow harvested, and HSCs collected by fluorescence-activated cell sorting (FACS). The resulting HSC population typically represents approximately 0.005% of cells found in the bone marrow.

2. Materials

2.1. Isolation of Murine Bone Marrow

1. For each experiment, harvest bone marrow from a minimum of ten C57BL/6 mice between the ages of 6–8 weeks of age. Older animals may be used if necessary, and this protocol may also be utilized for other strains of mice. Typically, sex-mismatched transplants are performed using male mice as donors (to use the Y chromosome as a marker of donor engraftment).
2. Minimal essential medium alpha medium (α-MEM, Invitrogen, Carlsbad, CA, USA).
3. 60 mm polystyrene tissue culture dishes.
4. Sterile gloves.
5. Sterile scissors to remove femurs and tibias.
6. Sterile gauze sponges to remove muscle tissues from bones.
7. Sterile 1-ml syringes with 25-G needles to expel bone marrow from medullary cavities.
8. Sterile 5-ml syringe with a 22-G needle to collect flushed marrow.

2.2. Elutriation

1. Elutriator: we currently use a Beckman J-6M centrifuge with a JE-6B elutriator rotor and a standard chamber. A master-flex peristaltic pump is used to regulate flow rate.
2. Stock solution of elutriation medium. Prepare stock solution of elutriation medium (30× concentrations) by adding 270 g of NaCl, 30 g of D-glucose, and 3 g of ethylenediaminetetraacetic acid (EDTA) together in 1 l of sterile distilled water. Combine reagents together using heat and sterile filter. This stock solution can be stored at room temperature.

3. Working solution of elutriation medium. Immediately before elutriation, prepare working solution by adding 16.7 ml of 30× elutriation media and 8.3 ml of 30% BSA to 500 ml of sterile distilled water.
4. Collection bottles.
5. Optional reagent: Erythrocyte lysis buffer such as ACK Lysis Buffer (Quality Biological, Gaithersburg, MD, USA).

2.3. Density Gradient Separation

1. Percoll medium (GE Healthcare, Piscataway, NJ, USA).
2. 10× and 1× phosphate-buffered saline (PBS).
3. Distilled water.
4. Density gradient beads (GE Healthcare).
5. Sterile transfer pipettes.
6. 50-ml conical centrifuge tubes.
7. Optional reagent: Erythrocyte lysis buffer such as ACK Lysis Buffer (Quality Biological).

2.4. Lineage Depletion (Immuno-Panning)

1. Lineage marker antibodies: GR-1 (granulocytes), B220 (B lymphocytes), CD5 (T lymphocytes), Mac-1 (macrophage/monocyte), AA4.1 (pre-B lymphocytes and early hematopoietic progenitors), and TER-119 (erythroid cells). Antibodies can be purchased from various suppliers.
2. 30% bovine serum albumin.
3. 0.05 M Tris-buffer, pH 9.0.
4. 60 mm polystyrene tissue culture dishes.
5. 15-ml centrifuge tubes.
6. Minimal essential medium alpha medium (without serum).
7. Or another established method to deplete differentiated cells using the antibodies listed above.

2.5. PKH26 Labeling and Injection of Cells

1. PKH26 red fluorescent linker kit (Product no. PKH26-GL Sigma Aldrich, St. Louis, MO, USA). Typically stored at room temperature. Note product instructions on storage and stability of the dye.
2. Minimal essential medium alpha medium (without serum).
3. Heat inactivated fetal bovine serum (heat inactivated at 56°C for 60 min and sterile filtered).
4. Lethally irradiated syngeneic recipient mice (1100 cGy—given in a single dose).
5. Sterile 1-ml syringes with 25-G needles.

2.6. Acquisition of 2-Day Homed Cells

1. Reagents required for isolation of bone marrow (*see* **Subheading 2.1.**).
2. Erythrocyte lysis buffer.
3. 15-ml centrifuge tubes.

4. 5-ml polystyrene round bottom tubes with cell strainer.
5. FACS machine.

3. Methods

Before beginning the experimental protocol, it is recommended that the investigator prepare stock and working solutions of elutriation media (*see* **Subheading 3.2.1.**) or density gradients (*see* **Subheading 3.3.1.**) and prepare lineage depletion plates (*see* **Subheading 3.4.**).

3.1. Isolation of Murine Bone Marrow

1. Euthanize mice by carbon dioxide inhalation.
2. Remove hind legs.
3. Aseptically separate the femurs and tibias. Sterile gauze can be used to remove the soft tissue surrounding the bones.
4. Cut ends of bones and expel bone marrow by flushing α-MEM through the medullary cavities using a 25-G needle and 1-ml syringe into a Petri dish.
5. Dissociate the cells into a single cell suspension by repeatedly passing the suspension through the needle.
6. Use the 5-ml syringe and 22-G needle to collect the expelled marrow for further processing.
7. Avoid contaminating the cells throughout the processing procedure as the cells will be transplanted into lethally irradiated recipients.

3.2. Physical Separation of Whole Bone Marrow: Elutriation

1. To elutriate, use a rotor speed of $1260 \times g$ and a temperature of 25°C.
2. After whole marrow has been added to the elutriation chamber, collect a minimum of 200 ml of media at a flow rate of 15 ml/min. Discard this fraction.
3. Keeping the rotor speed constant, adjust flow rate to 25 ml/min and collect 200–400 ml of media. This fraction contains cells with long-term repopulating ability and will be further enriched for HSC by the 2-day homing protocol. Retain this fraction for further processing.
4. Transfer elutriation media containing fraction 25 cells to sterile 50-ml conical tubes. Centrifuge at $400 \times g$ for 10 min at 4°C.
5. Re-suspend and combine all cell pellets in α-MEM to prepare for lineage depletion.
6. An optional step in this protocol is to remove contaminating erythrocytes using lysis buffer (*see* **Note 1**). After centrifugation, re-suspend the cell pellet in 5 ml of erythrocyte lysing buffer. Incubate for only 2–3 min. After lysing, fill the tube with α-MEM and centrifuge at $400 \times g$ for 10 min at 4°C. Re-suspend cell pellet in α-MEM for lineage depletion.

3.3. Physical Separation of Whole Bone Marrow: Density Gradient Separation

3.3.1. Preparation of Density Gradients

1. Prepare Percoll solutions of the following densities (1.064, 1.075, 1.081, and 1.087 g/ml) by diluting Percoll medium with 10× PBS and distilled water according to the manufacturer's one-step procedure.
2. To create the discontinuous gradient, layer 10 ml volumes of each density sequentially in a 50-ml conical centrifuge tube starting with the heaviest solution and continuing with decreasing concentrations.
3. Store gradients at 4°C and warm to room temperature before use.

3.3.2. Collection of High-Density Bone Marrow Cells

1. Suspend whole bone marrow in 8–10 ml of α-MEM and gently layer onto the preformed gradient (*see* **Note 2**).
2. To calibrate the gradient, suspend density gradient beads corresponding to the appropriate densities in a similar volume of α-MEM and layer on a separate gradient (run in parallel).
3. Centrifuge gradient at 400 × g at 20°C for 30 min (*see* **Note 3**). Turn centrifuge brake off.
4. Handle carefully after centrifugation as not to disturb the gradient.
5. Using a sterile transfer pipette, aspirate 5 ml of media at the 1.081/1.087 density interface and place in a 15-ml tube (*see* **Fig. 1** and **Note 4**).
6. Wash cell fraction with 1× PBS and centrifuge at 400 × g for 10 min at 4°C.
7. Re-suspend the cell pellet in α-MEM to prepare for lineage depletion.
8. An optional step in this protocol is to remove contaminating erythrocytes using lysis buffer (*see* **Note 1**). After centrifugation, re-suspend the cell pellet in 5 ml of erythrocyte lysing buffer. Incubate for only 2–3 min. After lysing, fill the tube with α-MEM and centrifuge at 400 × g for 10 min at 4°C. Re-suspend cell pellet in α-MEM for lineage depletion.

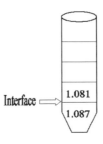

Fig. 1. Schematic drawing of the discontinuous gradient used for whole bone marrow separation. The arrow indicates the interface where high-density bone marrow cells are collected.

3.4. Lineage Depletion (Immuno-Panning)

1. Prepare lineage depletion plates by adding 1.5 ml of Tris-buffer (0.05 M, pH 9.0) and 100 µg of each antibody to a Petri dish. Completely cover the surface of each plate and incubate overnight at 4°C. Wash plates twice with α-MEM and store at 4°C.
2. Before use, block lineage depletion plate at room temperature by adding 3 ml of 30% BSA for 15–30 min. Depletion can be carried out using one or two plates.
3. After blocking, wash plate twice with α-MEM.
4. Remove media from plate and add the cell suspension. Incubate at 4°C for 90 min. Remember to agitate the plate at the midpoint of the incubation.
5. After depletion, remove the supernatant carefully from the plate as not to disturb the adherent cells. A small volume of media can be used to gently wash the plate; however, care must be taken to not disturb attached cells.
6. Wash the cells in a 15-ml tube with α-MEM by centrifuging at $400 \times g$ for 10 min at 4°C.

3.5. PKH26 Labeling and Transplantation into Primary Recipients

1. Re-suspend cell pellet in 1–2 ml of the diluent provided with the PKH26 label.
2. Prepare dye immediately before staining using a working concentration of 10 µmol/l. Equal volumes of working dye and cell suspension must be used.
3. Slowly add the cell suspension to the dye and quickly pipette the mixture to evenly distribute the label. Incubate the cells with PKH26 at room temperature for 2–5 min. Gently agitate the cells (tube may be placed on a rocker). Do not vortex (*see* **Note 5**).
4. To stop the staining reaction add and equal volume of FBS and incubate for 1 min.
5. Wash the cell suspension by completely filling the tube with α-MEM and centrifuge at $400 \times g$ for 10 min at 4°C.
6. Discard the supernatant and re-suspend the cell pellet in α-MEM. Transfer to a new tube and wash with media. Two additional wash steps are recommended.
7. Stained cells can be observed by detecting a pink color to the cell pellet.
8. Re-suspend the cell pellet in α-MEM and determine the number of cells available for transplantation. Typically $10–20 \times 10^6$ cells are injected per animal.
9. Prepare cell dosages by re-suspending the cells in the appropriate volume of α-MEM. Suggested volumes are 400–500 µl of media per injection.
10. Prepare transplant recipients by providing 1100cGy of total body irradiation as a single dose (*see* **Note 6**). This step may be performed up to 1 h before injection of cells.
11. Inject cells peripherally into lethally irradiated recipient mice either by tail vein or by retro-orbital injection (*see* **Note 7**).

3.6. Acquisition of 2-Day Homed Cells

1. Euthanize mice between 40–48 h post-transplant of PHK26 labeled cells.
2. Collect bone marrow aseptically from the femur and tibias as described in **subheading 3.1**. Samples may be pooled if multiple animals were injected.

3. Centrifuge cells at $400 \times g$ for 10 min at 4°C.
4. Remove contaminating erythrocytes using lysis buffer.
5. Re-suspend cells in a small volume of media for cell sorting (typically 500 µl to 1 ml) and transfer to a 5-ml tube with cell strainer.
6. For cell acquisition, an unstained sample is recommended to define the background level of autofluorescence. Bone marrow collected from a lethally irradiated control recipient mouse (if available) or an untreated mouse may be used as a negative control.
7. Acquire a low forward scatter, PKH26 bright population by FACS. The PKH26 label will be visible in the FL2 channel (*see* **Note 8 and Fig. 2**).
8. Collect into a small volume of media.

The Two Day Homing Protocol

After 48 Hours Harvest Bone Marrow and Obtain HSCs by FACS

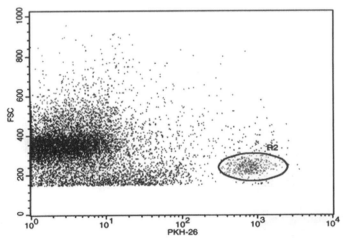

Fig. 2. Flow chart illustrating the 2-day homing protocol. The fluorescence-activated cell sorting (FACS) profile demonstrates the collection 2-day homed cells (low-forward scatter, PKH bright cells).

4. Notes

1. Erythrocyte lysis is not a necessary step in the 2-day homing protocol; however, it can be incorporated before immuno-panning to achieve optimal depletion of mature cells and to improve PKH26 labeling of cells.

2. For density gradient separation, a maximum of 130×10^6 cells per gradient are typically used. A distinct separation of the cells at various densities may not be observed if the gradient is overloaded.

3. Although density gradients can be stored at 4°C, ensure that the gradients are warmed up to room temperature before use and that the centrifuge is at 20°C for the initial 30-min spin. Centrifugation at a lower temperature may require a longer period of time for the cells to settle at the appropriate densities. It is also important to centrifuge the gradients immediately after the cell suspension has been added to the gradient, as clotting may occur if gradients are left for an extended period of time (greater than 20 min) before centrifugation.

4. A visible layer of cells may not always be present at the interface of 1.081/1.087 after density gradient separation (if low numbers of bone marrow cells were loaded onto the gradient). Aspirate the media at the interface regardless of whether you can detect a layer.

5. Aggregation of the cells during the staining procedure is a common problem. To avoid clumping of the cells, slowly add cells to the dye and triturate to distribute the stain quickly and evenly. The tube should be agitated several times during the staining procedure. Note that it is not uncommon to see pink clumps forming at this time. Also, ensure that the FBS has been warmed to room temperature before use, as the addition of cold FBS may also result in the clumping of cells. A trouble-shooting guide is provided with the PKH26 dye.

6. Young mice, 6–8 weeks of age, are typically used as primary recipients. We have noticed a decrease in HSC recovery after homing when aged mice are used as recipients. Avoid using animals older than 8 weeks of age.

7. Rarely, mice have died immediately after injection of the cells. As cell aggregates can form during and after staining, ensure that cell clumps are not present by passing the fluid through the needle before injection.

8. Occasionally, low numbers of HSC cells are detected using flow cytometry after the 2-day homing protocol. Possible reasons for poor recovery of cells include inadequate injection of cells into the peripheral circulation, low numbers of cells injected, improper staining of cells, inadequate irradiation of recipient mice, or contaminated cells were injected (mice would appear sick).

References

1. Wiktor-Jedrzejczak,W., Szczylik, C., Gonas, P., Sharkis, S.J., and Ahmed, A. (1979) Different marrow cell number requirements for the haemopoietic colony formation and the curve of the W/Wv anemia. *Experientia* **35**, 546–7.

2. Ikuta, K. and Weissman, I.L. (1992) Evidence that hematopoietic stem cells express mouse c-kit but do not depend on steel factor for their generation. *Proc Natl Acad Sci USA* **89**, 1502–6.

3. Uchida, N., Tsukamoto, A., He, D., Friera, A.M., Scollay, R., and Weissman, I.L. (1998) High doses of purified stem cells cause early hematopoietic recovery in syngeneic and allogeneic hosts. *J Clin Invest* **101**, 961–6.

4. Kiel, M.J., Yilmaz, O.H., Iwashita, T., Yilmaz, O.H., Terhorst, C., and Morrison, S. J. (2005) SLAM family receptors distinguish hematopoietic stem and progenitor cells and reveal endothelial niches for stem cells. *Cell* **121**, 1109–21.

5. Jones, R.J., Wagner, J.E., Celano, P., Zicha, M.S., and Sharkis, S.J. (1990) Separation of pluripotent haematopoietic stem cells from spleen colony-forming cells. *Nature* **347**, 188–9.

6. Jones, R.J., Collector, M.I., Barber, J.P., Vala, M.S., Fackler, M.J., May, W.S., Griffen, C.A., Hawkins A.L., Zehnbauer, B.A., Hilton, J., Colvin, O.M., and Sharkis, S.J. (1996) Characterization of mouse lymphohematopoietic stem cells lacking spleen colony-forming activity. *Blood* **88**, 487–91.

7. Lanzkron, S.M., Collector, M.I., and Sharkis, S.J. (1999) Hematopoietic stem cell tracking in vivo: a comparison of short-term and long-term repopulating cells. *Blood* **93**, 1916–21.

8. Krause, D.S., Theise, N.D., Collector, M.I., Henegariu, O., Hwang, S., Gardner, R., Neutzel, S., and Sharkis, S.J. (2001) Multi-organ, multi-lineage engraftment by a single bone marrow-derived stem cell. *Cell* **105**, 369–77.

9. Cogle, C.R., Yachnis, A.T., Laywell, E.D., Zander, D.S., Wingard, J.R., Steindler, D.A., and Scott, E.W. (2004) Bone marrow transdifferentiation in brain after transplantation: a retrospective study. *Lancet* **363**, 1432–7.

10. Orlic, D., Kajstura, J., Chimenti, S., Jakoniuk, I. Anderson, S.M., Li, B., Pickel, J., McKay, R., Nadal-Ginard, B., Bodine, D.M., Leri, A., and Anversa, P. (2001) Bone marrow cells regenerate infarcted myocardium. *Nature* **410**, 701–5.

11. Jang, Y.Y., Collector, M.I., Baylin, S.B., Diehl, A.M., and Sharkis, S.J. (2004) Hematopoietic stem cells convert into liver cells within days without fusion. *Nat Cell Biol* **6**, 532–9.

12. Juopperi, T.A., Schuler, W., Yuan, X., Collector, M.I., Dang, C.V., and Sharkis, S.J. (2007) Isolation of bone marrow derived stem cells using density gradient separation. *Exp Hematol* **35**, 335–341.

3

Mobilization of Hematopoietic Stem and Progenitor Cells in Mice

Simon N. Robinson and Ronald P. van Os

Summary

Animal models have added significantly to our understanding of the mechanism(s) of hematopoietic stem and progenitor cell (HSPC) mobilization. Such models suggest that changes in the interaction between the HSPC and the hematopoietic microenvironmental 'niche' (cellular and extracellular components) are critical to the process. The increasing availability of recombinant proteins (growth factors, cytokines, chemokines), antibodies, drugs (agonists and antagonists), and mutant and genetically modified animal models [gene knock-in (KI) and knock-out (KO)] continue to add to the tools available to better understand and manipulate mobilization processes.

Key Words: Mouse models; hematopoietic stem and progenitor cell (HSPC); mobilization; methodologies; endpoints.

1. Introduction

1.1. Rationale

The goal of this chapter is to provide a guide to the use of mice in the investigation of mobilization processes rather than to specifically discuss mechanism(s) of mobilization. Ultimately, a better understanding of the mechanism(s) of hematopoietic stem and progenitor cell (HSPC) mobilization may allow more effective (more rapid and higher HSPC yield) clinical mobilization strategies to be developed both for patients (autologous transplantation) and for healthy donors (allogeneic transplantation).

From: *Methods in Molecular Biology, vol. 430: Hematopoietic Stem Cell Protocols*
Edited by: K. D. Bunting © Humana Press, Totowa, NJ

1.2. Mobilizing Agents

A large number of factors can induce HSPC mobilization (summarized in **Table 1**). These include growth factors, cytokine, chemokines, enzymes, antibodies, and drugs. Whereas some factors appear to share common mechanisms of HSPC mobilization, others appear to utilize unique mechanisms.

Table 1
Factors that can Induce Hematopoietic Stem and Progenitor Cell (HSPC) Mobilization[*]

Growth factors, cytokines and chemokines
    Granulocyte colony-stimulating factor (G-CSF) *(9,29,30)*
    Granulocyte-macrophage colony stimulating factor (GM-CSF) *(11)*
    Stem cell factor (SCF)/c-kit ligand *(17,18)*
    Interleukin (IL)-1 *(6)*, IL-3 *(12)*, IL-7 *(15)*, IL-8 *(2)*, IL-11 *(10)*,
    IL-12 *(13)* and IL-17 *(16)*
    Macrophage inflammatory protein (MIP)-1α *(1)*
    Chemokine growth-related oncogene protein (GRO)-β *(3,4)*
    Erythropoietin (EPO) *(14)*
    Fms-like tyrosine kinase 3 ligand (Flt-3L) *(19–21)*
Enzymes
    Dipeptidylpeptidase IV (CD26) *(31,32)* (inhibited by Diprotin A/Val-Pyr)
Antibodies
    Disrupting the interaction between specific integrins and their ligands e.g.
    (a) Very late antigen (VLA)-4/CD49d and vascular cell adhesion molecule
    (VCAM)-1 *(22–24)*
    (b) Leukocyte function-associated antigen (LFA)-1 and intra-cellular adhesion
    molecule (ICAM)-1 *(33)*
Drugs
    β2 Adrenergic agonist *(34)*
    Lipopolysaccharide (LPS) *(7)*
    Sulfated polysaccharide *(35)*
    Chemotherapy agents *(9,36)*
    CXCR-4 agonist (CTCE-0021) *(37)*
    CXCR-4 antagonist (AMD3100) *(5,8)*

[*]These include growth factors, cytokines, chemokines, enzymes, antibodies, and drugs. Whereas some factors share common mechanisms, others appear to utilize unique pathways. Factors differ in both the magnitude and the speed of HSPC mobilization. Some factors stimulate HSPC mobilization after a single injection, and elevated levels of HSPC are evident in the blood within minutes or hours, whereas others require once- or twice-daily injection for multiple days before elevated levels of HSPC are detectable in the blood. The precise mechanism(s) by which the different factors elicit HSPC mobilization remain to be determined as do qualitative comparisons of the HSPC mobilized by the different factors.

Furthermore, whereas some factors can induce rapid HSPC mobilization (within minutes) *(1–4)* and follow only a single administration, others induce HSPC mobilization more slowly [over hours *(5–8)* or days *(9–24)*] and only after repeated once- or twice-daily administration over several days. The complexity of the mobilization process(es) is further increased when mobilizing agents are combined with the goal of stimulating more effective HSPC mobilization (larger and more rapid HSPC yield).

Experiments will usually fall into one of two types:

1. Experiments that are designed to investigate the mechanism(s) of HSPC mobilization. These might investigate (a) changes in gene-, protein-, and adhesion molecule-expression by HSPC and/or cells of the hematopoietic microenvironment following the administration of a given factor; (b) the impact (positive or negative) of a given molecule in the mechanism(s) of HSPC mobilization through the use of (i) mutant or genetically engineered [knock-out (KO) or knock-in (KI)] mice or (ii) the administration of neutralizing antibodies to reduce levels of activity or (iii) the administration of recombinant protein to increase levels of activity; and/or (c) novel putative mobilizing agents initially in wild-type mice, then in mutant or genetically engineered (KO or KI) mice, to characterize mobilization profiles and to determine whether the molecule has similarities/differences when compared with HSPC mobilization by G-CSF.

2. Experiments that are aimed at optimizing mobilization protocols for future clinical use. These studies often involve the use of combinations of factors often including G-CSF.

G-CSF is currently the best-characterized mobilization agent, and it is widely used in the clinic for the purpose of HSPC mobilization in patients and healthy donors. Although the precise mechanism(s) of HSPC mobilization by G-CSF still remain unclear, it can be considered as the "gold standard" positive control against which the mobilization properties of other agents can be compared. The inclusion of G-CSF as a positive control in mobilization experiments, especially those using novel agents or novel delivery strategies, should be given serious consideration.

In addition to trying to understand the mechanism(s) by which a factor elicits HSPC mobilization, a basic understanding of the pharmacokinetic profile of the molecule under investigation is important. Such knowledge enables the administration of a factor to be optimized for HSPC mobilization. For example, is it possible to determine how rapidly the molecule is cleared from the plasma? More rapid clearance from the plasma will require more frequent administration of the molecule to sustain plasma levels sufficient to yield HSPC mobilization. Enzyme-linked immunosorbant assay (ELISA) of plasma samples can be used to address such pharmacokinetic issues, but this approach is limited by the availability of antibodies to the molecule of interest.

Conversely, where the pharmacokinetic profile of a molecule (e.g. G-CSF) is well established, attempts can be made to alter its profile to biological and clinical advantage. For example, PEGylation *(25)* of G-CSF markedly reduces its clearance from the plasma. In mouse models, by sustaining levels of G-CSF at biologically relevant levels for an increased period of time in the plasma, a single injection of PEG-G-CSF ["Neulasta®" (pegfilgrastim), Amgen Inc., Thousand Oaks, CA, USA] can match the HSPC mobilization currently achieved only after 4 days of single, or twice daily, G-CSF injection. However, the safety and efficacy of Neulasta® for HSPC mobilization has not yet been evaluated in patient studies, and as a result, its current use is primarily for the treatment of chemotherapy-associated neutropenia.

To this end, PEGylation and the development of sustained release formulations *(21,26)* are strategies that can be used to alter the pharmacokinetic profiles of mobilizing agents with the goal of yielding potentially more effective HSPC mobilization strategies. These are examples of the types of studies that require the use of established mouse HSPC mobilization models and appropriate controls.

1.3. Endpoints

Although the hematopoietic system of the mouse can be considered to consist of three compartments: *(1)* bone marrow (BM), *(2)* spleen (*See* **Subheading 1.4.2.**), and *(3)* blood, studies of mobilization are focused primarily on the appearance of increased numbers of HSPC in the blood. As the blood is easily accessed and can be sampled repeatedly, specific hematopoietic and hematologic changes that occur with time can be followed. Mobilization is usually marked by an increase in white blood cell (WBC) cellularity. This can be measured using an electronic blood analyzer, an electronic particle counter, or manually using a hemacytometer. Electronic blood analyzers may also provide information on red blood cell and platelet numbers.

Endpoints used to measure hematopoiesis and which can be used to determine the speed and magnitude of HSPC mobilization in the blood are discussed elsewhere in this publication [e.g., in vitro assays: colony-forming unit (CFU), cobblestone area-forming cell (CAFC), long-term culture-initiating cell (LTC-IC) assays; and in vivo: repopulation and competitive repopulation studies]. At the end of mobilization experiments, the cellularities of blood, spleen, and BM can be determined and the redistribution of HSPC between these compartments with mobilization investigated.

A more mechanistic approach to mobilization might require that other analytical techniques also be considered as endpoints. These might include the isolation of plasma for ELISA and/or zymography, which might reveal the role of downstream factors (e.g., enzymes and/or secondary cytokines) in the mobilization process; flow cytometry of cells in the blood, spleen, and BM compartments of the hematopoietic system, which might reveal the presence of specific HSPC subsets; the up/downregulation of cell-surface antigens (e.g., adhesion molecules) and their cell-cycle kinetics; and analysis of mRNA from isolated cells, which might reveal changes in gene expression as a consequence of mobilization.

1.4. Caveats

1.4.1. Species Specificity of Factors and Strain Differences in Responses

Although the use of mouse models provides valuable insights into the in vivo action of specific factors, it carries with it a number of important *caveats*. Whereas certain factors and mechanisms may be conserved between species (demonstrating their evolutionary importance), other factors may be species-specific. Furthermore, the concentration of a given factor required to generate similar biologic responses between species may be markedly different. Even between strains of mice, there are differences in the magnitude of mobilization in response to a given mobilization stimulus (e.g., DBA2 > Balb/c > C57Bl/6) *(27)*. These differences may ultimately allow a better understanding of what genetic factors regulate mobilization and reveal what measures may be predictive of a good or poor mobilizer. Furthermore, genetically engineered, or mutant, mice are often generated in less well-characterized wild-type strains making rigorous controls all the more important

1.4.2. The Extramedullary Role of the Spleen in Mice

Under normal (non-pathologic) conditions in man, active hematopoiesis is limited to the bone cavities of the vertebrae, cranium, sternum, ribs, and pelvis. Other bone cavities are filled with fatty marrow. In contrast, under normal conditions in mice, the majority of BM is actively hematopoietic, and as a consequence, hematopoietic stress or stimulation results in extramedullary hematopoiesis (outside of the BM). This occurs primarily in the spleen, leading to splenomegaly. The spleen therefore has an important role in murine hematopoiesis. As a consequence, to better mimic the changes that may occur in the human hematopoietic system where the spleen is less important, some researchers use surgically splenectomized mice in their studies *(28)*.

1.4.3. The Use of Genetically Engineered or Mutant Mice

Genetically engineered and mutant mice provide the researcher with a powerful tool with which to investigate the mechanism(s) of HSPC mobilization. However, an important consideration is that while an experiment using available KO, or mutant, mice can be designed with a very specific question and targeted endpoint, nature may confound these efforts. As most KO and mutant mice are viable, healthy organisms, it is possible that the absence of the specific molecule of interest may be compensated by the upregulation of a molecule with similar, albeit not identical activity. This compensatory upregulation may ultimately obscure any impact of the deficient molecule. Consequently, although all evidence may point to the critical role of a specific molecule in the mobilization process in wild-type mice (e.g., absence of mobilization when treated with a specific neutralizing antibody), a genetically engineered KO population (specific molecule absent) might still mobilize as effectively as a wild-type cohort, leading to the conclusion that the molecule does not in fact play a critical role. The use of conditional KO models, where specific genes can be selectively inactivated by manipulation of the *Cre-loxP* system, may reduce the impact of compensatory mechanisms. However, the use of such models is limited by their availability.

2. Materials

2.1. Mice and Mouse Vendors

Established strains of mice are usually readily available from vendors (*The Jackson Laboratory*, 600 Main Street, Bar Harbor, ME. http://www.jax.org; *Taconic*, One Hudson City Centre, Hudson, NY. http://www.taconic.com; *Charles River Laboratories, Inc.*, 251 Ballardvale Street, Wilmington, MA. http://www.criver.com; *Harlan Bioproducts for Science, Inc.*, P.O. Box 29176 Indianapolis, IN. http://www.harlan.com). Splenectomized mice may also be available with the surgery performed by the suppliers for an additional cost. Vendors are constantly developing more exotic strains of mice and are making these available to researchers.

Less well-established mouse strains, including mutant mice and those generated by KO or KI technologies, might only be available in limited numbers from other research establishments, or directly from investigators. As an option, breeding pairs may be provided and the generation of sufficient similarly aged and sexed mice from these breeding animals and management of the breeding colony within available space limitations becomes your direct concern as does ensuring the genotype of offspring with respect to any mutation (*see* **Note 1-4**).

When using genetically modified mice, careful attention should be paid to the selection of appropriate wild-type controls as differences in the magnitude of mobilization in response to a given mobilization stimulus exist between different mouse strains.

Established mouse strains are sold by age, sex, and weight and are usually readily available from established vendors. More exotic strains might require special ordering. Mice should be between 8 and 12 weeks of age (unless age is a variable under evaluation) at the time of experimentation. Upon arrival, mice will usually require 1 week of acclimation, which should be taken into account if a specific age at the time of treatment is required. Unless sex is one variable to be investigated, male or female mice should be chosen. Female mice are generally more docile than male mice, and although, alpha females may barber, clip fur of, subordinate females in a group, alpha males may fight with and severely injure and/or mutilate subordinate males. Alpha males should be identified and separated as soon as possible as persistent injuries can lead to stress and infection and adversely influence carefully planned experiments.

2.2. Mobilization Agents

Mobilizing agents that are used clinically (e.g., G-CSF—Neupogen Filgrastim, Amgen, Inc., http://www.amgen.com) are available to researchers through, for example, hospital pharmacies. Most research-grade growth factors, cytokines, and chemokines can be purchased from established vendors (e.g., *R&D Systems*, 614 McKinley Place NE, Minneapolis, MN. http://www.rndsystems.com, or *CellGenix*, 602 Hillside Avenue, Antioch, IL. http://www.cellgenix.com). Factors are usually provided in a lyophilized form. Because factors are usually supplied as 10- to 50-µg aliquots, special and/or bulk orders obtained through direct contact with companies may be required to minimize costs and to ensure sufficient factor(s) are available to complete the proposed study. The cost of sufficient factor(s) to complete a study is an important consideration, especially for those factors that are short-lived in vivo and/or require repeated administration. The availability of sufficient specialized and less readily available reagents (e.g., specialized antibodies, inhibitors and drugs) should also be an important consideration when planning experiments.

3. Methods

3.1. Training

Possibly the most important part of working confidently with animals is training in the techniques that will be used [e.g., methods of handling and restraint and administration of agents [subcutaneous (SC), intraperitoneal (IP), or intramuscular (IM)]. Most animal facilities will provide rudimentary training in handling and restraint skills. It is up to the individual researcher to learn and develop expertise in the procedures required.

3.2. Treatment Groups

3.2.1. Treatment Groups and Controls

Animals of the same sex and age should be approximately the same weight. Any significant outliers (alpha dominant mice or runts) should be excluded from experiments. Unless the amount of the factor under evaluation is limiting, 5 mice *per* group is a good target, although more mice *per* group improves statistical power as does the performance of replicate experiments. Appropriate controls of the same age, sex, and weight must be treated identically to the treatment group, including the injection of an identical amount of diluent, such that the only difference is the administration of the test agent.

In an experiment where the ability of a factor (Factor X) to induce HSPC mobilization is under investigation, identical mice receiving only diluent (no Factor X) provide a good control. A simple statistical analysis between the hematologic and hematopoietic indices of the two groups (steady-state and treated with Factor X) will yield important information about HSPC mobilization by Factor X.

In performing experiments with genetically modified mice, it is important to realize that there are a number of additional controls to consider. As above, in the experimental group, whereas genetically modified mice would receive Factor X, one control would be identical genetically modified mice receiving only diluent (no Factor X). A simple statistical analysis between the hematologic and hematopoietic indices of the two groups (steady-state and treated with Factor X) will yield important information about HSPC mobilization by Factor X in the genetically modified mice. However, an additional important consideration is the use of a wild-type control. These wild-type controls will be treated identically to the genetically modified groups. One group will receive Factor X and one will receive diluent (no Factor X). Wild-type controls reveal information about the specific impact of the genetic modification during both steady-state hematopoiesis and following Factor X administration. The inclusion of this important control immediately doubles the size of the proposed experiment.

The selection of appropriate controls when using genetically modified mice (mutant, KO or KI) requires careful consideration. Ideally, litter mates not carrying the genetic modification are the best control for comparison with those carrying the genetic modification. However, once a specific genetic modification has been backcrossed for more than 10 generations, the original wild-type strain can be used as a control.

3.2.2. Calculation of Dose

The majority of mobilization protocols administer agents on a weight of agent *per* unit body weight of recipient. Each group of mice should be weighed and a mean ± standard deviation (SD) calculated. The SD should be small indicating that the individuals within the group are closely matched. The mean dose of factor to be administered *per* individual in the group can then be calculated and appropriate stock and working solutions prepared. Ensure that the concentration of the working solution is such that the volume of material to be administered is compatible with the route of administration.

3.2.3. Preparation of Stock and Working Solutions

Information provided with the agent of interest by the supplier should be thoroughly reviewed. Use of an appropriate diluent is important. This diluent will also be used to inject control animals. The presence of protein (e.g., serum albumin) in the diluent is sometimes indicated to avoid adsorption to plastic. The use of polypropylene rather than polystyrene plastic ware will also reduce adsorption. Some lyophilized materials may already contain protein and salts and only require the addition of sterile water. Some factors that are supplied as a liquid may precipitate if diluted in saline. Maximum recommended dilutions are often indicated, and repeated freeze/thaw cycles of stock solutions are to be avoided. Stock solutions should be prepared, aliquoted, clearly labeled, and frozen. Observe manufacturer's instructions regarding the storage of materials provided as a liquid (e.g., G-CSF Neupogen Filgrastim, Amgen). Such materials will only usually require refrigeration. Be careful to maintain sterility of stock solutions that are used repeatedly.

3.2.4. Administration of Factor

Mobilizing agents can be administered IP or SC in mice as single injections, or as once- or twice-daily injections for multiple days. IM injections are rare in the mouse as muscles are small. Mobilizing agents are rarely administered IV. Alternatively, mini-osmotic pumps (Alzet Osmotic Pumps, Cupertino, CA, USA) can be loaded with a factor of interest and surgically implanted SC to deliver sustained levels of a factor over 7–14 days, or factors can be administered SC in sustained release gel formulations *(21,26)*.

3.2.5. IP Injection

For IP injection, the animal should be restrained and held on its back in a slightly head down position. This allows peritoneal viscera to slide away from the injection site into the peritoneal cavity. Held perpendicular to the skin surface, a preloaded 1-ml syringe with a 0.75-inch, 25-G needle can then be used to puncture the skin and peritoneum of the abdomen. The appropriate volume of the drug contained in diluent can then be delivered and the needle withdrawn. Relatively large volumes of drug and diluent can be delivered by this route. Only penetration sufficient to enter the peritoneum is required. Too deep may cause damage to underlying viscera and/or organs. If repeated injections are to be administered over time, the injection site can be alternated between left and right of the midline to avoid excessive irritation to any one injection site.

3.2.6. SC Injection

There are a number of sites recommended for the SC administration of agents. With the mouse restrained and on its back, an area of loose skin is evident where the hind limb meets the abdomen. Repeated injections should be alternated between right and left hind limbs. With a pre-loaded syringe and 0.75-inch, 25-G held at a shallow angle (almost parallel) to the skin, the skin can be punctured and agent gently delivered. The volume administered SC should be less than 200 µl as an abscess is being generated. Withdrawal of the needle is often associated with the efflux of a small volume of the drug from the injection site.

3.2.7. IM Injection

IM injections in mice are difficult, as there are no large muscle groups. If required, low volume (e.g., 10 µl) IM injections can be made into the muscles (quadriceps) surrounding the femur. Excessive muscle damage may lead to local bleeding. Repeated injections should be alternated between right and left limbs.

3.2.8. Post-injection

After injection, mice should be gently placed to a new cage to avoid accidental mixing of injected and non-injected mice. Mice should be monitored immediately after injection and should return to normal behavior (e.g., grooming) within a short time. Generally, the administration of agents widely investigated for stem cell mobilization is not associated with acute or chronic adverse events.

3.3. In vivo Assay During Mobilization

The extent and rapidity of HSPC mobilization can be monitored by analysis of the number of HSPC in the blood. Obtaining blood before and after administration of a mobilizing agent allows for the dynamic measurement of hematologic (RBC, WBC, and platelet counts) and hematopoietic indices (flow cytometry and hematopoietic assays). It is important not to overbleed mice, as this will itself induce hematopoietic stress. Between 100 and 200 μl of blood (approximately 5–10% total blood volume depending on the age and strain of the mice) can be drawn at one time on a weekly basis. Alternatively, smaller volumes (20–40 μl) can be drawn on a daily or every-other-day basis, to follow changes in WBC cellularity and hematopoiesis with time after injection of factor(s).

Samples of plasma for analysis by ELISA or zymography can be obtained by centrifugation of larger volumes of blood (100–200 μl) and can be aliquoted and frozen until assayed. ELISA and/or zymographic analysis of plasma may reveal changes in the levels of downstream molecules following the administration of a mobilizing agent. RNA can also be isolated from isolated WBC to enable analysis of gene expression during the mobilization process using established molecular biology protocols. If sufficient cells are available, flow cytometry may also reveal specific phenotypic changes (e.g., expression of cell adhesion molecules) occurring during mobilization.

3.3.1. Obtaining Blood

Some researchers cut the tip of the tail using a sharp scalpel blade to release blood for collection into heparinized capillary tubes. However, this is a painful procedure for the mouse, and bleeding continues after the blood has been collected. "Vetbond" (clinical grade superglue) can be used to rapidly seal the open wound and assist with healing. An effective alternative is to lightly anesthetize the mouse using an inhaled anesthetic (e.g., halothane or isoflurane) and to draw blood from the retro-orbital plexus. Mice are sufficiently anesthetized when a toe-pinch does not elicit a pain withdrawal reflex of the limb.

Using a sterilized microcapillary tube, Drummond Scientific Company has a range of calibrated micropipets Catalog no. 2-000-001 to 2-000-200). The eyeball of the anesthetized mouse can be gently displaced and the retro-orbital venous plexus gently disrupted by a gentle rotation of the microcapillary tube. Blood flow is usually rapid and blood is collected into a known volume of heparin (Drummond Scientific Company provides an aspirating tube to expel the blood from the microcapillary tube). Once sufficient blood (10–200 μl) has been withdrawn, allowing the eyeball gently back into its socket rapidly staunches the blood flow. It is important to alternate eyes over repeated bleeds to avoid excessive injury to any one retro-orbital plexus. The mouse will rapidly regain consciousness after anesthesia and should soon be observed to groom and return to normal activity after the procedure. Antibiotic gel can be applied to the eye to reduce opportunistic infection and to avoid excessive drying; however, in the authors' experience, the material is rapidly removed and consumed by the mouse as it grooms.

3.3.2. Analysis of Blood

3.3.2.1. HEMATOLOGIC INDICES

An electronic cell counter with veterinary settings (e.g., Careside H-2000, Serono Baker Hematology Series Cell Counter, or Sysmex Automated Hematology Analyzer) and the use of appropriate calibration controls allows a rapid, reproducible measure of RBC, WBC, and platelet numbers using a relatively small volume of blood. Some machines may also provide a WBC differential count (granulocyte, monocyte, and lymphocyte) to allow the measurement of changes in WBC populations with mobilization. A Coulter Counter (Beckman Coulter: http://www.beckmancoulter.com) can provide an alternative for electronic particle counting providing WBC data. If an electronic cell counter is not available, WBC can be enumerated by dilution of blood in a 2% acetic acid solution that lyses RBC (crystal violet can be added to the solution as a counter stain). In addition, blood smears can be prepared and stained histochemically (e.g., Diff-Quik stain set, Dade Behring, Newark, DE, USA) for manual WBC differential counts.

3.3.2.2. HEMATOPOIETIC INDICES

A known volume of whole blood is collected into a known volume of heparin and dilution factor noted, or else whole blood is collected using heparinized microcapillary tubes (available through Drummond Scientific Company). WBC cellularity is determined and the CFU content of heparinized blood determined by adding 10 µl of heparinized blood to a 3-ml aliquot of methylcellulose-based culture medium supplemented with hematopoietic growth factors (StemCell Technologies, http://www.stemcell.com; StemCell Technologies offers a variety of pre-aliquoted methylcellulose culture media supplemented with different growth factor cocktails. For example, MethoCult GF M3434 contains SCF, IL-3, IL-6, and EPO and supports the proliferation of murine BFU-E, CFU-GM, CFU-G, CFU-M, CFU-GEMM). One milliliter aliquots of MethoCult GF M3434 containing the cells from approximately 3.3 µl of blood are plated *per* 30-mm culture dish and cultured as *per* manufacturer's protocol (The details of the CFU assay are discussed elsewhere in this publication.) The number of CFU *per* dish (i.e., *per* 3.3 µl blood) can be determined and converted to a number of CFU/ml blood to reveal the magnitude of HSPC mobilization.

If larger volumes of blood can be drawn (e.g., 100–200 μl) and/or individual samples pooled for a given treatment group, blood can be processed to yield sufficient WBC to perform a variety of hematopoietic assays, e.g., CFU, CAFC, and/or LTC-IC at a given time point within the mobilization process. The frequency of HSPC from the isolated WBC can be related back to the whole blood to reveal the frequency of HSPC/ml.

Another technique that can be used to measure HSPC frequencies in blood involves the use of hematocrit tubes. A standard 75-mm hematocrit tube contains approximately 75 μl of blood when full. Therefore approximately 75 μl of blood containing anticoagulant can be collected into non-heparinized hematocrit tubes, or 75 μl of fresh blood can be collected into heparinized hematocrit tubes and the blood-filled hematocrit tubes centrifuged using a hematocrit microcentrifuge. After centrifugation, the blood is separated into RBC, buffy coat (WBC), and plasma fractions. By carefully etching the glass just below the buffy coat, the hematocrit tube can be broken such that the buffy coat is retained together with the plasma in one part of the hematocrit tube and the majority of RBC are retained in the other. The WBC comprising the buffy coat can be flushed from the hematocrit tube and assayed for the presence of HSPC activity. These data can be related back to the original 75-μl volume of blood from which they were derived. The frequency of HSPC from the isolated WBC can be related back to the whole blood to reveal the frequency of HSPC/ml.

3.4. Termination of Experiment

As much information as possible needs to be collected from the mice at the end of the experiment. Consider end-points carefully. Outside of immediate hematologic and hematopoietic assays, decide whether it is feasible/warranted to preserve any tissues or parts of tissues for histology or immunohistochemistry. This will require knowledge of appropriate fixation protocols. Generation of a worksheet will assist the recording of data. Gross measures, e.g., total body weight, might reveal changes in the condition of the mice with treatment when compared with that of the controls. Caution: use of CO_2 asphyxia to terminate animals (preferred in some institutions) may impact hematologic and hematopoietic measures. Furthermore, cervical dislocation may impair the collection of blood by cardiac puncture. Instead, animals should be deeply anesthetized (inhalation anesthesia) and fur and skin thoroughly wetted with a sterilizing agent, e.g., 70% ethanol, before processing.

3.4.1. Blood

1. A maximum volume (1–2 ml) of blood should be drawn into heparin in a 15-ml centrifuge tube using sterile technique, either through cardiac puncture (e.g., 3-ml

syringe fitted with a 22-G needle) or retro-orbital bleeding. Measure the volume of blood drawn.
2. RBC, WBC, and platelet counts should be obtained by the use of an electronic blood analyzer; otherwise, manual hemacytometer counts will reveal WBC cellularity. Blood smears can also be prepared for manual WBC differential counting.
3. Plasma can be isolated from the blood by centrifugation and frozen in small aliquots for future assay (e.g., ELISA, zymography, etc.).
4. RBC in the blood can be removed by

(a) Hypotonic lysis
(b) Density gradient separation

However, neither technique will remove nucleated RBC.

3.4.2. Removal of RBC by Hypotonic Lysis

3.4.2.1. AMMONIUM CHLORIDE-BASED RBC-LYSING BUFFER

RBC-lysing buffer contains 0.83% ammonium chloride in 0.01 M Tris (e.g., Sigma R7757 – specifically intended for the removal of RBC from mouse blood).

3.4.2.1.1. Procedure.

1) Add 1 ml of the lysis buffer to the cell pellet and mix gently for 1 min.
2) Lysis is halted by the addition of 15–20 ml of medium (to restore isotonicity).
3) WBC are isolated by centrifugation (e.g. $300 \times g$, 10 min).
4) If RBC lysis is incomplete steps 1–3 can be repeated. WBC cellularities are then determined before hematopoietic assay.

3.4.2.2. HYPOTONIC SHOCK USING COLD, STERILE H_2O

1) Add 9 ml of cold sterile water (e.g., autoclaved ultrapure/mQ water) to a cell pellet, or whole blood, using a 10-ml pipette and thoroughly mix for *only* 5–10 s by drawing the cell suspension repeatedly (e.g., 3 times) through the pipette (*see* **Note 5**).
2) Lysis is halted by the restoration of isotonicity. Add 1 ml of room temperature 10× Dulbecco's phosphate buffered saline and thoroughly mix.
3) To remove RBC from BM or spleen samples, the addition of 4.5 ml of water and restoration of isotonicity with 0.5 ml 10× DPBS is usually sufficient. WBC cellularities are then determined before hematopoietic assay.
4) WBC can be isolated by centrifugation (e.g., $300 \times g$, 10 min).
5) If RBC lysis is incomplete, steps 1–3 can be repeated; however, a small volume of serum should be added after lysis to increase the viscosity of the now isotonic solution before centrifugation.

3.4.3. Density Gradient Separation

This technique utilizes the fact that different components of blood (RBC, granulocytes, lymphocytes, monocytes, and HSPC) have different densities. Use of separation media of appropriate densities allows removal of RBC, granulocytes, platelets, and dead cells from blood and allows the isolation of mononuclear cells containing the HSPC activity. This technique is best suited to the processing of larger volumes of blood (>1 ml). For smaller volumes of blood and to remove RBC from BM and spleen samples, lysis is the method of choice.

1) Dilute blood (containing anticoagulant) two-fold using DPBS or medium.
2) Carefully overlay diluted blood over an equal volume of room temperature Histopaque® 1083 (Sigma). The density of Histopaque® 1083 is 1.083 g/ml at 25°C.
3) Centrifuge at $400 \times g$ for 20 min at room temperature with brake off. Note: During centrifugation: (a) mononuclear cells (lymphocytes, macrophages, and HSPC) will collect at the interface between the 1.083 g/ml Histopaque® lower fraction and diluted plasma upper fraction, (b) RBC will aggregate in the presence of polysucrose (a component of Histopaque) and rapidly sediment to form a pellet, (c) granulocytes are more dense than the mononuclear cells and are driven into the Histopaque, and (d) platelets are less dense than the mononuclear cell fraction and remain in the upper layer.
4) Carefully aspirate the majority of upper layer to waste. This removes the majority of platelets and plasma proteins.
5) Collect the mononuclear cells at the interface into a new tube. Note: Avoid collecting too much of the Histopaque® (lower) fraction when harvesting the interface, as this will increase granulocyte contamination of the mononuclear cells.
6) Wash the mononuclear cell fraction twice to remove residual Histopaque® (Washing consists of the dilution of cells in medium, centrifugation at $300 \times g$ for 10 min and removal of supernatant to waste).
7) After the final wash, cells are resuspended in a known volume of medium and cellularity determined before hematopoietic assay (*see* **Note 6**).

3.4.4. Measurement of HSPC Activity Contained in Blood

Once isolated, WBC should be refrigerated or held on ice until processed further. WBC cellularity is determined (electronically or manually) and HSPC activity measured using hematopoietic assays of choice (in vitro and/or in vivo) and/or phenotypic analysis performed using flow cytometry. Once assays are completed, calculations should be performed to reveal the absolute frequencies of the various hematopoietic progenitors *per* milliliter of blood. An example of how to calculate the number of HSPC *per* milliliter of blood is:

$$\frac{\text{Total number of cells isolated} \times \text{Number of HSPC measured at assay}}{\text{Number of cells plated in HSPC assay} \times \text{Original blood volume collected}}$$

For reference, during steady-state hematopoiesis, 10–100 CFU-GM may be circulating *per* milliliter of blood; however, this is very strain-dependent.

3.4.5. Spleen

1. Spleen should be dissected, weighed, and a single cell suspension generated by gently squashing the spleen in a small amount of cold medium (e.g., using a sterile plunger from a 3-ml syringe and a 30- to 60-mm Petri dish) so as to rupture the spleen capsule and release its contents. All procedures should be performed in a sterile environment. Upon gross examination, splenomegaly associated with the administration of mobilization agents can often be dramatic and is indicative of a transition of the spleen from a lymphoid organ to a site of active hematopoiesis.
2. Cells can be passed gently through a 22-G needle to generate a single cell suspension in a known volume of medium. Use of a 70-μm mesh filter (e.g., 352350, BD Falcon, Bedford, MA, USA) will remove remnants of the capsule and debris from the cell suspension.
3. RBC contamination of spleen cell suspensions does not usually warrant lysis. If required, it can be performed as for blood.
4. Cell suspensions should be refrigerated or held on ice until processed further. The cellularity of the spleen cell suspension can be determined using either an electronic cell counter or a hemacytometer, and total spleen cellularity calculated and hematopoietic measures undertaken (e.g., CFU, CAFC, flow cytometry).
5. Calculations should be performed to reveal spleen cellularities, the frequencies of the various hematopoietic progenitors, and the absolute numbers of the various hematopoietic progenitors *per* spleen.

3.4.6. Bone Marrow

1. BM either from femora or from combined tibia and femora can be isolated. Long bones are removed by dislocation of the hip joint and dissected free of muscle. Denuded bones will desiccate if left open to air. To avoid desiccation, bones should be covered with a small volume of sterile medium in small (e.g., 30 mm), sterile, uniquely labeled Petri dishes and stored refrigerated, or on ice, until processed.
2. Marrow can be obtained from the bone cavities by one of two techniques either by (i) flushing of the marrow cavities or (ii) by grinding of the bones using a sterile mortar and pestle. All procedures should be performed in a sterile environment.

3.4.7. Flushing of the Marrow Cavities

1. The knee joint is dislocated and femur and tibia separated.
2. The ball joint is removed using a scalpel blade and the femur cavity entered through the cartilage-coated area of bone that comprises the knee joint using a 1-ml syringe containing a known volume of cold medium and fitted with a 22-G needle. A gentle drilling action achieved by rolling the syringe and needle between

finger and thumb while holding the bone firmly in a pair of sterile forceps will enable the needle to pop through the cartilage.

3. Flushing medium through the femoral cavity will displace the femoral marrow often as a plug, into a collection tube. (A similar process will remove tibial marrow.) Marrow cavities should be flushed repeatedly to ensure all marrow is removed. Bones will become pale in color as the marrow is removed.

4. Gentle drawing of the marrow plug, or marrow fragments, through the needle and syringe will produce a single cell suspension.

5. Cell suspensions should be refrigerated or held on ice until processed further. RBC contamination of BM cell suspensions does not usually warrant lysis. If required, it can be performed as for blood.

6. BM cellularities can be determined and hematopoietic measures undertaken (e.g., CFU, CAFC, flow cytometry).

7. Calculations should be performed to reveal femur, or lower limb (femur + tibia), cellularities, the frequencies of the various hematopoietic progenitors, and the absolute numbers of the various hematopoietic progenitors *per* femur, or lower limb (femur + tibia).

3.4.8. Grinding of the Bones

1. This technique is especially useful if many bones from one mouse, or bones from many mice, are to be processed. Mortar and pestle apparatus can be sterilized by autoclaving. A small amount of water can be added to the apparatus to generate steam during the sterilization process and the whole wrapped in aluminum foil. The integrity of the aluminum foil over wrap will confirm sterility of the apparatus.

2. For processing, bones are placed in the mortar in a small volume of cold medium and gently but firmly ground using the pestle. Once sufficiently ground, the BM cells can be collected by rinsing the collected debris into a known volume of cold medium using a pipette. Bone fragments will become white once separated from the BM. The grinding action of the pestle is usually sufficient to break aggregates of BM, thereby generating a single cell suspension.

3. Bone fragments can be removed by filtration of the cell suspension through a sterile cell strainer.

4. Aggregates of BM can be dispersed by gentle drawing of the cell suspension through a 22-G needle and syringe.

5. Cell suspensions should be refrigerated or held on ice until processed further. RBC contamination of BM cell suspensions does not usually warrant lysis. If required, it can be performed as for blood. BM cellularities can be determined and hematopoietic measures made (e.g., CFU, CAFC, flow cytometry).

6. To avoid cellular cross-contamination between individual samples using the mortar and pestle, they can be cleaned during use by removal of bone debris to waste using sterile gauze and thorough rinsing with 70% ethanol, then thorough rinsing with sterile cold medium.

7. Calculations should be performed to reveal femur, or lower limb (femur + tibia) cellularities, the frequencies of the various hematopoietic progenitors and the

absolute numbers of the various hematopoietic progenitors *per* femur, or lower limb (femur+tibia).

3.5. Data acquisition and analysis

Experiments will generate a large amount of data from control and treatment groups. To investigate the repopulating potential of the mobilized cells, more advanced experimental measures can also be developed (*see* **Notes 7–8**).

Keep all raw data safe (consider protection against water and fire and back-up electronic data) and process data in a timely manner. For each mouse, there will be total body and spleen weights, hematologic (RBC, WBC, and platelet counts) and hematopoietic data (spleen and BM cellularities, frequencies and absolute numbers of the various hematopoietic progenitors in the blood, spleen, and BM), and flow cytometric data. Subsequent to these data, there will be the possibility of data from the analyses of plasma and isolated RNA. As data accrue, appropriate statistical analyses should be performed between control and treatment groups for each variable. Replicate experiments will add power to your analyses.

4. Notes

1. Discuss your proposed experiment(s) with colleagues especially those that are knowledgeable in the field and are familiar with the use of animals. They are a valuable resource and can usually offer suggestions and guidance. Early critical peer-review of your scientific rationale and experimental design will make your study more effective ensuring that (i) appropriate controls are included, (ii) the number and size of treatment groups and number of replicate experiments provide sufficient statistical power; and (iii) appropriate endpoints are proposed. These considerations may facilitate IACUC approval (*see* **Note 2**).

2. Authorization from the Institutional Animal Care and Use Committee (IACUC) (or similar) of your institution must be obtained before any ordering of animals or experimentation. This step allows critique of proposed study by veterinary and facilities management staff. Approval will be dependent on your demon-strating the use of appropriate approved techniques and confirm that the studies you are proposing meet animal care and use criteria. An IACUC Amendment system allows for the modification of originally described experimental proce-dures. Amendments are subject to approval by the IACUC and are appended to the original file. The complete IACUC documentation (original submission and amendments) should be kept up-to-date and accurately reflect experimental procedures. Annual review and IACUC renewal may be required.

3. All research staff, technicians, and assistants involved in animal studies must receive appropriate training from the animal facility in the handling of animals

and demonstrate proficiency in the techniques proposed for the study. Training is usually certified and monitored by the administration of the animal facility.

4. Genetically engineered mutants are now becoming more widely available. Specific genes can now be modified (KO or KI or conditionally expressed). Provided that the mutations induced are not lethal, viable heterozygous and homozygous offspring can be generated. Breeding may sometimes be more successful between heterozygous parents in which case genotyping of offspring becomes important. The genotype [wild-type (+/+), heterozygous (+/–), or homozygous (–/–)] of individuals can be confirmed by the use of the polymerase chain reaction (PCR), and the presence/absence of functional proteins may be confirmed by Western blot and ELISA if detection antibodies exist, or zymography if the molecule is an enzyme. Mutations may not generate marked phenotypic differences from wild-type emphasizing the need for careful genotyping of individuals. Once genetically typed, individuals can be "labeled" by use of a unique ear punch code, numeric ear tags, or individual radiometric tags. Although genetically engineered mutants provide powerful tools with which to address questions about the specific role(s) of molecules in biological processes, it is important to appreciate that dynamic biological systems are the product of a complex balance of positive and negative influences. As such, it is possible that the phenotypic impact of a genetically modified gene (KO or KI) may be obscured, at least in part, by changes in the compensatory up/downregulation of molecule(s) with similar (albeit not identical) activity.

5. Lysis of RBC is marked by a change from a cloudy RBC "suspension" to a clear solution of hemoglobin.

6. An alternative density separation media designed specifically to isolate mononuclear cells from mouse blood is: Lympholyte®-Mammal (e.g., CL5110), (Cedarlane Labs, Ltd., Hornby, Ontario, Canada; www.cedarlanelabs.com). The procedure is similar to that described for Histopaque® 1083 with the following exceptions: (a) Lympholyte® has components that may separate with storage and should be mixed vigorously before use, (b) it has a density of 1.086 g/ml at 22°C, and (c) centrifugation is performed at $800 \times g$.

7. Transplantation of mobilized WBC into lethally irradiated syngeneic recipients can also be performed. Competitive repopulation studies can also be undertaken, e.g., between BM- and PB-derived WBC. Following hematopoietic reconstitution in vivo in these mice will provide qualitative and quantitative data about the HSPC mobilized by the agent under investigation.

8. RNA can also be extracted from the WBC from the blood, spleen, and BM and compared with that of control samples to yield information about changes in gene-expression that occur during the mobilization process.

Acknowledgments

The authors thank the animal handling, technical, and veterinary staff that assisted with these studies. SNR thanks Dr. J. Graham Sharp, University of

Nebraska Medical Center, Omaha, Nebraska, for his support and advice with the preparation of this chapter.

References

1. Lord, B.I., Woolford, L.B., Wood, L.M., Czaplewski, L.G., McCourt, M., Hunter, M.G., and Edwards, R.M. (1995) Mobilization of early hematopoietic progenitor cells with BB-10010: a genetically engineered variant of human macrophage inflammatory protein-1 alpha. *Blood* **85**, 3412–3415.
2. Laterveer, L., Lindley, I.J., Hamilton, M.S., Willemze, R., and Fibbe, W.E. (1995) Interleukin-8 induces rapid mobilization of hematopoietic stem cells with radio-protective capacity and long-term myelolymphoid repopulating ability. *Blood* **85**, 2269–2275.
3. Pelus, L.M., Horowitz, D., Cooper, S.C., and King, A.G. (2002) Peripheral blood stem cell mobilization. A role for CXC chemokines. *Crit. Rev. Oncol. Hematol.* **43**, 257–275.
4. Pelus, L.M., Bian, H., King, A.G., and Fukuda, S. (2004) Neutrophil-derived MMP-9 mediates synergistic mobilization of hematopoietic stem and progenitor cells by the combination of G-CSF and the chemokines GRObeta/CXCL2 and GRObetaT/CXCL2delta4. *Blood* **103**, 110–119.
5. Flomenberg, N., Devine, S.M., Dipersio, J.F., Liesveld, J.L., McCarty, J.M., Rowley, S.D., Vesole, D.H., Badel, K., and Calandra, G. (2005) The use of AMD3100 plus G-CSF for autologous hematopoietic progenitor cell mobilization is superior to G-CSF alone. *Blood.* **106**, 1867–1874.
6. Hestdal, K., Jacobsen, S.E., Ruscetti, F.W., Dubois, C.M., Longo, D.L., Chizzonite, R., Oppenheim, J.J., and Keller, J.R. (1992) In vivo effect of interleukin-1 alpha on hematopoiesis: role of colony-stimulating factor receptor modulation. *Blood.* **80**, 2486–2494.
7. Molendijk, W.J., van Oudenaren, A., van Dijk, H., Daha, M.R., and Benner, R. (1986) Complement split product C5a mediates the lipopolysaccharide-induced mobilization of CFU-s and haemopoietic progenitor cells, but not the mobilization induced by proteolytic enzymes. *Cell Tissue Kinet.* **19**, 407–417.
8. Flomenberg, N., DiPersio, J., and Calandra, G. (2005) Role of CXCR4 chemokine receptor blockade using AMD3100 for mobilization of autologous hematopoietic progenitor cells. *Acta Haematol.* **114**, 198–205.
9. Neben, S., Marcus, K., and Mauch, P. (1993) Mobilization of hematopoietic stem and progenitor cell subpopulations from the marrow to the blood of mice following cyclophosphamide and/or granulocyte colony-stimulating factor. *Blood* **81**, 1960–1967.
10. Mauch, P., Lamont, C., Neben, T.Y., Quinto, C., Goldman, S.J., and Witsell, A. (1995) Hematopoietic stem cells in the blood after stem cell factor and interleukin-11 administration: evidence for different mechanisms of mobilization. *Blood* **86**, 4674–4680.

11. Socinski, M.A., Cannistra, S.A., Elias, A., Antman, K.H., Schnipper, L., and Griffin, J.D. (1988) Granulocyte-macrophage colony stimulating factor expands the circulating haemopoietic progenitor cell compartment in man. *Lancet* **1**, 1194–1198.

12. Lie, A.K., Rawling, T.P., Bayly, J.L., and To, L.B. (1996) Progenitor cell yield in sequential blood stem cell mobilization in the same patients: insights into chemotherapy dose escalation and combination of haemopoietic growth factor and chemotherapy. *Br. J. Haematol.* **95**, 39–44.

13. Jackson, J.D., Yan, Y., Brunda, M.J., Kelsey, L.S., and Talmadge, J.E. (1995) Interleukin-12 enhances peripheral hematopoiesis in vivo. *Blood* **85**, 2371–2376.

14. Kessinger, A., Bishop, M.R., Jackson, J.D., O'Kane-Murphy, B., Vose, J.M., Bierman, P.J., Reed, E.C., Warkentin, P.I., Armitage, J.O., and Sharp, J.G. (1995) Erythropoietin for mobilization of circulating progenitor cells in patients with previously treated relapsed malignancies. *Exp. Hematol.* **23**, 609–612.

15. Grzegorzewski, K., Komschlies, K.L., Mori, M., Kaneda, K., Usui, N., Faltynek, C.R., Keller, J.R., Ruscetti, F.W., and Wiltrout, R.H. (1994) Administration of recombinant human interleukin-7 to mice induces the exportation of myeloid progenitor cells from the bone marrow to peripheral sites. *Blood* **83**, 377–385.

16. Schwarzenberger, P., Huang, W., Oliver, P., Byrne, P., La, R., V, Zhang, Z., and Kolls, J.K. (2001) Il-17 mobilizes peripheral blood stem cells with short- and long-term repopulating ability in mice. *J. Immunol.* **167**, 2081–2086.

17. Andrews, R.G., Bensinger, W.I., Knitter, G.H., Bartelmez, S.H., Longin, K., Bernstein, I.D., Appelbaum, F.R., and Zsebo, K.M. (1992) The ligand for c-kit, stem cell factor, stimulates the circulation of cells that engraft lethally irradiated baboons. *Blood* **80**, 2715–2720.

18. Bodine, D.M., Seidel, N.E., Zsebo, K.M., and Orlic, D. (1993) In vivo administration of stem cell factor to mice increases the absolute number of pluripotent hematopoietic stem cells. *Blood* **82**, 445–455.

19. Brasel, K., McKenna, H.J., Morrissey, P.J., Charrier, K., Morris, A.E., Lee, C.C., Williams, D.E., and Lyman, S.D. (1996) Hematologic effects of flt3 ligand in vivo in mice. *Blood* **88**, 2004–2012.

20. Robinson, S., Mosley, R.L., Parajuli, P., Pisarev, V., Sublet, J., Ulrich, A., and Talmadge, J. (2000) Comparison of the hematopoietic activity of flt-3 ligand and granulocyte-macrophage colony-stimulating factor acting alone or in combination. *J. Hematother. Stem Cell Res.* **9**, 711–720.

21. Robinson, S.N., Chavez, J.M., Pisarev, V.M., Mosley, R.L., Rosenthal, G.J., Blonder, J.M., and Talmadge, J.E. (2003) Delivery of Flt3 ligand (Flt3L) using a poloxamer-based formulation increases biological activity in mice. *Bone Marrow Transplant.* **31**, 361–369.

22. Craddock, C.F., Nakamoto, B., Andrews, R.G., Priestley, G.V., and Papayannopoulou, T. (1997) Antibodies to VLA4 integrin mobilize long-term repopulating cells and augment cytokine-induced mobilization in primates and mice. *Blood* **90**, 4779–4788.

23. Papayannopoulou, T. and Nakamoto, B. (1993) Peripheralization of hemopoietic progenitors in primates treated with anti-VLA4 integrin. *Proc. Natl. Acad. Sci. U.S.A.* **90**, 9374–9378.

24. Kikuta, T., Shimazaki, C., Ashihara, E., Sudo, Y., Hirai, H., Sumikuma, T., Yamagata, N., Inaba, T., Fujita, N., Kina, T., and Nakagawa, M. (2000) Mobilization of hematopoietic primitive and committed progenitor cells into blood in mice by anti-vascular adhesion molecule-1 antibody alone or in combination with granulocyte colony-stimulating factor. *Exp. Hematol.* **28**, 311–317.

25. Molineux, G., Kinstler, O., Briddell, B., Hartley, C., McElroy, P., Kerzic, P., Sutherland, W., Stoney, G., Kern, B., Fletcher, F.A., Cohen, A., Korach, E., Ulich, T., McNiece, I., Lockbaum, P., Miller-Messana, M.A., Gardner, S., Hunt, T., and Schwab, G. (1999) A new form of Filgrastim with sustained duration in vivo and enhanced ability to mobilize PBPC in both mice and humans. *Exp. Hematol.* **27**, 1724–1734.

26. Robinson, S.N., Chavez, J.M., Blonder, J.M., Pisarev, V.M., Mosley, R.L., Sang, H., Rosenthal, G.J., and Talmadge, J.E. (2005) Hematopoietic progenitor cell mobilization in mice by sustained delivery of granulocyte colony-stimulating factor. *J. Interferon Cytokine Res.* **25**, 490–500.

27. Kessinger, A., Mann, S., Murphy, B.O., Jackson, J.D., and Sharp, J.G. (2001) Circulating factors may be responsible for murine strain-specific responses to mobilizing cytokines. *Exp. Hematol.* **29**, 775–778.

28. Molineux, G., Pojda, Z., and Dexter, T.M. (1990) A comparison of hematopoiesis in normal and splenectomized mice treated with granulocyte colony-stimulating factor. *Blood* **75**, 563–569.

29. Hoglund, M., Smedmyr, B., Simonsson, B., Totterman, T., and Bengtsson, M. (1996) Dose-dependent mobilisation of haematopoietic progenitor cells in healthy volunteers receiving glycosylated rHuG-CSF. *Bone Marrow Transplant.* **18**, 19–27.

30. Hoglund, M., Smedmyr, B., Bengtsson, M., Totterman, T.H., Cour-Chabernaud, V., Yver, A., and Simonsson, B. (1997) Mobilization of CD34+ cells by glycosylated and nonglycosylated G-CSF in healthy volunteers–a comparative study. *Eur. J. Haematol.* **59**, 177–183.

31. Christopherson, K.W., Cooper, S., and Broxmeyer, H.E. (2003) Cell surface peptidase CD26/DPPIV mediates G-CSF mobilization of mouse progenitor cells. *Blood* **101**, 4680–4686.

32. Christopherson, K.W., Cooper, S., Hangoc, G., and Broxmeyer, H.E. (2003) CD26 is essential for normal G-CSF-induced progenitor cell mobilization as determined by CD26-/- mice. *Exp. Hematol.* **31**, 1126–1134.

33. Pruijt, J.F., van Kooyk, Y., Figdor, C.G., Lindley, I.J., Willemze, R., and Fibbe, W.E. (1998) Anti-LFA-1 blocking antibodies prevent mobilization of hematopoietic progenitor cells induced by interleukin-8. *Blood* **91**, 4099–4105.

34. Katayama, Y., Battista, M., Kao, W.M., Hidalgo, A., Peired, A.J., Thomas, S.A., and Frenette, P.S. (2006) Signals from the sympathetic nervous system regulate hematopoietic stem cell egress from bone marrow. *Cell* **124**, 407–421.

35. Sweeney, E.A., Priestley, G.V., Nakamoto, B., Collins, R.G., Beaudet, A.L., and Papayannopoulou, T. (2000) Mobilization of stem/progenitor cells by sulfated polysaccharides does not require selectin presence. *Proc. Natl. Acad. Sci. U.S.A.* **97**, 6544–6549.
36. Schwartzberg, L.S., Birch, R., Hazelton, B., Tauer, K.W., Lee, P., Jr., Altemose, R., George, C., Blanco, R., Wittlin, F., and Cohen, J. (1992) Peripheral blood stem cell mobilization by chemotherapy with and without recombinant human granulocyte colony-stimulating factor. *J. Hematother.* **1**, 317–327.
37. Pelus, L.M., Bian, H., Fukuda, S., Wong, D., Merzouk, A., and Salari, H. (2005) The CXCR4 agonist peptide, CTCE-0021, rapidly mobilizes polymorphonuclear neutrophils and hematopoietic progenitor cells into peripheral blood and synergizes with granulocyte colony-stimulating factor. *Exp. Hematol.* **33**, 295–307.

4

Transmigration of Human CD34⁺ Cells

Seiji Fukuda and Louis M. Pelus

Summary

Understanding mechanisms responsible for engraftment of hematopoietic stem cells (HSC) is important to achieve successful HSC transplantation. Homing of HSC to the bone marrow niche is believed to be a crucial step for engraftment. However, the molecular mechanisms that regulate HSC homing are not understood well. Migration of HSC in response to cytokines and chemokines is believed to be one of the critical steps of HSC homing and mobilization. Evaluating the migration of HSC will help to understand the mechanisms responsible for their homing and/or mobilization. In this chapter, we will describe the methodology utilized in our laboratory to evaluate migration of human CD34⁺ cells that contain HSC.

Key Words: CD34⁺ cells; chemotaxis assay; SDF1/CXCR4; transmigration.

1. Introduction

Hematopoietic stem cells (HSC) require an appropriate marrow niche to differentiate and self renew *(1,2)*, and only HSC that home to marrow contribute to long-term repopulation *(3,4)*. Homing and mobilization of hematopoietic stem and progenitor cells (HSPC) are regulated by hematopoietic cytokines *(5–7)*, chemokines *(8–10)*, and adhesion molecules *(11,12)*; however, the regulatory interactions mediated through these molecules are poorly understood. The total number of HSC within the hematopoietic graft is limited, and only pluripotential HSC that lodge to the appropriate bone marrow niche contribute to long-term hematopoiesis. Thus, one way to obtain superior engraftment is to enhance homing of HSC to the marrow microenvironment. Increasing

From: *Methods in Molecular Biology, vol. 430: Hematopoietic Stem Cell Protocols*
Edited by: K. D. Bunting © Humana Press, Totowa, NJ

the efficacy of engraftment is clinically relevant in order to minimize the leukocytopenia and thrombocytopenia that follows HSC transplantation as a consequence of myelosuppressive preconditioning. Enhanced homing is also important for umbilical cord blood cell transplantation where the HSC number is limiting, particularly for adult recipients. To increase the engraftment efficacy of HSC, it is imperative to understand the mechanism responsible for HSC trafficking.

1.1. Role of SDF1α/CXCR4 in CD34⁺ Cell Trafficking and Migration

The G-protein-coupled chemokine receptor CXCR4 is expressed on primitive HSPC, and its ligand CXCL12 (SDF1) is expressed on cells within the marrow microenvironment, including stromal cells *(13)* and osteoblasts *(14,15)*. Interaction between SDF1 and CXCR4 is believed to play an important role in retention, homing, mobilization, and survival of HSPC *(6,10,16–18)*. Although redundancy between chemokines and their receptors occurs for the majority of chemokine/chemokine receptor interactions, CXCR4 is the only receptor for SDF1. SDF1 can attract primitive hematopoietic cells that express CXCR4 to the marrow microenvironment *(6,16)*, whereas disruption of SDF1/CXCR4 interaction within marrow can under appropriate circumstances facilitate their mobilization to the peripheral circulation *(10,18)*. Other chemokine receptors expressed within the primitive hematopoietic compartment are CCR3, CCR9 *(19)*, CCR5, CXCR1, CXCR2 *(20)*, and CXCR3 *(21)*. However, there is no clear evidence that these chemokine receptors are involved in HSPC migration. Enhanced in vitro transmigration of CD34⁺ cells to SDF1 is associated with hematopoietic recovery *(22)*. SDF1 also activates the adhesion receptors, very late antigen-4 (VLA-4) and lymphocyte function associated antigen-1 (LFA-1) on HSPC, which can contribute to the homing process *(23)*. Studies have shown that SDF1/CXCR4 signaling and migration of CD34⁺ cells are regulated by hematopoietic cytokines such as stem cell factor *(16)* and Flt3 ligand *(24)*. In contrast, other reports suggest that the SDF1/CXCR4 axis may not be absolutely required for engraftment and homing by all populations of HSC *(25–27)*. Nevertheless, migration of HSPC to SDF1 is believed to be one of the major mechanisms for homing and understanding their in vitro migratory behavior, and the components involved in migration of transplantable HSC will facilitate development of novel strategies to enhance HSPC homing.

Chemoattractants may be present in bone marrow and in peripheral blood at different concentration, creating a concentration gradient. Depending on the concentration, distribution, or the combination of the chemoattractants, cell migratory behavior may vary. The presence of two or more chemoattractants in the local area may synergistically increase or decrease the migratory behavior of the cells, whereas the presence of chemoattractants in the opposite direction may be antagonistic and inhibit migration. The presence of a chemoattractant gradient appears to be required for initiating cell migration. To test migratory behavior in vitro, different gradients of chemoattractant are created using transwell systems, i.e., zero (+/+), negative (+/–), and positive (–/+) gradients. This chapter describes the routine techniques used in our laboratory for evaluating migration of CD34⁺ cells or other populations of human or mouse HSPC.

2. Materials

1. Human cord blood, bone marrow, or mobilized peripheral blood (*see* **Note 1**)
2. Ficoll-Paque™ PLUS (GE Health Care, Piscataway, NJ)
3. PBS
4. 0.5% BSA, 2 mM EDTA, in PBS
5. Direct CD34⁺ cell isolation kit (Milteny Biotec, Auburn, CA)
6. MS or LS magnetic column (Milteny Biotec)
7. MACS™ Separator (MiniMACS, OctoMACS or MidiMACS from Milteny Biotec)
8. 0.5% BSA in RPMI-1640
9. 0.25% Triton-X 100; 1% BSA in PBS
10. Transwells (6.5 mm diameter, 5.0 μm pore size, Costar 3421, Corning, NY)
11. Calcein/AM (Molecular Probes, Eugene, OR)
12. Polycarbonate membranes (5.7 mm diameter, 5 μm pore size; NeuroProbe, Gaithersburg, MD)

3. Methods

3.1. Isolation of CD34⁺ Cells from Umbilical Cord Blood

The most common sources of cells for HSC transplantation are bone marrow, umbilical cord blood (UCB), and mobilized blood. The use of UCB CD34⁺ cells for migration assay may be the easiest option owing to their availability (*see* **Note 1**).

1. Dilute cord blood with PBS at 1:3–1:4. Gently overlay 35 ml diluted blood onto 15 ml Ficoll-Paque™ PLUS in a 50-ml centrifuge tube (do not disturb the Ficoll and blood layer).
2. Centrifuge tubes at $400 \times g$ at room temperature (RT) in a swinging bucket rotor without brake for 30 min.
3. Carefully aspirate off the top layer and collect the mononuclear layer at the interface (*see* **Note 2**).
4. Wash monolayer cells with 0.5% BSA/2 mM EDTA in PBS twice to remove Ficoll. If cells are still floating in the supernatant after centrifugation, collect supernatant in a separate tube and re-spin (*see* **Note 3**).

5. Spin down and collect the cells in 15 ml conical tubes (*see* **Note 3**).
6. Aspirate the supernatant and re-suspend the cell pellet in 300 µl of 0.5% BSA/2 mM EDTA in PBS. (Optional: cell number may be counted at this step.)
7. Add Fc blocking solution (100 µl/10^8 cells) supplied in the direct CD34$^+$ cell isolation kit. Mix well.
8. Add anti-CD34 antibody (supplied: 100 µl/10^8 cells). Mix well and incubate tubes at 6–12 °C for 15 min (on ice or placing the tubes in the refrigerator also works fine in our experience).
9. Wash cells with 10 ml of 0.5% BSA/2 mM EDTA in PBS and centrifuge for 10 min at 450 g. If cells are still floating in the supernatant after centrifugation, collect supernatant in a separate tube and spin again. Aspirate supernatant and re-suspend cell pellet in 500 µl of 0.5% BSA, 2 mM EDTA in PBS.
10. Option: Cells may be filtered using MACS pre-separation filters to remove clumps. This prevents cells from clogging the columns during separation but may cause loss of some cells (*see* **Note 5**).
11. Set the MS or LS column on a MACS™ Separator (magnet) and wash columns once with 1 or 5 ml of 0.5% BSA/2 mM EDTA in PBS, respectively (*see* **Note 6**).
12. Apply cells onto the column (*see* **Note 5**).
13. Rinse columns with 0.5 or 2.5 ml (MS and LS column, respectively) of 0.5% BSA/2 mM EDTA in PBS three times (*see* **Note 7**).
14. Remove columns from the magnet and place into appropriate tubes. Elute cells with 0.5% BSA/2 mM EDTA in PBS (1 or 3 ml for MS and LS column, respectively). Using a plunger flush columns again with the same volume of buffer to increase the recovery.
15. To maximize the purity of CD34$^+$ cells, sequential column purification is recommended. Collected cells should be processed in the same way using a new column (*see* **Note 8**). Apply collected cells to new columns pre-rinsed with 0.5% BSA/2 mM EDTA in PBS.
16. Count cell number using a hemocytometer and Trypan blue.
17. The purity of separated CD34$^+$ cells should be evaluated by staining with fluorescent labeled antibody against CD34 (*see* **Note 9**).

3.2. Transmigration Assay for CD34$^+$ Cells Using Transwell Chamber

Transmigration of CD34$^+$ cells is normally determined using transwells placed in 24-well plates. Cells loaded onto the transwells and migrated to the bottom chamber are quantitated (**Fig. 1**). The results can be expressed as percentage migration or converted to migration index for comparative studies.

1. Wash cells once with 0.5% BSA/RPMI-1640 (chemotaxis buffer: *see* **Note 10**) once and re-suspend at 10^5–10^6 cells/ml in 0.5% BSA/RPMI-1640. The cell concentration may vary and depends on the number of CD34$^+$ cells recovered and the scale of the experiments.

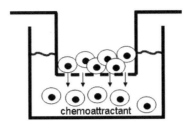

Fig. 1. Transwell chamber for migration assay: Cells are placed in the top chamber of the transwell. Chemoattractant is placed either in the bottom, top or both chambers. Cells migrate through the small pores in the trans-membrane in response to chemoattractant.

2. Add 0.5 ml of 0.5% BSA/RPMI-1640 medium in the bottom chamber (24-well plates) with appropriate cytokines, chemokines, or medium alone.
3. Load 0.1×10^5 to 1×10^6 cells/100 μl of 0.5% BSA/RPMI-1640 onto the transwells (chemokines or cytokines may be added to top wells, bottom wells, or both). Place the same amount of cells (100 μl) in tubes containing 0.4 ml 0.5% BSA/RPMI-1640 and keep them in a humidified incubator at 37°C, 5% CO_2. These will be used later to enumerate input cells.
4. Carefully place transwells into the 24-well plate (*see* **Notes 11–13**).
5. Incubate the plates and the tubes containing input cells in a humidified tissue culture incubator for 4 h at 37°C, 5% CO_2 (*see* **Note 12**).
6. Following migration, carefully remove transwells.
7. Resuspend migrated cells in the bottom chamber using a pipette and transfer to a 5 ml culture tube to count cells using flow cytometry. Cells may be fixed by adding 100 μl of 5% paraformaldehyde (final 1% paraformaldehyde). Make sure to add exactly the same amount of paraformaldehyde solution to all the samples because the cell count depends on the volume.
8. Count cell events for 20–60 s at the highest speed on the flow cytometer (*see* **Note 14**).
9. Similarly, count the number of events for the input cells (*see* **Note 15**).
10. Percentage migration may be calculated by dividing the number of events migrated to the bottom well by the total input events multiplied by 100. The degree of migration can also be expressed as Migration Index: fold change over background migration in the absence of any chemoattractants. This will be given by dividing the migrated cell events by the background migration (*see* **Note 16**).

3.3. Analysis of Migration Using Fluorescence-Tagged CD34$^+$ Cells

Chemotaxis assays can be performed using 96-well chemotaxis chambers and fluorescence-tagged cells *(28)*. This method allows one to quantitate migrated cells faster and easier based on fluorescence signal and minimizes the use of chemokines and cells. This is especially beneficial for analyzing rare populations, such as CD34$^+$ cells. However, this protocol is not practical for performance of CFU-assay or to analyze intracellular or cell-surface markers of migrated cells. In addition, a microplate spectrofluorometer is required. Although this is not a routine protocol in our hands, it may be convenient for readers who perform chemotaxis assays on a regular basis.

1. Add 300 µl of chemotaxis buffer with or without chemokines (i.e., SDF1) but without phenol red in the lower chamber of 96-well flat bottom tissue culture plates.
2. Place 20,000 fluorescence-tagged (4 µg/ml of Calcein/AM; Molecular Probes) CD34$^+$ cells in 50 µl of medium to the upper side of the membrane (NeuroProbe: 5.7 mm diameter, 5 µm pore size, polycarbonate membrane).
3. Incubate chamber at 37°C for 4 h.
4. Measure fluorescence (excitation, 485 nm; emission, 530 nm) of both migrated cells and input cells on a microplate spectrofluorometer.
5. Calculate the number of cells using samples with known cell concentration and fluorescence and determine percent migration by dividing the number of cells in the lower well by the total cell input multiplied by 100.

3.4. Evaluating Migration of CD34$^+$ Cells for Prolonged Period of Time

In some experiments, evaluating cell migration over 24 h may be necessary. The chemokine gradient normally reaches plateau after 24 h owing to the diffusion of the chemokines from bottom to the top wells or desensitization of the receptor. However, if chemokinetic activity will be determined, cells can be incubated for longer than 24 h in the transwell, as diffusion of the chemoattractant does not affect chemokinesis. CD34$^+$ cell migration for greater than 48 h can be determined as we reported using Flt3 ligand and/or SDF1 (*24*).

1. Place CD34$^+$ cells in the transwell as described (chemokines or cytokines may be included with the cells). Cells can be re-suspended in medium containing serum to prevent excessive cell death; however, this normally increases background migration.
2. Place transwells including cells into 24-well plates containing chemokines or cytokines.
3. Incubate input cells with the same concentration of chemoattractant as controls during the migration time frame (*see* **Note 17**).
4. After migration, collect migrated cells and input cells.
5. Count cell events for migrated and input cells using flow cytometry.
6. Calculate viable cells in migrated and input cells based on the integrity obtained by the forward and side scatter profiles.
7. Calculate percent migration.

3.5. Evaluating Migration of Partially Purified CD34⁺ Cells

If partially purified CD34$^+$ cells (i.e., isolated using one magnetic column purification) are to be used for migration assays, migrated and input cells can be stained with anti-CD34 antibody and the specific percentage migration of CD34$^+$ cells can be calculated.

1. Perform migration assays using partially purified CD34$^+$ cells as described above.
2. Count cells using flow cytometry for 30–60 sec.
3. Collect migrated and input cells in 1.5-ml tubes. Split samples into two replicates and stain them with anti-CD34 (*see* **Note 18**) antibody or an appropriate isotype control.
4. Determine the percentage of CD34$^+$ cells in input and migrated cells.
5. Calculate the events for CD34$^+$ cells in input and migrated cells by multiplying total cell events by the percentage of CD34$^+$ cells and divide by 100. Calculate percentage migration by dividing the CD34$^+$ events migrated to the bottom well by the input CD34$^+$ events multiplied by 100.

3.6. Analysis of Migration of Subpopulations of CD34⁺ Cells Using Cell-Surface Molecules or Intracellular Proteins as Markers

The CD34$^+$ cell fraction is heterogenous with respect to hematopoietic cell populations, differentiation markers, growth factor receptors, intracellular molecule expression, and cell cycle status. To identify migration of subpopulations of CD34$^+$ cells, cells can be sorted by fluorescence-activated cell sorting (FACS) based on these parameters followed by their use in migration assays. This strategy can be used for subpopulations defined by cell-surface markers; however, there is always substantial cell loss during isolation. Therefore, this strategy is feasible but may not always be practical. As an alternative strategy, CD34$^+$ cells can be stained with specific markers after migration. This allows one to examine migration of CD34$^+$ cells with different cell-surface markers, DNA content, intracellular cell cycle markers, expression of intracellular molecules, or phosphorylation, provided that the appropriate validated antibodies are available *(24)*.

1. Perform migration assay and count migrated and input cells as described.
2. Transfer cells to 1.5-ml microtubes and centrifuge for 2 min.
3. Wash cells once with chilled 0.5% BSA in PBS.
4. If intracellular makers are to be stained, skip steps 5–9 and go to steps 10–15.
5. Incubate input and migrated cells with Fc blocking reagent on ice for 10 min.
6. Split cells into two replicates and stain with cell-surface antibodies or an appropriate isotype control in 0.5% BSA in PBS on ice for 30 min.
7. Wash cells once with 0.5% BSA in PBS.
8. If secondary staining is necessary, stain cells with secondary Ab for 20–30 min on ice.

9. Wash cells once with 0.5% BSA in PBS and re-suspend in PBS or fix with 1% paraformaldehyde in PBS.
10. If intracellular makers will be stained, fix cells with 1% paraformaldehyde on ice for 30 min or longer (*see* **Note 19**).
11. Wash cells once with 0.25% Triton-X 100/1% BSA in PBS (*see* **Note 20**) and incubate on ice with the same buffer.
12. Incubate cells with Fc blocking reagent on ice for 10 min.
13. Split cells into two replicates and stain with specific antibodies or an appropriate isotype control in 0.25% Triton-X 100/1% BSA in PBS on ice for 30 min.
14. If secondary staining is necessary, stain cells with secondary Ab in 0.25% Triton-X 100/1% BSA in PBS for 20–30 min on ice.
15. Wash cells once with 0.25% Triton-X/1% BSA in PBS and resuspend in 0.25% Triton-X 100/1% BSA in PBS (PBS or 1% paraformaldehyde is fine).
16. Determine the percentage of positive or negative cells for the selected marker in migrated and input cells using flow cytometry.
17. Calculate the number of positive or negative cells for the marker in input and migrated cells by multiplying the number of total cells (migrated and input) by the percentage of positive or negative cells.
18. Calculate percentage migration.

3.7. Analysis of Cell Cycle of Migrated CD34+ Cells Using DNA Staining

Migrated cells can be stained with propidium iodide (PI) or 7-AAD to evaluate DNA content and cell cycle. This analysis will determine the cell cycle of migrated and input cells and address whether migration takes place during particular stages of the cell cycle.

1. Fix input cells and migrated cells with 1% paraformaldehyde at 4 °C for 30 min or longer (*see* **Note 19**). Alternatively, cells may be fixed in 70% ethanol overnight at −20 °C.
2. Wash cells with 0.25% Triton-X 100/1% BSA in PBS or PBS if cells were fixed with 70% ethanol.
3. Add 1 μg/ml of PI or 7-AAD in 0.5 ml of 0.25% Triton-X 100/1% BSA in PBS (or in PBS if cells were fixed with ethanol) for 30 min at RT in the dark.
4. Analyze DNA content by flow cytometry using FL2A or FL3A channel (PI or 7-AAD, respectively) (*see* **Notes 21** and **22**) and determine cell cycle in migrated and input cells.
5. Calculate the cell number in each phase of the cell cycle for migrated and input cells.
6. Calculate percent migration of cells in different phase of the cell cycle by dividing the number of migrated cells in each phase of the cell cycle by the number of the input cells in the same cell cycle phase and multiply by 100.

3.8. Evaluating Migration of GFP⁺ CD34⁺ Cells

CD34⁺ cells can be electroporated with plasmid vectors (*see* **Note 23**). GFP is a powerful marker for cells expressing a gene of interest if the gene is linked with GFP through internal ribosomal entry site (IRES). Cells can also be co-transfected with GFP vector plus a plasmid containing a gene of interest, although equivalent transfection cannot be automatically assumed. One can investigate the involvement of the protein produced by the gene of interest in cell migration by monitoring GFP expression. Cells can be sorted using FACS before quantitating migration or subjected to migration assay without cell sorting followed by FACS analysis for cell count and GFP quantitation. If cells are sorted before the migration assay, migration can be evaluated as described above. If unsorted cells are used, migration of GFP⁺ cells is evaluated as follows:

1. After 24- to 48-h transfection, perform migration assay as described.
2. Count cell events for GFP⁺ cells in migrated and input cells using flow cytometry. This can be done in a single tube without any sorting or staining.

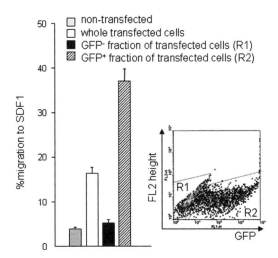

Fig. 2. Migration of Ba/F3 cells transfected with active-Ras–EGFP construct in response to SDF1. Mouse hematopoietic Ba/F3 cells were electroporated with plasmid containing constitutively active Ras–IRES–EGFP. IRES allows simultaneous expression of a gene of interest with EGFP, which serves as a marker for transfected cells. After 24-hour electroporation using a BioRad electroporator, migration in whole cells, GFP⁻ cells or GFP⁺ cells in response to 100 ng/ml SDF1 was evaluated for 4 hours. Migration of GFP⁻ cells was equivalent to untransfected cells, whereas GFP⁺ cells that express constitutive active Ras showed significantly higher migration. Whole cell (including GFP⁺ and GFP⁻ cells) migration was intermediate. The results indicate that activated Ras increases migration to SDF1.

Table 1
Calculation of Percent Migration of Ba/F3 cells Transfected with Activated Ras–IRES–EGFP Construct in Response to SDF1

	Total cells	% GFP$^+$	GFP$^+$ cells	% GFP$^-$	GFP$^-$cells
Input cells	100,000	40%	40,000	60%	60,000
Migrated cells	17,000	85%	14,400	15%	2,600
% migration	17%		36%		4%

After 24 hour transfection with activated Ras–IRES–EGFP plasmid, Ba/F3 cells were subjected to migration to SDF1 and input cells and migrated cell number were counted. The GFP$^+$ and GFP$^-$ fractions of input cells and migrated cells were determined using flow cytometry. The number of GFP$^+$ cells in input cells was calculated as the number of total input cells (100,000) multiplied by %GFP$^+$ (40%) divided by 100 (40,000). The number of migrated GFP$^+$ cells was determined as the number of migrated cells (17,000) multiplied by %GFP$^+$ in migrated cells (85%) divided by 100 (14,400). Percentage migration of GFP$^+$ cells was obtained by dividing the number of migrated GFP$^+$ cells (14,400) with GFP$^+$ input cells (40,000) $\times$ 100 (36%). Migration of GFP$^-$ cells was determined in the same manner (4%). The results indicate that the GFP$^+$ cells migrated better than GFP$^-$ cells, indicating that activated Ras increases migration to SDF1.

3. Calculate percent migration of GFP$^+$ cells by dividing the GFP$^+$ events migrated to the bottom well by the number of GFP$^+$ input cells multiplied by 100 (*see* **Note 24**).

As an example, migration and calculation of mouse BaF3 cells transfected with activated H-Ras in IRES–EGFP plasmid *(29)* is shown in **Fig. 2** and **Table 1**.

3.9. Migration of CFU in CD34$^+$ Cells

The CD34$^+$ cell population is enriched for hematopoietic stem and progenitor cells (HSPC) but is still heterogeneous. Migration of hematopoietic progenitor cells can be evaluated by quantitating colony-forming units (CFU) in the migrated CD34$^+$ cells using semisolid culture.

- Perform CD34$^+$ cell migration assay as described above.
- Wash migrated cells well to remove any cytokines/chemokines present in the samples.
- Centrifuge migrated and input cells (*see* **Note 25**), and re-suspend them in 1.1% methylcellulose or 0.3% agar containing hematopoietic growth factors that support CFU proliferation, e.g., stem cell factor, GM-CSF, Epo.
- Score colonies 2 weeks later.
- Percent migration of CFU is calculated by dividing the number of CFU in the lower chamber by the number of input CFU multiplied by 100.

3.10. Checkerboard Assay

Checkerboard assay is useful to evaluate chemotactic versus chemokinetic activity of a chemoattractant. This method was first described by Zigmond and Hirsch *(30)* where the gradient across the transwell is positive in certain chambers, i.e., concentration increasing across the transwell from top to the bottom, the gradient is negative in others, or there is no gradient, but the absolute concentration of the attractant in which the cells are migrating varies from chamber to chamber (*see* **Table 2**).

1. Prepare CD34⁺ cells as described.
2. Prepare chemotaxis buffer containing different concentrations of chemoattractant, i.e., 0.1, 1, 10, 50, 100, and 500 ng/ml for SDF1 or 0.1, 1, 10, 50, and 100 ng/ml for Flt3 ligand or Stem Cells Factor *(16,24)*.
3. Aliquot chemotaxis buffers containing chemoattractant prepared above into the bottom chambers of the plates (either 96-well or 24-well format).
4. Aliquot cells and add chemoattractant at the concentration required, i.e., 0.1, 1, 10, 50, 100, and 500 ng/ml for SDF1 or 0.1, 1, 10, 50, and 100 ng/ml for Flt3 ligand or Stem Cells Factor *(16,24)*.
5. Load cells into the top chamber of the transwell and transfer them to the wells containing chemoattractant.
6. Count cells after 4 h as described above.
7. Calculate percent migration.

If migration is observed exclusively in a positive gradient where the gradient is increasing from the top chamber to bottom chamber, it suggests that the chemoattractant has chemotactic activity, whereas if migration occurs in a zero or negative gradient, it suggests that it has chemokinetic activity.

Table 2
Checkerboard Assay to Study Cell Migration Induced by Various Gradient of Chemoattractant

	Chemoattractant in the bottom chamber (ng/ml)				
Chemoattractant in the top chamber (ng/ml)	1	10	100	500	1000 ng/ml
1					
10					
100					
500					
1000					

Cells are subjected to migration assay in various gradients of chemoattractant present in the top and/or bottom chamber of transwells.

3.11. Migration of CD34⁺ Cells Using Matrigel or ECM-Coated Transwells

Homing, migration, and mobilization of hematopoietic cells are considered to take place through the extracellular matrix and endothelial cell layer. Hematopoietic cell trafficking is regulated not just by chemokines but also by adhesion molecules expressed in the matrix or on endothelial cells *(11,12)*. Firm adhesion to the vascular endothelium is the first step in homing, followed by trans-endothelial migration *(31)*. Therefore, migration in the presence of endothelial cells or extracellular matrix may more closely represent physiological conditions. To evaluate migration through endothelial cells or extracellular matrix, transwells are coated with endothelial cells, adhesion molecules, such as VCAM-1 or fibronectin or matrix such as Matrigel™ (BD Bioscience, San Diego, CA), which contains basement membrane components (i.e., collagens, laminin, and proteoglycans) and some growth factors. The optimal concentration of extracellular matrix has to be determined in each researcher's hand. A good rule of thumb for coating transwells with extracellular matrix is 1–10 µg/ml for human fibronectin and 0.1–5 µg/ml for recombinant VCAM-1 in PBS for 1 h at 37°C or overnight at 4°C. After coating with matrix, transwells should be washed briefly with PBS. Treating wells with 2% BSA in PBS for 30 min at RT may help to inhibit non-specific binding of the cells. Transwells may be treated with Matrigel™ in a similar manner as follows:

3.11.1. Preparation of Transwells Coated with Matrigel

1. Thaw Matrigel™ at 4 °C overnight.
2. Dilute Matrigel™ in serum-free ice cold medium (the dilution factor has to be determined empirically, but 25–40 µg/well is commonly used *(32,33)* (*see* **Note 26**).
3. Incubate wells for 0.5–1 h at 37 °C to solidify Matrigel™.
4. Wash gelled wells gently with serum-free medium.
5. Load cells onto wells and perform the migration assay as described above.

3.12. Transendothelial Migration of CD34⁺ Cells

For transendothelial migration assay, various endothelial-derived cells have been used, i.e., HUVEC *(34)*, BMEC-1 cells *(35)*, STR-10, STR-12, and LE1SVO cells *(36)*.

The number of cells to be plated onto the transwells has to be determined experimentally and varies depending on the cell types and culture conditions. The following information may be useful to determine the optimal cell types or conditions. BMEC-1 cells (5×10^5) were seeded on transwells and incubated for 3 days before migration assay. The monolayers achieved full confluency following 3 days and were suitable for transmigration studies. In this report, the transwell inserts with the monolayers were placed in a 6-well tissue culture plate (3 μm transwell micropores) *(36)*. In other reports, 2×10^5, mouse STR-10, STR-12, and LE1SVO cells were seeded onto 24-well transwells (5 μm pores) and incubated for 4 days *(36)*. In both reports, integrity of the endothelial monolayer was maintained. In other experiments, transwells were coated with 25 μg/ml fibronectin before seeding endothelial cells, which prevents endothelial cells from detaching *(23)*. It is important that the endothelial cell monolayer is confluent at the time of assay. The formation of confluent monolayers on transwell filters can be verified by visual inspection using an inverted microscope. In addition, the ability of these monolayers to act as a barrier may be tested by measuring permeability to ^{125}I-labeled human serum albumin (^{125}I-HSA) or ^{14}C-albumin. Briefly, wash confluent monolayers on transwell filters and place aliquots of ^{125}I-HSA (1.8 mBq/ml, GE Health Care, Piscataway, NJ) in assay medium in the upper chambers and load medium or PBS only in the bottom chamber. Incubate at 37°C for the expected time frame to be used for the migration assay. Collect the contents of the lower chamber at the end of the incubation period and measure radioactivity. If the equilibration of ^{125}I-HSA is less than 10% across endothelium-covered Transwell filters over the migration period, the integrity of the layer can be considered to be intact *(34,35)*.

3.13. Migration of Mouse HSPC Cells

In this section, transmigration of mouse bone marrow or mobilized peripheral blood-derived HSPC will be described. The principle of the assay is the same as described for human CD34⁺ cells.

3.14. Evaluating Migration of HSPC Using CFU Assay

To determine migration of marrow-derived CFU, whole marrow cells or lineage-depleted marrow cells are placed in the transwell. The pore size and the diameter for the transwell is the same as used for CD34⁺ cells (5 μm pore size, 6.5 mm diameter).

1. Place 5×10^5 whole marrow cells or mononuclear cells in 100 μl of chemotaxis buffer (0.5% BSA/RPMI-1640) in the top chamber of the transwell.
2. Place the transwell containing cells into the 24-well plate containing appropriate chemokines in chemotaxis buffer and incubate for 4 h at 37°C (*see* **Note 27**).
3. After the migration assay, lift transwells carefully and collect migrated cells.
4. Wash cells twice with the chemotaxis buffer to remove chemokines or cytokines that could otherwise affect CFU proliferation and centrifuge.

5. Add 3–4 ml of 0.3% agar in 1× McCoy/15% fetal bovine serum (FBS) to the cell pellet and plate 1 ml/dish with appropriate hematopoietic growth factors (i.e., 10 ng/ml rm-GM-CSF and 50 ng/ml of rm-Stem Cell Factor). Alternatively, add 5 ml of 1% methylcellulose in IMDM/30% FBS and plate 1 ml/dish with appropriate hematopoietic growth factors (i.e., 10 ng/ml rm-GM-CSF, 50 ng/ml of rm-Stem Cell Factor plus 1 U/ml of rhEpo) (*see* **Note 28**).

6. Seed 2×10^4–5×10^4 input cells/plate (It is important to know the number of CFU in the input cells in order to calculate % migration.)

7. Score specific type of colonies (i.e., CFU-GEMM, CFU-GM, and BFU-E) after 7 days culture at 37°C 5% CO_2, 5% O_2 in humidified air.

8. Calculate percent migration by dividing the number of migrated CFU by the input CFU multiplied by 100. (For example, if you get 120 colonies from the 5×10^4 input cells, the number of CFU in the input was 1200, because you subjected 10-fold higher (5×10^5) cells for migration/well. If you found 30/25/28 colonies in three plates in the migrated cells (assuming migrated cells were re-suspended in 5 ml in agar or methylcellulose and plated 1 ml/dish into three plates), the percentage of migration will be calculated as $100 \times$ (30, 25, or 28) $\times 5/1200 =$ 12.5, 10.4, or 11.7%. Average percent migration and calculate standard deviation.

3.15. Evaluating Migration of Phenotypically Defined Hematopoietic Progenitor Cells

In addition to evaluating migration of total CFU, migration of specific populations of hematopoietic progenitor cells defined by their cell-surface markers can be examined. HPC, such as c-kit$^+$, lineage$^-$ cells (KL) have a frequency of approximately 1% in mouse bone marrow. Although HPC can be sorted by FACS based on the surface markers first and subjected to migration assay, cells will be lost during the sorting that will limit overall cell number, making it impractical to sort the HPC before performing the migration assay. We prefer to deplete lineage positive cells, subject lin^{neg} cells to migration, count migrated and input cells, and subsequently stain cells with surface markers for specific HPC. Lineage depletion may not be absolutely required but will help reduce background cells that can impede HPC migration.

1. Deplete lineage positive cells by Magnetic column separation according to the manufactures' instruction (Milteny Biotech, Auburn, CA). The lineage antibodies are biotinylated, enabling detection using fluorochromes conjugated with streptavidin (i.e., streptavidin PE-Cy7 or APC).

2. Re-suspend cells in chemotaxis buffer and determine cell number.

3. Place 1–5×10^5 cells/0.1 ml in the top chamber of the transwell. The cell number may vary depending on the cell number recovered after magnetic column separation.

4. Perform migration assay using appropriate chemokines or cytokines (normally 4 h at 37 °C). Be sure to keep input cells at 37°C for subsequent staining.

5. After migration, collect migrated cells.
6. Determine cell number for migrated and input cells using flow cytometry.
7. Centrifuge cells and split samples for staining.
8. Stain cells with antibodies for c-kit and/or Sca-1, which define mouse HPC (or other markers), along with PE-Cy7 or APC streptavidin. Stain replicate cells with appropriate isotype control antibodies to define the negative gate for each marker.
9. Analyze the percentage of c-kit and/or Sca-1⁺, lin⁻ cells in migrated and input cells.
10. Calculate the number of c-kit and/or Sca-1⁺, lin⁻ cells in input and migrated cells by multiplying the number of total cells (migrated and input) by the percentage of c-kit and/or Sca-1⁺, lin⁻ and dividing by 100.
11. Calculate percent migration by dividing the Sca-1⁺ and/or c-kit⁺, lin⁻ (KSL or KL) events migrated to the bottom well by input KSL or KL cells multiplied by 100 as described for CD34⁺ cell migration.

3.16. Migration of CFU in Mobilized Blood

Because the cloning efficiency of CFU in mobilized blood is 10-fold lower than bone marrow, even in mice mobilized by G-CSF, it is necessary to deplete lineage positive cells before performing the migration assay to avoid loading too many unnecessary cells in the transwell. The procedures are the same as described for mouse marrow HPC and CFU.

4. Notes

1. Fresh cord blood (within 24 h after delivery) will give the best CD34⁺ cell yield. Use heparinized collection tubes or syringes for blood harvest. Mix blood and heparin thoroughly to prevent clotting. Blood clots, even microclots, can reduce the overall CD34⁺ cell recovery. Alternatively, CD34⁺ cells may be purchased from commercial sources.
2. The mononuclear cell (MNC) fraction that contains CD34⁺ cells will appear intermediate between the Ficoll and plasma layers. Collect MNC using an automatic pippetor or a Pasteur pipette. It is important to minimize Ficoll contamination, which hampers subsequent cell collection following centrifugation.
3. Do not discard the supernatant. Cells remaining in the supernatant should also be collected. Centrifuge supernatant for an additional 20 min and combine recovered cells with the original pellet. This will increase the yield of CD34⁺ cells.
4. Do not lyse red blood cells (RBC) even if the pellet appears to be red. RBC lysis can be harmful for CD34⁺ cells.
5. If cells are re-suspended in a larger volume, it will help prevent cells from clogging up the column. One or two milliliter of buffer may be used to re-suspend cells if the cell number is too high or concentrated. Loading Ab-labeled cells onto the magnet column using a syringe fitted with a 26-G needle may be helpful in preventing clumps from forming.

6. An LS column can be used if the cell number is greater than 10^8.

7. When too many cells are loaded or cell clumps are present, it is likely that the columns will clog. If this occurs, remove columns from the magnet and flush with buffer using a plunger (supplied) and collect the cells into a new tube. Restart the procedure by refluxing through a 26-G needle and applying cells to a new pre-rinsed column.

8. If the cells are purified using just a single column, the purity of $CD34^+$ cells will vary between 50 and 70%. In contrast, sequential column purification normally gives >95% purity.

9. As the antibody for isolation is QBEND/10, we use RPE-Cy5-conjugated anti-CD34 antibody (clone: BIRMA-K3, DAKO) and RPE-Cy5-conjugated mouse IgG1 as an isotype control. This fluorochrome-conjugated antibody provides a very bright signal that can be easily distinguished from the isotype control. Stain $0.1–0.5 \times 10^5$ cells, which should be sufficient to determine the purity using flow cytometry. Although results fluctuate from sample to sample, we routinely obtain 0.5–2 million $CD34^+$ cells with >95% purity following sequential column separation starting with 40–50 ml of freshly isolated cord blood.

10. It is extremely important that no gradient exists in the chemotaxis buffer except that produced by the chemokines or cytokines to be evaluated. Cells need to be washed thoroughly with chemotaxis buffer to remove potential contamination from BSA or serum used in the isolation steps. Use the same batch of chemotaxis buffer throughout the experiment.

11. When transwells are placed into the wells in the plates, hold the plates at an angle. This will help to prevent bubbles from forming beneath the wells that can impede cell migration. Do not shake the plates.

12. The most well-characterized chemotactic chemokine for $CD34^+$ cells is SDF1(CXCL12). The optimal concentration for chemotaxis is usually between 50 and 200 ng/ml *(16,24)*. Ten ng/ml each of SCF and Flt3 ligand provide a maximal migratory stimulus *(16,24)*. If multiple wells are to be prepared with the same conditions, chemokines/cytokines should be prepared in the medium as a pre-mixture to minimize fluctuation between wells. If various gradients of chemokines are to be investigated, add chemokines or cytokines to the cells, mix and load the cells to the top chamber of the transwells immediately (*see* **Subheading 3.10**, *Checkerboard Assay*). Incubation for 4h works best in our experience; however, incubation time may vary and researchers should determine the appropriate incubation time depending on the cytokines or chemokines used or the cells to be analyzed (*see* **Subheading 3.4**, *Transmigration Assay for a Prolonged Period of Time*).

13. Because cell migration varies from well to well, it is highly recommended to prepare multiple wells with the same treatment in order to obtain the appropriate number of data points for statistical analysis.

14. Enumeration of migrated cells can be done using a hemacytometer if the migrated cell concentration is high enough to obtain reliable counts. Counting migrated cells and input cells using flow cytometry gives a more accurate cell number. Flow

cytometry permits measurement of cell events/time for both migrated and input cells. In addition, dead cells can be excluded based on forward and side scatter.

15. If enumeration of the absolute number of migrated cells is required, prepare tubes containing known cell numbers per milliliter pre-determined using a hemacytometer or other methods. Make serial dilutions and count cell events for 30–60 sec. This will give the ratio of cell number per milliliter to events/time, which allows calculating cells per milliliter and the absolute number of migrated cells. Alternatively, migrated and input cells can be mixed with a pre-determined number of polystyrene beads (15 fxm; Polysciences, Inc., Warrington, PA) before counting using flow cytometry. It has been shown that the beads and cells are easily distinguishable on a plot of side-scatter versus forward scatter. This counting does not need time acquisition but requires the events for beads and the cells, which allows the researcher to calculate the ratio of beads to cells and thereby calculate the total number of cells migrated to the bottom well. This counting method was confirmed to accurately represent the number of cells present by testing it with known numbers of cells and beads *(37)*.

16. Generally, 100 ng/ml of SDF stimulates maximal migration of CD34⁺ cells, usually on the order of approximately 10–20% in the presence of 0.5% BSA for 4 hours. At lower or higher SDF1 concentration, migration is normally lower. Some investigators may use serum in the chemotaxis buffer; however, serum contains factors that normally increase cell migration. If the effect of the chemokines or cytokines is low, the presence of serum may mask its effect. If this is the case, avoid using serum in the chemotaxis buffer.

    In short-term migration assays, SCF and Flt3 ligand synergistically increase migration of CD34⁺ cells to a positive gradient of SDF1 *(16,24)*. If CD34⁺ cells are cultured in medium containing serum or with hematopoietic growth factors, this may enhance migration in response to SDF1. Prolonged incubation with some hematopoietic growth factors, such as SCF or IL6, can increase surface CXCR4 expression and migration to SDF1 in CD34⁺ cells *(6)*, whereas Flt3 ligand can decrease CXCR4 expression and reduce migration *(24)*.

17. Because prolonged incubation with or without chemoattractant can affect cell viability and cell number, which in turn affects migration, input cells should be incubated with or without chemoattractant at the concentration to be used for migration assays in order to monitor cell proliferation or survival. If the cell number increases or decreases in the presence or absence of chemoattractant, the number of migrated cells should be normalized by dividing the fold change of input cells during the incubation period.

18. Because the antibody for isolation is QBEND/10, use an antibody that recognizes a different CD34 epitope, i.e., RPE-Cy5-conjugated anti-CD34 antibody (clone: BIRMA-K3, DAKO) and its isotype RPE-Cy5-conjugated mouse IgG1.

19. Cells can be fixed in 1% paraformaldehyde for at least 1 week at 4 °C.

20. Permeabilizing cells with 0.25% Triton-X 100/1% BSA in PBS works for most applications; however, commercial reagents are also available.

21. Cell cycle distribution can be calculated using ModFit Software (Beckton Dickinson).
22. As an alternative strategy, cells may be stained with other cell cycle markers, such as Ki-67 *(24)* or cyclins (*see* **Subheading 3.6**, *Analysis of migration of subpopulations of CD34⁺ cells using cell-surface molecules or intracellular proteins as markers*).
23. Amaxa GmbH (Koeln, Germany) or Amaxa, Inc. (Gaithersburg, MD) offers electroporation solutions and an electroporator specifically designed for CD34⁺ cell transfection. In our experience, as high as 70% of freshly isolated human cord blood CD34⁺ cells can be transfected using this system (Fukuda and Pelus: unpublished observation, 2003).
24. GFP⁻ cells can be used as a negative control, because these cells do not express the gene of interest, when using an IRES-GFP construct.
25. CFU cloning efficiency for cord blood CD34⁺ cells in methylcellulose assay is normally approximately 50%. Therefore, if migration efficiency is approximately 20% and 10,000 or 100,000 CD34⁺ cells were subjected to migration/well, the expected number of CFU migrated to the lower chamber would be 2,000 or 20,000, respectively. Therefore, dilute cells by 10- or 100-fold before plating so that the density of colonies falls within a range where individual colonies can be enumerated without crowding and overlap, which effects counting efficiency. If migration efficiency is lower, the dilution may not be necessary. It is recommended that both original and diluted cells are plated.
26. Keep Matrigel™ on ice at all times. As it solidifies relatively quickly at RT, using chilled pipette tips will facilitate coating the transwells.
27. For chemotaxis using SDF1, 100 ng/ml normally provides the optimal migration stimulus for mouse bone marrow-derived CFU in 0.5 ml of 0.5% BSA/RPMI-1640.
28. The percentage migration of marrow CFU in 0.5% BSA with 100 ng/ml SDF1 is normally less than 20%; therefore, it is not necessary to dilute the migrated cells to obtain accurate CFU counts if 5×10^5 cells are subjected to migration. However, if expected migration is higher than 50%, or if the input cells exceed 1×10^6 cells per transwell, dilute cells by fivefold and plate for CFU along with the original concentration in order to maximize the chance to obtain a reliable colony count. (Note: the cloning efficiency of mouse marrow cells is approximately 0.1–0.3%)

References

1. Taichman, R.S. (2005) Blood and bone: two tissues whose fates are intertwined to create the hematopoietic stem cell niche. *Blood* **105**, 2631–9.
2. Arai, F., Hirao, A., Ohmura, M., Sato, H., Matsuoka, S., Takubo, K., Ito, K., Koh, G.Y., and Suda, T. (2004) Tie2/angiopoietin-1 signaling regulates hematopoietic stem cell quiescence in the bone marrow niche. *Cell* **118**,149–61.

3. Nibley, W.E. and Spangrude, G.J. (1998) Primitive stem cells alone mediate rapid marrow recovery and multilineage engraftment after transplantation. *Bone Marrow Transpl.* **21**, 345-54.

4. Lanzkron, S.M., Collector, M.I., and Sharkis, S.J. (1999) Hematopoietic stem cell tracking in vivo: a comparison of short-term and long-term repopulating cells. *Blood* **93**, 1916–21.

5. Pelus, L.M. and Fukuda, S. (2006) Peripheral blood stem cell mobilization: the CXCR2 ligand GROβ rapidly mobilizes hematopoietic stem cells with enhanced engraftment properties. *Exp. Hematol.* **34**, 1010–20.

6. Peled, A., Petit, I., Kollet, O., Magid, M., Ponomaryov, T., Byk, T., Nagler, A., Ben-Hur, H., Many, A., Shultz, L., Lider, O., Alon, R., Zipori, D., and Lapidot, T. (1999) Dependence of human stem cell engraftment and repopulation of NOD/SCID mice on CXCR4. *Science* **283**, 845–8.

7. Molineux, G., McCrea, C., Yan, X.Q., Kerzic, P., and McNiece, I. (1997) Flt-3 ligand synergizes with granulocyte colony-stimulating factor to increase neutrophil numbers and to mobilize peripheral blood stem cells with long-term repopulating potential. *Blood* **89**, 3998–4004.

8. King, A.G., Horowitz, D., Levin, R., Farese, AM., MacVittie, T.J., and Pelus, L.M. (2001) Rapid mobilization of murine hematopoietic stem cells with enhanced engraftment properties and evaluation of hematopoietic progenitor cell mobilization in rhesus monkeys by a single injection of SB-251353, a specific truncated form of the human CXC chemokine GROβ. *Blood* **97**, 1534–42.

9. Pelus, L.M., Bian, H., King, A.G., and Fukuda, S. (2004) Polymorphonuclear neutrophil proteases mediate hematopoietic stem cell mobilization induced by granulocyte-colony stimulating factor (G-CSF) and the CXCR2 selective ligands GROβ (CXCL2) and a specific truncated variant GROβ$_T$/CXCL2$_{Δ4}$. *Blood* **103**, 110–9.

10. Levesque, J.P., Hendy, J., Takamatsu, Y., Simmons, P.J., and Bendall, L.J. (2003) Disruption of the CXCR4/CXCL12 chemotactic interaction during hematopoietic stem cell mobilization induced by G-CSF or cyclophosphamide. *J. Clin. Invest.* **111**,187–96.

11. Katayama, Y., Hidalgo, A., Furie, B.C., Vestweber, D., Furie, B., and Frenette, P.S. (2003) PSGL-1 participates in E-selectin-mediated progenitor homing to bone marrow: evidence for cooperation between E-selectin ligands and α4 integrin. *Blood* **102**, 2060–67.

12. Papayannopoulou, T., Priestley, G.V., Nakamoto, B., Zafiropoulos, V. and Scott, L.M. (2001) Molecular pathways in bone marrow homing: dominant role of α4β1 over β2-integrins and selectins. *Blood* **98**, 2403–11.

13. Nagasawa, T., Kikutani, H., and Kishimoto, T. (1994) Molecular cloning and structure of a pre-B-cell growth-stimulating factor. *Proc. Natl. Acad. Sci. U.S.A.* **91**, 2305–9.

14. Semerad, C.L., Christophe, M.J., Liu, F., Short, B., Simmons, P.J., Winkler, I., Levesque, J.P., Chappel, J., Ross, F.P., and Link, D.C. (2005) G-CSF potently

inhibits osteoblast activity and CXCL12 mRNA expression in the bone marrow. *Blood* **106**, 3020–7.

15. Jung, Y., Wang, J., Schneider, A., Sun, Y.X., Koh-Paige, A.J., Osman, N.I., McCauley, L.K., and Taichman, R.S. (2006) Regulation of SDF-1 (CXCL12) production by osteoblasts; a possible mechanism for stem cell homing. *Bone* **38**, 497–508.

16. Kim, C.H. and Broxmeyer, H.E. (1998) In vitro behavior of hematopoietic progenitor cells under the influence of chemoattractants: stromal cell-derived factor-1, steel factor, and the bone marrow environment. *Blood* **91**, 100–10.

17. Christopherson, K.W., 2nd, Hangoc, G., Mantel, C.R., and Broxmeyer, H.E. (2004) Modulation of hematopoietic stem cell homing and engraftment by CD26. *Science* **305**, 1000–3.

18. Liles, W.C., Broxmeyer, H.E., Rodger, E., Wood, B., Hubel, K., Cooper, S., Hangoc, G., Bridger, G.J., Henson, G.W., Calandra, G., and Dale, D.C. (2003) Mobilization of hematopoietic progenitor cells in healthy volunteers by AMD3100, a CXCR4 antagonist. *Blood* **102**, 2728–30.

19. Wright, D.E., Bowman, E.P., Wagers, A.J., Butcher, E.C., and Weissman, I.L. (2002) Hematopoietic stem cells are uniquely selective in their migratory response to chemokines. *J. Exp. Med.* **195**, 1145–54.

20. Rosu-Myles, M., Khandaker, M., Wu, D.M., Keeney, M., Foley, S.R., Howson-Jan, K., Yee, I.C., Fellows, F., Kelvin, D., and Bhatia, M. (2000) Characterization of chemokine receptors expressed in primitive blood cells during human hematopoietic ontogeny. *Stem Cells* **18**, 374–81.

21. Jinquan, T., Quan, S., Jacobi, H.H., Jing, C., Millner, A., Jensen, B., Jensen, B., Madsen, H.O., Ryder, L.P., Svejgaard, A., Malling, H.J., Skov, P.S., and Poulsen, L.K. (2000) CXC chemokine receptor 3 expression on CD34(+) hematopoietic progenitors from human cord blood induced by granulocyte-macrophage colony-stimulating factor: chemotaxis and adhesion induced by its ligands, interferon gamma-inducible protein 10 and monokine induced by interferon gamma. *Blood* **96**, 1230–8.

22. Voermans, C., Kooi, M.L., Rodenhuis, S., van der Lelie, H., van der Schoo,t C.E., and Gerritsen, W.R. (2001) In vitro migratory capacity of CD34+ cells is related to hematopoietic recovery after autologous stem cell transplantation. *Blood* **97**, 799–804.

23. Peled, A., Kollet, O., Ponomaryov, T., Petit, I., Franitza, S., Grabovsky, V., Slav, M.M., Nagler, A., Lider, O., Alon, R., Zipori, D., and Lapidot, T. (2000) The chemokine SDF-1 activates the integrins LFA-1, VLA-4, and VLA-5 on immature human CD34(+) cells: role in transendothelial/stromal migration and engraftment of NOD/SCID mice. *Blood* **95**, 3289–96.

24. Fukuda, S., Broxmeyer, H.E., and Pelus, L.M. (2005) Flt3-ligand and the Flt3 receptor regulate hematopoietic cell migration by modulating the SDF-1α(CXCL12)/CXCR4 axis. *Blood* **105**, 3117–26.

25. Wiesmann, A. and Spangrude, G.J. (1999) Marrow engraftment of hematopoietic stem and progenitor cells is independent of Gαi-coupled chemokine receptors. *Exp. Hematol.* **27**, 946–55.

26. Kawabata, K., Ujikawa, M., Egawa, T., Kawamoto, H., Tachibana, K., Iizasa, H., Katsura, Y., Kishimoto, T., and Nagasawa, T. (1996) A cell-autonomous requirement for CXCR4 in long-term lymphoid and myeloid reconstitution. *Proc. Natl. Acad. Sci. U.S.A.* **96**, 5663–7.

27. Bonig, H., Priestley, G.V., and Papayannopoulou, T. (2006) Hierarchy of molecular pathway usage in bone marrow homing and its shift by cytokines. *Blood* **107**, 79–86.

28. Levesque, J.P., Leavesley, D.I., Niutta, S., Vadas, M., and Simmons, P.J. (1995) Cytokines increase human hemopoietic cell adhesiveness by activation of very late antigen (VLA)-4 and VLA-5 integrins. *J. Exp. Med.* **181**,1805–15.

29. Fukuda, S. and Pelus, L.M. (2006) Internal tandem duplication of Flt3 modulates chemotaxis and survival of hematopoietic cells by SDF1α but negatively regulates marrow homing in vivo. *Exp. Hematol.* **34**,1041–5.

30. Zigmond, S.H. and Hirsch, J.G. (1973) Leukocyte locomotion and chemotaxis. New methods for evaluation, and demonstration of a cell-derived chemotactic factor. *J. Exp. Med.* **137**, 387–410.

31. Sackstein, R. (2005) The lymphocyte homing receptors: gatekeepers of the multistep paradigm. *Curr. Opin. Hematol.* **12**, 444–50.

32. Reca, R., Mastellos, D., Majka, M., Marquez, L., Ratajczak, J., Franchini, S., Glodek, A., Honczarenko, M., Spruce, L.A., Janowska-Wieczorek, A., Lambris, J.D., and Ratajczak, M.Z. (2003) Functional receptor for C3a anaphy-latoxin is expressed by normal hematopoietic stem/progenitor cells, and C3a enhances their homing-related responses to SDF-1. *Blood* **101**, 3784–93.

33. Rao, Q., Zheng, G.G., Lin, Y.M., and Wu, K.F. (2004) Production of matrix metalloproteinase-9 by cord blood CD34+ cells and its role in migration. *Ann. Hematol.* **83**, 409–13.

34. Yong, KL., Watts, M., Thomas, N.S., Sullivan, A., Ings, S., and Linch, D.C. (1998) Transmigration of CD34$^+$ cells across specialized and nonspecialized endothelium requires prior activation by growth factors and is mediated by PECAM-1 (CD31). *Blood* **91**,1196–205.

35. Mohle, R., Moore, M.A., Nachman, R.L., and Rafii, S. (1997) Transendothelial migration of CD34+ and mature hematopoietic cells: an in vitro study using a human bone marrow endothelial cell line. *Blood* **89**, 72–80.

36. Imai, K., Kobayashi. M., Wang, J., Ohiro, Y., Hamada, J., Cho, Y., Imamura, M., Musashi, M., Kondo, T., Hosokawa, M., and Asaka, M. (1999) Selective transendothelial migration of hematopoietic progenitor cells: a role in homing of progenitor cells. *Blood* **93**,149–56.

37. Campbell, J.J., Qin, S., Bacon, K.B., Mackay, C.R., and Butcher, E.C. (1996) Biology of chemokine and classical chemoattractant receptors: differential requirements for adhesion-triggering vs chemotactic responses in lymphoid cells. *J. Cell Biol.* **134**, 255–66.

5

Flow Cytometry-Based Cell Cycle Measurement of Mouse Hematopoietic Stem and Progenitor Cells

Hongmei Shen, Matthew Boyer, and Tao Cheng

Summary

The balance between the quiescent hematopoietic stem cell (HSC) and the highly prolif-erative hematopoietic progenitor compartments maintains homeostasis in the hematopoietic system. Therefore, the entry of HSCs into the cell cycle and the rate of proliferation of hematopoietic progenitor cells are fundamental aspects in the field. This chapter describes two intracellular staining methods for DNA and RNA in conjunction with membrane staining for multiple hematopoietic cell-surface markers, and subsequent flow cytometric analysis to determine the cell cycle characteristics of primitive hematopoietic cells. First, the DNA stain Hoechst 33342 and the RNA dye Pyronin Y are used in combination with cell-surface markers to identify the proportion of cells in G_0 and G_1 in hematopoietic stem and progenitor cells. The second details the staining of bromodeoxyuridine incor-porated into replicating DNA as a measure for the cycling cell fraction within a specific hematopoietic cell subset.

Key Words: Hoescht 33342; Pyronin Y; bromodeoxyuridine; multi-color flow cytometry.

1. Introduction

In humans, blood cell turnover requires the production of tens of billions of cells per day, increasing significantly during times of physiological stress, to maintain homeostasis within the hematopoietic system. Maintenance of cell production requires cytokine-responsive hematopoietic progenitor cell (HPC) pools with robust proliferative capacity, and a smaller population of hematopoietic stem cells (HSCs) intermittently yielding daughter cells

From: *Methods in Molecular Biology, vol. 430: Hematopoietic Stem Cell Protocols*
Edited by: K. D. Bunting © Humana Press, Totowa, NJ

into the proliferative HPC compartments. The HSC pool itself is relatively quiescent and cytokine resistant, a state that appears to be necessary for the prevention of premature depletion under times of stress *(1–4)*. Limited proliferation of HSCs does occur *(5)*, but in a highly regulated fashion *(6)* and appears to be development-dependent *(7)*. HSC proliferation has been directly measured by bromodeoxyuridine (BrdU)-labeling experiments, and cell cycle lengths have been estimated at approximately 30 days in small adult rodents *(8)*. Similar analyses using population kinetics have estimated that HSCs replicate once per 10 weeks in cats *(9)* and once per year in higher order primates *(10)*. A precise analysis based on a single cell culture system suggested that most human HSCs, as measured by the assay for the long-term culture-initiating cell (LTC-IC), reside in the quiescent G_0 subcompartment of $CD34^+$ cells *(11,12)*. Exiting from quiescence might result in premature exhaustion of adult HSCs under stress conditions as evidenced in the absence of the cyclin-dependent kinase inhibitor p21 *(13)* or the polycomb protein Gfi-1 *(14)*. In contrast, an essential feature of the HPC population is that it undergoes a number of rapid cell divisions, operating as a cellular amplification machine, and irreversibly develops into mature cells. In short, the dichotomy of resistance to proliferative signals by HSCs and the brisk responsiveness by HPCs is a central feature of the blood-forming system, and therefore, cell cycle analysis is a critical parameter for studies on HSCs and HPCs.

HSCs and HPCs have been best defined in mice as compared with other species. Different strategies have been employed for the prospective isolation of murine HSCs and HPCs *(15–20)*. While cycling cells can be functionally examined by treatment with a S-phase-specific inhibitor such as 5-fluorouracil (5-FU), which blocks thymidine synthesis during DNA replication *(21)* (*see* **Note 1**), a complete snapshot of the cell cycle profile can be more definitively assessed in a phenotypically defined HSC or HPC subset with flow cytometry (FCM). There are many available dyes for determining G_0/G_1 versus S and G_2/M phases based on the DNA content measured by FCM. However, a DNA dye is not able to distinguish cells residing in G_0 or G_1 phases. This can be achieved by quantifying RNA content, which rises during G_1 and remains high during mitosis, with Pyronin Y (PY) in conjunction with the DNA dye, Hoechst 33342 (HO) *(22)*. This approach has been used, along with multiple surface markers, to identify the primitive quiescent G_0 stem-like cells in human hematopoietic tissues *(11)*. In this chapter, we describe a protocol to measure DNA and RNA contents in conjunction with multiple cell-surface markers on living cells by multi-color FCM techniques. This protocol will allow investigators to distinguish cells in the G_0 phase of the cell cycle from those in G_1 within a given HSC or HPC subset. In addition, we also provide a FCM-based method employing BrdU incorporation to track the proliferative kinetics of HSCs and HPCs under steady-state or stress conditions *in vivo*.

2. Materials

2.1. DNA or RNA Dyes for Cell Cycle

1. Hoechst 33342 (HO): 1 mg/ml in distilled H_2O stock solution. Store at –20°C.
2. PY: 10 mg/ml in DMSO stock solution. Store at 4°C.

2.2. Antibodies and Conjugation (see Note 2)

1. PE or Alexa Fluor 700-conjugated anti-Sca-1 (D7).
2. APC-conjugated anti-c-kit (2B8).
3. FITC-conjugated or biotinylated CD34 (RAM34).
4. Lineage markers: PE-Cy7-conjugated anti-CD3 (CT-CD3), anti-CD4 (CT-CD4), anti-CD8 (CT-CD8a), anti-CD45R (RA3-6B2), anti-CD11b (M1/70.15), anti-Gr-1 (RB6-8C5), and anti-TER-119 (TER-119).
5. FITC-conjugated isotype control for anti-CD34 (if using FITC-conjugated CD34) or PE-Texas Red-conjugated streptavidin (if using biotinylated CD34).

2.3. BrdU Incorporation

1. BrdU Flow Kit (BD Pharmingen): FITC-conjugated Anti-BrdU antibody, BD Cytofix/Cytoperm Buffer, 10× BD Perm/Wash Buffer, BD Cytoperm Plus Buffer, 10 mg/ml BrdU, and 1 mg/ml DNase.
2. BrdU: 10 mg/ml in PBS stock solution. Store at –80°C.

2.4. Buffers or Medium

1. IMEM+ medium: Iscove's Modified Eagle's Medium, 2% FBS, 10 mM HEPES–HCl, pH 7.2–7.5.
2. Hank's balanced salt solution (HBSS+) buffer: Hanks Balanced Salt Solution, 2% FBS, 10 mM HEPES–HCl, pH 7.2–7.5.
3. Staining buffer: PBS, 2% heat-inactivated FBS, 0.09% (w/v) sodium azide (NaN_3).

2.5. Other Reagents

1. Verapamil: 5 mM in 95% ethanol stock solution. Store at –20°C.
2. Propidium iodide (PI): 10 mg/ml in distilled H_2O stock solution. Store at –20°C. 200 µg/ml in PBS working solution. Store at 4°C.

2.6. Flow Cytometer

1. An analytical cytometer or cell sorter with multiple-laser excitation is required for these protocols. A 488-nm laser is required for FITC. A 488-nm or a green laser (~530 nm) can be used for PE, PE-Texas Red, PI, PE-Cy5.5, PE-Cy7, and PY excitation. A red laser (633, 635, or 647 nm) is required for APC. HO requires UV (~350 nm) excitation.

3. Methods

3.1. Cell Cycle Analysis of Living Mouse HSCs/HPCs by Combining HO, PY, and Cell-Surface Marker Staining

3.1.1. Staining Procedure for Multi-Color Flow Cytometric Analysis

1. Harvest cells from mouse femurs and tibias or hematopoietic organs (*see* **Note 1**).
2. Suspend 2×10^7 cells in 20 ml of pre-warmed (37°C) IMEM+ medium.
3. Add HO to a final concentration of 5–10 μM and verapamil to 50 μM (*see* **Note 3**).
4. Mix well and incubate at 37°C for 45 min in the dark.
5. Add PY to a final concentration of 3–5 μM.
6. Mix well and continue to incubate at 37°C for another 45 min in the dark.
7. Centrifuge at $300 \times g$ for 10 min at 4°C. Aspirate off the supernatant and leave about 50 μl of medium (*see* **Note 4**).
8. Add the following reagents to the cells (*see* **Note 2**):

   - FITC-conjugated CD34 antibody
   - Alexa Fluor 700-conjugated Sca-1 antibody
   - APC-conjugated c-kit antibody
   - Lineage cocktail containing PE-Cy7-conjugated CD3, CD4, CD8, CD45R, CD11b, Gr-1, and TER-119 antibodies

9. Mix well and incubate on ice for 15 min in the dark.
10. Wash cells with 3–5 ml of ice-cold HBSS+ buffer and centrifuge at $300 \times g$ for 10 min. Aspirate off the supernatant.
11. Suspend the cells in 0.5 ml of ice-cold HBSS+ buffer.
12. Add PI to a final concentration of 1 μg/ml right before running samples for FCM for discrimination of dead cells (*see* **Note 5**).

For instrument set-up, unstained cells as well as single positive controls for each fluorochrome should be included. Using an isotype control for anti-CD34-FITC in **step 8** above, instead of anti-CD34-FITC itself, is recommended for identifying long-term repopulating HSCs (LT-HSCs) (*see* **Note 6**).

3.1.2. Data Analysis

In mice, HSCs are highly enriched by Sca-1 and c-Kit expression and the absence of lineage marker expression (Lin$^-$Sca-1$^+$c-Kit$^+$, LKS) *(15)*. On the basis of functional bone marrow transplantation results, LKS cells can be divided into LT-HSCs, which lack CD34 expression and short-term repopulating HSCs (ST-HSCs), which express CD34 *(17)*. **Figure 1** shows the HO/PY-staining profiles for HSCs and HPCs.

3.2. Measurement of Cycling Mouse HSCs/HPCs by Incorporation of the Thymidine Analog BrdU

3.2.1. BrdU Administration

For kinetic measurements of cycling cells, BrdU is administered continuously to mice through their drinking water at a concentration of 0.5 mg/ml for 13 weeks. For a single pulse experiment, an intraperitoneal injection of 100 µg/g of BrdU is administered 16 h before harvesting bone marrow (BM) cells.

3.2.2. Staining Procedure for Multi-Color Flow Cytometric Analysis

1. Harvest BM from BrdU-treated mice and collect 2×10^7 cells.
2. Add anti-CD34-biotinylated antibody.
3. Incubate on ice for 15 min.
4. Wash with 3–5 ml of staining buffer. Centrifuge at $300 \times g$ for 10 min. Aspirate off the supernatant.

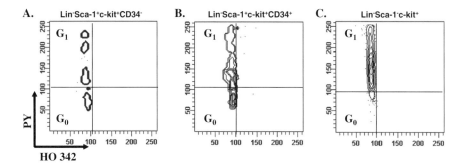

Fig. 1. HO/PY staining in hematopoietic stem cells (HSCs) and hematopoietic progenitor cells (HPCs). Bone marrow (BM) cells stained with the nucleic acid dyes HO and PY, anti-Sca-1, anti-c-Kit and for lineage markers were analyzed by flow cytometry. Plots show the profile of HO and PY staining in LT-HSCs (A), ST-HSCs (B), and HPCs (C).

5. Repeat **step 4** for a total of two washes. Leave approximately 50 μl of staining buffer after aspirating off the second supernatant.

6. Add the following reagents:

   - PE-Texas Red-conjugated streptavidin.
   - Lineage cocktail containing PE-Cy7-conjugated anti-CD3, anti-CD4, anti-CD8, anti-CD45R(B220), anti-CD11b, anti-Gr-1, and anti-TER-119 antibodies.
   - PE-conjugated Sca-1 antibody.
   - APC-conjugated c-kit antibody.

7. Mix well and incubate on ice for 15 min.

8. Wash with 3–5 ml of staining buffer once. Centrifuge for 10 min at $300 \times g$ at 4°C. Aspirate off supernatant.

9. Resuspend cells in 200 μl of Cytofix/Cytoperm Buffer and incubate on ice for 20 min.

10. Add 2 ml of staining buffer and store at 4°C overnight. This step can be eliminated if the rest of the staining procedures and Flow Cytometry acquisition can be finished in the same day.

11. Centrifuge at $300 \times g$ for 10 min at 4°C and remove the supernatant.

12. Resuspend cells with 200 μl of Cytoperm plus buffer and incubate on ice for 10 min.

13. Add 2 ml of 1× Perm Wash Buffer and centrifuge at $300 \times g$ for 10 min at 4°C.

14. Remove the supernatant. Add 200 μl of Cytofix/Cytoperm Buffer and incubate at room temperature for 5 min.

15. Add 2 ml of 1× Perm/Wash Buffer and centrifuge at $300 \times g$ for 10 min to remove the supernatant.

16. Remove the supernatant. Resuspend cells with 100 μl of 300 mg/ml DNase. Incubate at 37°C for 1 h.

17. Add 2 ml of 1× Perm/Wash Buffer and centrifuge at $300 \times g$ for 10 min.

18. Remove the supernatant. Resuspend the cells in 50 μl of 1:50 diluted FITC conjugated anti-BrdU diluted in 1× Perm/Wash Buffer and incubate at room temperature for 20 min.

19. Add 2 ml of 1× Perm/Wash Buffer and centrifuge at $300 \times g$ for 10 min.

20. Remove the supernatant. Add 0.3 ml of staining buffer for flow cytometric analysis.

21. For instrument set-up, unstained cells as well as single positive controls for each fluorochrome need to be included. Using an isotype control for CD34-biotin in **step 2** above, instead of anti-CD34-biotin, is recommended for identifying LT-HSCs (*see* **Note 5**).

3.2.3. Data Analysis

The number of cycling cells that are BrdU-positive can be calculated in each gated population (*see* **Fig. 2**). For kinetic BrdU uptake, the average

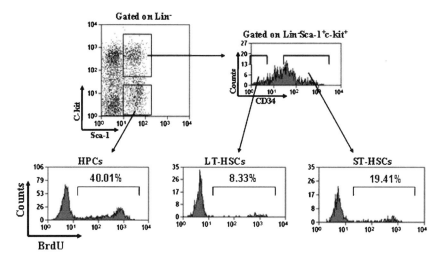

Fig. 2. BrdU incorporation in cycling hematopoietic stem cells (HSCs)/ hematopoietic progenitor cells (HPCs). Mice were administered 100 µg/g of BrdU through an intraperitoneal injection. Sixteen hours later, bone marrow (BM) cells were then stained with anti-BrdU, anti-Sca-1, anti-c-Kit, and lineage markers. Data were acquired and analyzed by using flow cytometry. This representative result shows the frequency of BrdU-positive cells in gated HPCs (Lin⁻Sca-1⁻c-Kit⁺), LT-HSCs (Lin⁻Sca-1⁺c-Kit⁺CD34⁻), or ST-HSCs (Lin⁻Sca-1⁺c-Kit⁺CD34⁺).

turnover time of HSCs can be calculated according to linear regression analysis based on the plot of the $\log_{10}$ of BrdU-negative cells as a function of time *(8)*. Because of the hierarchy of the hematopoietic system, the cellular phenotype may change after primitive cells proliferate or differentiate into more mature cells. Therefore, a higher BrdU content in relatively mature cells resulting from continuing BrdU uptake during differentiation does not necessarily reflect a greater percentage of cycling cells in that mature cell population.

4. Notes

1. 5-FU can be employed to enrich for LT-HSCs for both *in vitro* and *in vivo* assays. A single 200 mg/kg dose of 5-FU can be administered intravenously 24 h before collecting HSCs for the long-term culture initiating cell assay in which limiting dilutions of cells are plated in replicates to determine the frequency of stem cell functionality in the population. The higher degree of cycling cells, the greater proportion killed by 5-FU treatment, and the lower stem cell activity in downstream assays per input cell. A similar strategy can be applied to the colony-forming cell (CFC) assay to measure the proportion of HPCs in S-phase *(13)*.

Alternatively hydroxyurea, which inhibits ribonucleotide reductase thereby killing cells in S-phase and synchronizing the cell cycle status of those that survive, can be used *(8)*.

2. It is important to use the same antibody clone (indicated in parentheses) to generate reproducible results. The fluorochrome conjugation of antibodies can be chosen as dictated by the flow cytometry instrument configuration. The final concentration for each individual antibody during the staining could be different because of the source of antibodies. Titration of the antibodies may be needed.

3. For DNA content analysis of live mouse BM HSCs by HO, verapamil must be added at the same time as HO staining. HSCs in mouse BM expressing ABC-transporters such as Bcrp-1 efflux HO dye generating a lower intensity signal than that of the G_0/G_1 peak. This phenomenon was reported as a side population (SP) when HO emission was collected in blue (~450 nm) and red (~670 nm) channels *(18)* and will affect the DNA and RNA content profile. Verapamil serves as an inhibitor to completely block HO efflux from these cells.

4. All the procedures after HO and PY staining should be done at 4°C or on ice to prevent HO and PY leakage from the cells.

5. Because of the toxicity of HO and PY, there is always a certain percentage of dead cells after staining. To obtain the most accurate results, dead cells should be discriminated by PI uptake in order to remove the "noise" generated by these cells on the DNA/RNA profile or by non-specific antibody binding.

6. Because both ST- and LT-HSCS are separated based on CD34 expression or absence, respectively, in LKS cells and the majority of LKS cells express CD34 antigen, there are not distinct positive and negative populations. In multi-color flow cytometric analysis a "Fluorescence Minus One" (FMO) control, which includes all other markers or dyes except CD34, will serve as the best control for gating during analysis.

References

1. Mauch, P., J. Ferrara, and S. Hellman. (1989) Stem cell self-renewal considerations in bone marrow transplantation. *Bone Marrow Transplant* **4**, 601–607.
2. Mauch, P., and S. Hellman. (1989) Loss of hematopoietic stem cell self-renewal after bone marrow transplantation. *Blood* **74**, 872–875.
3. Mauch, P., L. Constine, J. Greenberger, W. Knospe, J. Sullivan, J.L. Liesveld, and H.J. Deeg. (1995) Hematopoietic stem cell compartment: acute and late effects of radiation therapy and chemotherapy. *Int J Radiat Oncol Biol Phys* **31**, 1319–1339.
4. Gardner, R.V., C.M. Astle, and D.E. Harrison. (1997) Hematopoietic precursor cell exhaustion is a cause of proliferative defect in primitive hematopoietic stem cells (PHSC) after chemotherapy. *Exp Hematol* **25**, 495–501.
5. Cheshier, S.H., S.J. Morrison, X. Liao, and I.L. Weissman. (1999) In vivo proliferation and cell cycle kinetics of long-term self-renewing hematopoietic stem cells. *Proc Natl Acad Sci USA* **96**, 3120–3125.

6. Pawliuk, R., C. Eaves, and R.K. Humphries. (1996) Evidence of both ontogeny and transplant dose-regulated expansion of hematopoietic stem cells in vivo. *Blood* **88**, 2852–2858.

7. Bowie, M.B., K.D. McKnight, D.G. Kent, L. McCaffrey, P.A. Hoodless, and C.J. Eaves. (2006) Hematopoietic stem cells proliferate until after birth and show a reversible phase-specific engraftment defect. *J Clin Invest* **116**, 2808–2816.

8. Bradford, G.B., B. Williams, R. Rossi, and I. Bertoncello. (1997) Quiescence, cycling, and turnover in the primitive hematopoietic stem cell compartment. *Exp Hematol* **25**, 445–453.

9. Abkowitz, J.L., S.N. Catlin, and P. Guttorp. (1996) Evidence that hematopoiesis may be a stochastic process in vivo. *Nat Med* **2**, 190–197.

10. Mahmud, N., S.M. Devine, K.P. Weller, S. Parmar, C. Sturgeon, M.C. Nelson, T. Hewett, and R. Hoffman. (2001) The relative quiescence of hematopoietic stem cells in nonhuman primates. *Blood* **97**, 3061–3068.

11. Gothot, A., R. Pyatt, J. McMahel, S. Rice, and E.F. Srour. (1997) Functional heterogeneity of human CD34(+) cells isolated in subcompartments of the G0 /G1 phase of the cell cycle. *Blood* **90**, 4384–4393.

12. Gothot, A., R. Pyatt, J. McMahel, S. Rice, and E.F. Srour. (1998) Assessment of proliferative and colony-forming capacity after successive in vitro divisions of single human CD34+ cells initially isolated in G0. *Exp Hematol* **26**, 562–570.

13. Cheng, T., N. Rodrigues, H. Shen, Y. Yang, D. Dombkowski, M. Sykes, and D.T. Scadden. (2000) Hematopoietic stem cell quiescence maintained by p21(cip1/waf1). *Science* **287**, 1804–1808.

14. Hock, H., M.J. Hamblen, H.M. Rooke, J.W. Schindler, S. Saleque, Y. Fujiwara, and S.H. Orkin. (2004) Gfi-1 restricts proliferation and preserves functional integrity of haematopoietic stem cells. *Nature* **431**, 1002–1007.

15. Spangrude, G.J., S. Heimfeld, and I.L. Weissman. (1988) Purification and characterization of mouse hematopoietic stem cells. *Science* **241**, 58–62.

16. Morrison, S.J., and I.L. Weissman. (1994) The long-term repopulating subset of hematopoietic stem cells is deterministic and isolatable by phenotype. *Immunity* **1**, 661–673.

17. Osawa, M., K. Hanada, H. Hamada, and H. Nakauchi. (1996) Long-term lympho-hematopoietic reconstitution by a single CD34- low/negative hematopoietic stem cell. *Science* **273**, 242–245.

18. Goodell, M.A., K. Brose, G. Paradis, A.S. Conner, and R.C. Mulligan. (1996) Isolation and functional properties of murine hematopoietic stem cells that are replicating in vivo. *J Exp Med* **183**, 1797–1806.

19. Kiel, M.J., O.H. Yilmaz, T. Iwashita, C. Terhorst, and S.J. Morrison. (2005) SLAM family receptors distinguish hematopoietic stem and progenitor cells and reveal endothelial niches for stem cells. *Cell* **121**, 1109–1121.

20. Adolfsson, J., R. Mansson, N. Buza-Vidas, A. Hultquist, K. Liuba, C.T. Jensen, D. Bryder, L. Yang, O.J. Borge, L.A. Thoren, K. Anderson, E. Sitnicka, Y. Sasaki, M. Sigvardsson, and S.E. Jacobsen. (2005) Identification of flt3(+) lympho-myeloid

stem cells lacking erythro-megakaryocytic potential a revised road map for adult blood lineage commitment. *Cell* **121**, 295–306.

21. Van Zant, G. (1984) Studies of hematopoietic stem cells spared by 5-fluorouracil. *J Exp Med* **159**, 679–690.

22. Shapiro, H.M. (1981) Flow cytometric estimation of DNA and RNA content in intact cells stained with Hoechst 33342 and Pyronin Y. *Cytometry* **2**, 143–150.

6

Analysis of Apoptosis in Hematopoietic Stem Cells by Flow Cytometry

William G. Kerr

Summary

Analysis of apoptosis can be used to assess aging and survival in the hematopoietic stem cell (HSC) compartment in the context of disease, therapeutic manipulation, or genetic mutation. Two different methods to assess the frequency of apoptosis in the HSC compartment are presented. The first method utilizes an intracellular TUNEL assay that detects DNA strand breaks, a late apoptotic event. The second method relies on an extracellular stain with recombinant AnnexinV that detects flipping of phosphatidylserine groups to the outer membrane leaflet, an early apoptotic event. Both methods involve an initial magnetic enrichment or sorting of hematopoietic stem/progenitor cells from whole bone marrow (BM). Magnetic sorting is followed by polychromatic antibody (Ab) stains that detect AnnexinV or TUNEL staining in the KFTLS or KTLS HSC phenotypes, respectively. Because of the intracellular detection required for the TUNEL assay, that procedure also includes cell fixation and permeabilization. Electronic gating strategies to assess the frequency of AnnexinV$^+$ or TUNEL$^+$ cells in KFTLS or KTLS HSC phenotypes are also described along with representative examples.

Key Words: HSC; TUNEL; AnnexinV; flow cytometry; Sca1; Flk2; Lin; *c-Kit*; Thy1; SH2-containing inositol phosphatase (SHIP).

1. Introduction

Apoptosis is a physiological process that controls tissue kinetics and homeostasis (*1*). Cells that form a tissue or compartment in the body receive constant input from the extracellular milieu. This input determines whether a cell will continue to survive or meet its demise through apoptosis. The extracellular inputs that influence this survival versus apoptosis decision include growth

From: *Methods in Molecular Biology, vol. 430: Hematopoietic Stem Cell Protocols*
Edited by: K. D. Bunting © Humana Press, Totowa, NJ

factors, extracellular matrix components, and hormones. Cells receive these inputs through receptors specific for each of these extracellular components, whereas intracellular signaling pathways coupled to these receptors enable the cells to interpret and integrate this diverse array of signals to determine the final outcome: death or survival.

Like other tissues or cell types in the body, the hematopoietic stem cell (HSC) compartment appears to regulate its numbers. The HSC compartment must achieve a balance between maintaining sufficient HSC numbers to sustain hematopoiesis, while also avoiding inappropriate expansion of the compartment that would predispose the host to leukemogenesis. The need for fine control of HSC survival versus apoptosis is exemplified by hematologic dysfunction and bone marrow (BM) failure that occurs with increased frequency in older humans and mice *(2–5)*. In recent years, genetic analysis has revealed intracellular signaling and apoptosis components that play a crucial role in determining the frequency of apoptosis in the HSC compartment. These include SH2-containing inositol phosphatase (SHIP) *(6)*, p16INK4a *(7)*, GATA1 *(8)*, Bcl2 *(9,10)*, and Mcl1 *(11)*. Osteopontin, an extracellular matrix protein expressed on BM stroma, has been shown to restrict HSC compartment size indicating extracellular factors in the BM niche can also play a pivotal role in the survival versus apoptosis decision of HSC *(12)*.

These studies of HSC survival or demise in disease and genetic mutation highlight the need for quantitative assays to assess the frequency of apoptosis in the HSC compartment. Below, we present two different approaches that enable the quantitation of apoptosis in HSC. These assays will facilitate the study of how genetic mutation, disease, or therapeutic manipulation impact survival in the HSC compartment.

2. Materials

2.1. Buffers, Reagents, and Materials Required for Both the TUNEL and the AnnexinV Assays

1. Tissue medium: RPMI 1640 medium supplemented with 3% fetal bovine serum (FBS) and 10 mM HEPES. Stored at 4 °C.
2. Red blood cell (RBC) lysis buffer: Purchased from E-bioscience, Inc. (San Diego, CA, USA).
3. Phosphate-buffered saline (PBS): Purchased from GIBCO, Inc. (Carlsbad, CA, USA).
4. Staining medium (SM): PBS supplemented with 3% FBS and 10 mM HEPES. Stored at 4 °C.
5. 70-μm cell strainer: Purchased from BD Biosciences, Inc. (Bedford MA, USA).
6. Hemacytometer: Purchased from Hausser Scientific, Inc. (Horsham PA, USA).

7. Trypan Blue: Purchased from SIGMA-Aldrich, Inc. (St Louis, MO, USA).
8. Fc receptor (FcR)-blocking Ab: Anti-CD16 (clone 2.4G2). Purchased from BD Biosciences, Inc. (San Diego, CA, USA).
9. Miltenyi buffer (MB): PBS supplemented with 0.5% BSA, 2 mM EDTA, and 10 mM HEPES. Stored at 4 °C.
10. MACS Separation Column: Purchased from Miltenyi Biotec, Inc. (Auburn, CA, USA).
11. AutoMACS: The AutoMACS separator from Miltenyi Biotec is an automated benchtop magnetic cell sorter.
12. BD FACSAria Cell-Sorting System: The FACSAria cell sorter is an easy-to-use benchtop system used for high-speed sorting and multicolor analysis.
13. All fluorochrome-conjugated antibodies used in the TUNEL and the AnnexinV assays can be purchased from BD Biosciences, Inc. or E-Biosciences, Inc.

2.2. TUNEL Assay to Measure Apoptosis in HSC by Flow Cytometry

1. Lineage antibodies conjugated to PE (anti-B220, anti-CD3, anti-CD5, anti-Gr1, anti-Mac1, anti-Ter119): Purchased from BD Biosciences, Inc.
2. Anti-PE Microbeads: Purchased from Miltenyi Biotec, Inc. (Cat. No. 130-048-801)
3. "In Situ Cell Death Detection Kit, Fluorescein" Purchased from Roche Applied Sciences (Indianapolis, IN, USA) (Cat. No. 11684795910), which includes TUNEL Enzyme Solution and TUNEL Labeling Solution.
4. Fixation solution: 4% paraformaldehyde in PBS, pH 7.4. Should be prepared immediately before use.
5. TUNEL Reaction Mixture: Mix 50 µl of TUNEL Enzyme Solution with 450 µl of TUNEL Labeling Solution, both provided in Roche kit.
6. DNAseI recombinant: Purchased from Roche Applied Sciences (Cat. No. 04536282001).
7. Permeabilization reagent: 0.1% Triton-X 100 in 0.1% sodium citrate.
8. Clear polyolefin sealing tape size 121 × 79 mm: Purchased from Fisher Scientific (Hampton, NH, USA)
9. V-bottom 96-well microplate: Purchased from Nunc, TM. Distributed by Fisher Scientific.

2.3. Gating Strategy for Analysis of TUNEL Flow Cytometry Data

1. FlowJo software: Purchased from Treestar, Inc. (Ashland, OR, USA)

2.4. AnnexinV Assay to Measure Apoptosis in HSC by Flow Cytometry

1. Anti-Sca1-biotin antibody: Purchased from BD Biosciences, Inc.
2. Anti-Biotin Microbeads: Purchased from Miltenyi Biotec, Inc. (Cat. No. 130-090-485).
3. AnnexinV stain: AnnexinV-Cy5.5 (BD Biosciences, Inc.) (Cat. No. 559935).

4. 1× AnnexinV-Binding Buffer: diluted from 10× AnnexinV-Binding Buffer (BD Biosciences, Inc.) (Cat. No. 556454).
5. DAPI stain in AnnexinV-binding buffer: Prepare "DAPI dilution 1" from a 375 ng/ml DAPI stock by diluting 2 μl into 2 ml of AnnexinV-Binding Buffer (DAPI dilution 1 = 5 μg/ml). To prepare the final DAPI stain in AnnexinV-binding buffer, place 60 μl of "DAPI dilution 1" in 3.94 ml of AnnexinV-binding buffer (Final DAPI concentration = 75 ng/ml).

2.5. Gating Strategy for Analysis of AnnexinV Flow Cytometry Data

1. FlowJo software: Purchased from Treestar, Inc.

3. Methods

We currently use two different assays to analyze the frequency of apoptosis in the HSC compartment: the TUNEL assay and AnnexinV staining. The TUNEL assay detects double-stranded breaks in chromosomal DNA that represent the final stage of apoptosis. These strand-breaks are detected by incorporation of dUTP covalently labeled with biotin by the enzyme terminal deoxynucleotide transferase (TdT). The AnnexinV assay detects an earlier event in apoptosis— "flipping" of phospholipids that contain a phosphatidylserine group from the inner leaflet in the phospholipid bilayer that constitutes the plasma membrane to the outer leaflet. Extracellular exposure of phosphatidylserine is then detected by staining with recombinant AnnexinV coupled to a fluorochrome. Because the TUNEL assay measures an intracellular apoptosis event, it must be performed on fixed and permeabilized cells. The AnnexinV assay can be performed on viable cells.

3.1. TUNEL Assay to Measure Apoptosis in HSC by Flow Cytometry

1. Isolate whole BM cells from humerus, tibia, femur, and vertebral column by flushing or crushing in tissue media (RPMI 1640, 3% FBS, 10 mM HEPES) (*see* **Note 1**).
2. Combine the single cell suspension prepared from the marrow of the above tissues in a single 50-ml conical tube by passing through a 70-μm mesh filter or cell strainer and pellet them at $300 \times g$ (*see* **Note 2**) for 5 min at 4 °C.
3. Decant supernatant and resuspend pellet in RBC lysis buffer (~4 ml/mouse in tube).
4. Allow RBC to lyse at room temperature (RT) for 2 min and then dilute with 10 vol of cold PBS to halt hypotonic lysis.
5. Pellet cells at $300 \times g$ for 5 min at 4 °C.
6. Decant supernatant and resuspend pellet in 15 ml SM.

7. Pass the cells through a 70-μm cell strainer to remove any clumps.
8. Perform a viable cell count on a hemacytometer:

   (a) Take a 20-μl aliquot and transfer to a microfuge tube.
   (b) To that same microfuge tube add 20 μl of PBS.
   (c) Then add 40 μl of Trypan Blue and mix well.
   (d) Take 9.5 μl of the mixture and transfer onto a hemacytometer.
   (e) Count the cells:

   – (X1 + X2 + X3 + X4)/4 × 10,000 × 4 = amount of cells/ml, where X is one of the four quadrants on the hemacytometer
   – Amount of cells/ml × 15 ml = total amount of cells

   (f) Resuspend at 1×10^6 cells/50 μl SM buffer for blocking and staining steps.

9. Once the cells are counted and resuspended at the designated concentration, set aside cells for control stains (*see* **Notes 3** and **4**).

   (a) Suggested controls for Tunnel assay:

   – No stain control: 1 million cells
   – *Negative control* for TUNEL stain: FMO-FITC stain with TUNEL Labeling Solution added: 2 million cells
   – FMO-PE: 2 million cells
   – FMO-PeCy7: 2 million cells
   – FMO-APC: 2 million cells
   – FMO-PeCy5: 2 million cells
   – *Positive control* for TUNEL reaction (DNAseI treatment) (stain): 2 million cells (removed later from Lin⁻ cells)

10. Take the remaining cells and prepare them for depletion of cells positive for lineage commitment marker panel using the Lin-PE antibody (Ab) cocktail below.
11. To prevent non-specific Ab staining because of Fc region binding, incubate the cells with FcR-blocking Ab (0.5 μg/1×10^6 cells) in SM (1×10^6 cells/50 μl SM).
12. Incubate at 4 °C for 15–20 min and then wash with 4 vol of SM, mix and pellet cells at $300 \times g$ for 5 min at 4 °C.
13. Prepare Lin-PE Ab cocktail for the cells. Lin-PE cocktail: anti-B220, anti-CD3, anti-CD5, anti-Gr1, anti-Mac1, anti-Ter119 (0.5 μg/1×10^6cells) in SM (1×10^6 cells/50 μl SM) (*see* **Note 5**).
14. Resuspend cell pellet in Lin-PE Ab cocktail and incubate at 4 °C for 15–20 min.
15. Wash with 2 vol of MB, pellet at $300 \times g$ for 5 min at 4 °C, decant and resuspend in Miltenyi Biotec anti-PE microbeads diluted in MB. Anti-PE microbeads should be used at 10 μl of beads/million cells in 90 μl MB/million cells. Optimally, the microbead stain should be done for 15–20 min in the refrigerator (4–8 °C)

(*see* **Note 6**). After incubation, dilute the cell/bead mixture with 4 vol of MB and pellet at $300 \times g$ for 5 min at 4 °C.

16. Resuspend the anti-PE bead-cell pellet at a concentration of 1×10^8 cells/500 µl MB and proceed with the AutoMACS depletion step.

17. For the AutoMACS depletion step, run sample through a fresh separation column on the program "DepleteS" (sensitive depletion)—keep the negative pass-through cell fraction (Lin⁻ fraction) and recount these cells for subsequent HSC and TUNEL staining (*see* **Note 7**).

18. Pellet the Lin⁻ cell fraction at $300 \times g$ for 5 min at 4 °C and count on a hemacytometer as before. Resuspend at 50 µl/million cells in SM.

19. Set aside 2×10^6 Lin⁻ cells for the TUNEL-positive control stain (DNAse treatment) mentioned earlier.

20. On the remaining cells, repeat the FcR block step (0.5 µg/1×10^6 cells) in SM for 15–20 min at 4 °C. Add 4 vol of SM to wash and centrifuge at $300 \times g$ for 5 min at 4 °C. Now proceed with the extracellular Ab stains for the HSC phenotype of choice. For the KTLS HSC phenotype *(13)*, perform the following extracellular stains:

    (a) Sca1-biotin (0.5 µg/1×10^6 cells) in SM (1×10^6 cells/50 µl). Stain at 4 °C for 15–20 min, then wash with 4 vol of SM. Pellet at $300 \times g$ for 5 min at 4 °C and remove the supernatant.

    (b) Secondary stain: Lineage-PE [B220, CD3, CD4, CD5, CD8, Gr1, Mac1, Ter119, NK1.1 (0.4 µg/1×10^6), SA-PE/Cy7 (0.4 µg/1×10^6), *c-Kit*-APC (0.5 µg/1×10^6), Thy1.2-PE/Cy5 (0.3 µg/1×10^6) in SM (1×10^6 cells/50 µl). Stain at 4 °C for 15–20 min.]

21. Add 4 vol of PBS, mix and pellet cells at $300 \times g$ for 5 min at 4 °C.

22. Wash with 4 vol of PBS and pellet cells at $300 \times g$ for 5 min at 4 °C. Repeat and resuspend cell pellet at 2×10^7 cells/ml in PBS.

23. Transfer the cell suspension at 100 µl/well into a V-bottom microplate (*see* **Note 9**). Also transfer the control cell samples, including the TUNEL-positive and TUNEL-negative control samples that were previously set aside, to the V-bottom plate for fixation and permeabilization.

24. Add 100 µl of freshly prepared fixation solution to each cell suspension so that the final concentration of paraformaldehyde is 2%. Resuspend well and incubate for 60 min at 15–25 °C on an orbital shaker (*see* **Note 8**). Gentle agitation is advised to avoid cell clumping during fixation.

25. Centrifuge microplate at $300 \times g$ for 8 min at 4 °C. Decant supernatant being careful to remove all excess supernatant.

26. Resuspend the fixed cell pellet with 200 µl/well of PBS and centrifuge at $300 \times g$ for 8 min at 4 °C.

27. Resuspend the fixed cell pellet in 200 µl/well of PBS and store in the dark at 4 °C until ready to proceed with the TUNEL reaction below (*see* **Note 10**).

28. Preparation of the TUNEL Reaction mixture: (a) Set aside 100 µl of TUNEL Labeling Solution provided in the Roche kit for the labeling of negative controls.

Add 50 μl of TUNEL Enzyme Solution to the remaining 450 μl of TUNEL Labeling Solution. Mix well to equilibrate. This should yield a total of 500 μl of TUNEL Reaction Mixture (enough for five reactions).

29. When ready to proceed with the TUNEL reaction, the fixed Lin⁻ cells in the microplate are spun at 300 × g for 10 min at 4°C and washed once more with PBS and pelleted.

30. Remove all excess supernatant from the fixed cell pellet and resuspend in 100 μl/well of permeabilization reagent and incubate for 2 min at 4°C. After a 2-min incubation, wash the cells by adding PBS (100 μl/well) and pellet the cells at 300 × g for 10 min at 4°C.

31. Add 250 μl/well of PBS and leave on ice until the TUNEL-positive control sample is ready to be centrifuged.

32. To prepare the TUNEL-positive control sample, first prepare DNaseI solution by adding 6 μl of 1 mg/ml DNAseI solution to 194 μl of Tris/BSA (50 mM Tris–HCl, pH 7.5, 1 mg/ml BSA). This yields a buffered DNAseI solution. Add 3 U/ml of DNaseI solution to the Lin⁻ cell sample set aside in **step 3.1.19** and incubate for 10 min at RT to induce DNA strand breaks.

33. Centrifuge the microplate including the experimental samples and the DNAse-treated positive control and negative control cell samples at 300 × g for 10 min at 4°C.

34. Add 200 μl PBS to no stain and compensation control cell samples from **step 3.1.9**.

35. Add 50 μl/well of TUNEL Reaction Mixture to the test sample(s) and the TUNEL-positive control sample.

36. Add 50 μl of the unmixed TUNEL Labeling Solution to the TUNEL negative control cell sample from **step 3.1.9**.

37. Incubate all samples from **steps 3.1.34–36** for 60 min at 37°C in the dark with a sealing polyolefin or parafilm cover over the microplate.

38. Centrifuge at 300 × g for 8 min at 4°C. Resuspend the cell pellet in 250 μl/well of PBS.

39. Centrifuge at 300 × g for 8 min at 4°C.

40. Resuspend in 300 μl/well of PBS. Analyze the samples on the FACS Aria. Collect data on 500,000–1,000,000 events per stain.

3.2. Gating Strategy for Analysis of TUNEL Flow Cytometry Data (see Note 13)

1. After collecting the raw list mode data on the flow cytometer apply a non-rectangular gate (*see* **Fig. 1**, GATE 1) based on Forward (FSC) and Side Scatter (SSC). This gate excludes cell debris and cell ghosts that have very low FSC and unusually large cells representing myeloid cells, stromal cells, or cell doublets. These events can have anomalous fluorescence properties or non-specific absorption of antibodies, and thus, this gate enhances the accuracy of fluorescent measurements in subsequent gates.

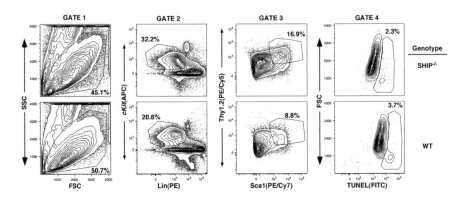

Fig. 1. Gating scheme to assess the frequency of TUNEL$^+$ cells in the hematopoietic stem cell (HSC) compartment. Sequence of electronic gates applied to list mode cytometer data collected for SH2-containing inositol phosphatase$^{-/-}$ (SHIP$^{-/-}$) and wild type (WT) BM cells following magnetic cell sorting to deplete Lin$^+$ cells as described above in **Subheading 3.1**. The percentage shown in each contour plot represents the fraction of events that fall within the indicated gate. All cytometry plots shown represent two-parameter contour plots (1% probability) with the last 1% of cells displayed as single dots. The flow cytometry analysis program FlowJo6.0 was used to generate the contour plots shown and to estimate the frequency of cells present in each gate. These examples show gates for the KTLS phenotype of Morrison and Weissman *(13)*.

2. Apply a non-rectangular c-Kit$^+$Lin$^-$ gate analogous to that shown in **Fig. 1** (GATE 2). This gate captures the fraction of BM that contains essentially all HSC along with less-primitive hematopoeitic progenitors.
3. Apply a non-rectangular Sca1$^+$Thy1$^+$ gate analogous to that shown in **Fig. 1** (GATE 3).
4. Apply a non-rectangular gate analogous to that shown in **Fig. 1** (GATE 4) to estimate the proportion of TUNEL$^+$cells present in the cKit$^+$Thy1$^+$Lin$^-$Sca1$^+$ HSC compartment.

3.3. AnnexinV Assay to Measure Apoptosis in HSC by Flow Cytometry

1. Isolate whole BM cells from humerus, tibia, femurs, and vertebral column by flushing or crushing in tissue media (RPMI 1640, 3% FBS, 10 mM HEPES).
2. Combine the single cell suspension prepared from the marrow of the above tissues into a single 50-ml conical tube after passing through a 70-μm mesh filter or cell strainer to remove any cell clumps. Pellet the cells at $300 \times g$ for 5 min at 4 °C.
3. Decant the supernatant and resuspend the cell pellet in 15 ml of SM (PBS, 3% FBS, 10 mM HEPES) (*see* **Note 11**).
4. Count the cells on a hemacytometer as described in **Subheading 3.1.8**.

5. After counting the cells, set aside cells for various control stains (*see* **Notes 3 and 4**).

   Suggested controls:
   No stain: 1 million cells
   FMO-FITC: 2 million cells
   FMO-PE: 2 million cells
   FMO-PECy7: 2 million cells
   FMO-APC: 2 million cells
   FMO-APCCy7: 2 million cells
   FMO-Cy5.5: 2 million cells
   Positive control (incubated at 95 °C for 2 min): 2 million cells

6. Take the remaining cells and prepare them for magnetic enrichment. First block the FcR on the cells with FcR-blocking Ab (0.5 µg/1 × 10^6cells) in SM at 1 × 10^6 cells/50 µl SM buffer. Incubate at 4 °C for 15–20 min. Then add 4 vol of SM, mix and pellet cells at 300 × g for 5 min at 4 °C.

7. To prepare the cells for magnetic enrichment, resuspend them in a cocktail of anti-Sca1-biotin (0.5 µg/1 × 10^6 cells) in SM (1 × 10^6 cells/50 µl SM). Incubate for 15–20 min at 4 °C.

8. Wash with 2 vol of MB, pellet at 300 × g for 5 min at 4 °C, decant and resuspend in Miltenyi Biotec anti-biotin microbeads diluted MB. Anti-biotin microbeads should be used at 10 µl beads/million cells in 90 µl MB/million cells. Optimally, the microbead stain should be done for 15–20 min in the refrigerator (4–8 °C) (*see* **Note 6**). After incubation dilute the cell/bead mixture with 4 vol of MB and pellet at 300 × g for 5 min at 4 °C.

9. Resuspend the bead–cell mixture at a concentration of 1 × 10^8 cells/500 µl MB and proceed with the AutoMACS enrichment. Run sample over a fresh Miltenyi separation column on the program "Posel S" (sensitive positive selection).

10. Keep the Sca1$^+$ fraction from the AutoMACS enrichment and recount these cells for staining (*see* **Note 12**).

11. After counting, pellet the cells and resuspend them at 50 µl/1 × 10^6 cells in SM.

12. Repeat FcR block stain (0.5 µg/1 × 10^6 cells) in 1 × 10^6 cells/50-µl SM buffer for 15–20 min at 4 °C. Wash with 4 vol of SM, mix and then pellet the cells at 300 × g for 5 min at 4 °C.

13. Proceed with the "KTLSFlk" HSC phenotype *(14)* stain on the Sca1$^+$-enriched cell fraction.

    (1) Primary stain: Anti-Sca1-biotin (0.5 µg/1 × 10^6 cells) in SM. Incubate at 4 °C for 15–20 min, then wash with 4 vol of SM. Pellet at 300 × g for 5 min at 4 °C and remove the supernatant.

    (2) Secondary stain: Lineage-FITC (B220, CD2, CD3, CD4, CD5, CD8, Gr1, Mac1, Ter119, NK1.1 (1 µg/1 × 10^6cells), Flk2-PE (0.4 µg/1 × 10^6cells), SA-PE/Cy7 (0.4 µg/1 × 10^6cells), cKit-APC/Cy7 (0.5 µg/1 × 10^6cells), Thy1-APC (0.5 µg/1 × 10^6) SM at 50 µl/1 × 10^6 cells. Stain at 4 °C for 15–20 min.

14. Add 4 vol of SM, mix and pellet cells at $300 \times g$ for 5 min at 4 °C.
15. Wash with 2 vol of PBS and pellet cells at $300 \times g$ for 5 min at 4 °C.
16. Resuspend the cell pellet of each sample in 100 µl DAPI stain in AnnexinV-binding buffer (*see* **Subheading 2.3.5**).
17. Add 10 µl AnnexinV-Cy5.5 to each sample. Gently vortex the cells and incubate for 15 min at RT in the dark.
18. Add an additional 400 µl of DAPI stain in AnnexinV-binding buffer to each sample and analyze by flow cytometry within 1 hour. Analyze the samples on the FACS Aria. Collect data on 500,000–1,000,000 events per stain.

3.4. Gating Strategy for Analysis of AnnexinV Flow Cytometry Data (see Note 13)

1. After collecting the raw list mode data on the flow cytometer, apply a non-rectangular gate (*see* **Fig. 2,** GATE 1) based on FSC and SSC. This gate excludes cell debris, cell ghosts that have very low FSC and unusually large cells representing myeloid cells, stromal cells, or cell doublets. Because these events can have anomalous fluorescence properties or non-specific absorption of antibodies, this gate typically enhances the accuracy of fluorescent measurements in subsequent gatings.

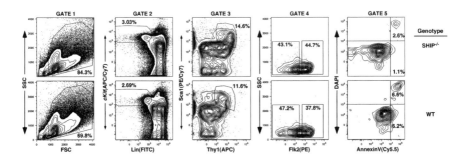

Fig. 2. Gating scheme to assess the frequency of AnnexinV$^+$ hematopoietic stem cells (HSC). Sequence of electronic gates applied to list mode cytometer data collected for SH2-containing inositol phosphatase$^{-/-}$ (SHIP$^{-/-}$) and WT BM cells following magnetic cell sorting of Sca1$^+$ stem/progenitor cells as described above in **Subheading 3.2**. Each contour plot represents the events contained in the previous electronic gate. The percentage shown in each contour plot represents the fraction of events that fall within the indicated gate. All flow cytometry plots shown represent two-parameter contour plots (1% probability) with the last 1% of cells displayed as single dots. The flow cytometry analysis program FlowJo6.0 was used to generate the contour plots shown and to estimate the frequency of cells present in each gate. These examples show gates for the KTFLS phenotype of Christensen and Weissman *(14)*.

2. Apply a non-rectangular c-Kit$^+$Lin$^-$ gate analogous to that shown in **Fig. 2** (GATE 2). This gate captures the fraction of BM that contains essentially all HSC along with less primitive hematopoeitic progenitors.
3. Apply a non-rectangular Sca1$^+$Thy1$^+$ gate analogous to that shown in **Fig. 2** (GATE 3).
4. Display SSC versus Flk2 contour plot on the cells from GATE 3 and then apply a rectangular Flk2$^-$ gate analogous to that shown in **Fig. 2** (GATE 4).
5. Apply a rectangular gate analogous to that shown in **Fig. 2** (GATE 5) to estimate the proportion of AnnexinV$^+$/Dapi$^-$ cells present in the cKit$^+$Thy1$^+$Flk2$^-$Lin$^-$Sca1$^+$ HSC compartment. The AnnexinV$^+$/Dapi$^-$ proportion of the HSC compartment is present in the lower right hand rectangular gate or quadrant.

4. Notes

1. Harvest of tissues: To obtain marrow from arm and leg bones, we flush the marrow out of the hollow limb bones with a 3-cc needle and syringe. The vertebral marrow is obtained by crushing the spinal column with a mortar and pestle.
2. The 300 × g centrifugal speed used to pellet single cell suspension in physiological buffers is typically ~1100–1200 rpm on low speed centrifuges designed for cell culture.
3. FMO = Fluorescence Minus One. FMO stains are highly advised as they help the investigator assess both the degree of autofluorescence in a given fluorescence channel and the amount of emission "spillover" into that channel from other fluorochromes used in a given poly-chromatic cytometry stain.
4. The suggested controls listed for each assay is a minimal set. It is recommended that you include additional controls as dictated by your specific experimental design.
5. In **Subheading 3.1**, **step 13**, the Lin-PE used is a "minimal" Lin panel and is only for the magnetic cell-sorting step used to deplete the majority of lineage-committed hematopoietic cells before the actual FACS stain for HSC. The Lin panel used for flow cytometric detection and quantitation of HSC numbers is a more extensive panel of Abs.
6. Microbead staining as described in **Subheading 3.1**, **step 15**, and **Subheading 3.3**, **step 8**, is optimal at 4–8 °C, and thus, this step is performed in a refrigerator rather than on ice.
7. After Lineage depletion of cells, we typically obtain 2–3 × 10^6 Lin$^-$ cells per mouse. Note that humeri, tibia, femurs, and vertebral column are harvested from each mouse for maximal cell numbers.
8. The shaker that we use is an orbital shaker set at speed 4–5 (100 rpm): Purchased from BELLCO Glass, Inc. (Vineland, NJ, USA).
9. The use of a V-bottomed plate is recommended in the TUNEL assay to minimize cell loss during labeling.

10. After **step 27**, **Subheading 3.1**, and before proceeding with TUNEL reaction, we find that the fixed Lin⁻ cells can be stored overnight at 4 °C in the dark.

11. For the AnnexinV assay, no RBC lysis is performed on whole BM cells to avoid disruption of cell integrity.

12. After enrichment for Sca1$^+$ cells, we typically obtain 2–3 × 10^6 Sca1$^+$ cells per mouse. Note that humeri, tibia, femurs, and vertebral column are harvested from each mouse as this maximizes cell yield.

13. We encourage users to consult a concise and helpful review of software-based approaches to analysis of flow cytometry data *(15)*.

Acknowledgments

The author thanks Caroline Desponts (now a post-doctoral fellow at Scripps Institute) who originally defined the protocols described here as part of her thesis work. The author is also particularly indebted to Amy Hazen who helped distill the above protocols from various laboratory notebooks. This work was supported in part by grants from NHLBI (R01 HL72523), NIDDK (R21 DK071872-01), the Pediatric Cancer Foundation, Susan Komen Breast Cancer Research Foundation, and the Philip Morris Lung Cancer Research Foundation. WGK is the Newman Family Scholar of the Leukemia and Lymphoma Society.

References

1. Kerr, J. F., Wyllie, A. H., and Currie, A. R. (1972) Apoptosis: a basic biological phenomenon with wide-ranging implications in tissue kinetics. *Br J Cancer* **26**, 239–257.

2. Ogden, D. A. and Micklem, H. S. (1976) The fate of serially transplanted bone marrow cell populations from young and old donors. *Transplantation* **22**, 287–293.

3. Morrison, S. J., Wandycz, A. M., Akashi, K., Globerson, A., and Weissman, I. L. (1996) The aging of hematopoietic stem cells. *Nat Med* **2**, 1011–1016.

4. Chen, J., Astle, C. M., and Harrison, D. E. (1999) Development and aging of primitive hematopoietic stem cells in BALB/cBy mice. *Exp Hematol* **27**, 928–935.

5. Liang, Y., Van Zant, G., and Szilvassy, S. J. (2005) Effects of aging on the homing and engraftment of murine hematopoietic stem and progenitor cells. *Blood* **106**, 1479–1487.

6. Desponts, C., Hazen, A. L., Paraiso, K. H., and Kerr, W. G. (2006) SHIP deficiency enhances HSC proliferation and survival but compromises homing and repopulation. *Blood* **107**, 4338–4345.

7. Janzen, V., Forkert, R., Fleming, H. E., Saito, Y., Waring, M. T., Dombkowski, D. M., Cheng, T., DePinho, R. A., Sharpless, N. E., and Scadden, D. T. (2006) Stem-cell ageing modified by the cyclin-dependent kinase inhibitor p16INK4a. *Nature* **443**, 421–426.

8. Rodrigues, N. P., Janzen, V., Forkert, R., Dombkowski, D. M., Boyd, A. S., Orkin, S. H., Enver, T., Vyas, P., and Scadden, D. T. (2005) Haploinsufficiency of GATA-2 perturbs adult hematopoietic stem-cell homeostasis. *Blood* **106**, 477–484.

9. Domen, J., Cheshier, S. H., and Weissman, I. L. (2000) The role of apoptosis in the regulation of hematopoietic stem cells: Overexpression of Bcl-2 increases both their number and repopulation potential. *J Exp Med* **191**, 253–264.

10. Orelio, C., Harvey, K. N., Miles, C., Oostendorp, R. A., van der Horn, K., and Dzierzak, E. (2004) The role of apoptosis in the development of AGM hematopoietic stem cells revealed by Bcl-2 overexpression. *Blood* **103**, 4084–4092.

11. Opferman, J. T., Iwasaki, H., Ong, C. C., Suh, H., Mizuno, S., Akashi, K., and Korsmeyer, S. J. (2005) Obligate role of anti-apoptotic MCL-1 in the survival of hematopoietic stem cells. *Science* **307**, 1101–1104.

12. Stier, S., Ko, Y., Forkert, R., Lutz, C., Neuhaus, T., Grunewald, E., Cheng, T., Dombkowski, D., Calvi, L. M., Rittling, S. R., and Scadden, D. T. (2005) Osteo-pontin is a hematopoietic stem cell niche component that negatively regulates stem cell pool size. *J Exp Med* **201**, 1781–1791.

13. Morrison, S. J. and Weissman, I. L. (1994) The long-term repopulating subset of hematopoietic stem cells is deterministic and isolatable by phenotype. *Immunity* **1**, 661–673.

14. Christensen, J. L. and Weissman, I. L. (2001) Flk-2 is a marker in hematopoietic stem cell differentiation: a simple method to isolate long-term stem cells. *Proc Natl Acad Sci USA* **98**, 14541–14546.

15. Herzenberg, L. A., Tung, J., Moore, W. A., Herzenberg, L. A., and Parks, D. R. (2006) Interpreting flow cytometry data: a guide for the perplexed. *Nat Immunol* **7**, 681–685.

III

In Vitro Assays and Differentiation

7

In Vitro Hematopoietic Differentiation of Murine Embryonic Stem Cells

Jinhua Shen and Cheng-Kui Qu

Summary

In recent years, the field of stem cells has become one of the most rapidly growing areas in biological and medical sciences. Embryonic stem (ES) cells differentiate efficiently in vitro and give rise to many different somatic cell types. The ability to generate a wide spectrum of differentiated cell types from ES cells in culture offers a powerful approach for studying lineage induction and specification and a promising source of progenitors for cell replacement therapy. Hematopoietic progenitors present within ES cell-derived embryoid bodies (EB) can be assayed by directly replating EB cells or by replating sorted cell populations into semisolid media with hematopoietic growth factors. The developmental kinetics of various hematopoietic lineage precursors within EBs and molecular and cellular studies of these cells have suggested that the sequence of events leading to the onset of hematopoiesis within EB is similar to that found within the mouse embryo. Thus, the in vitro differentiation model of ES cells to hematopoietic cells provides a unique opportunity to study onset mechanisms involved in hematopoietic development and to characterize hematopoietic lineage-specific gene expression. In this chapter, we attempt to be as comprehensive as possible and yet focus on what we perceive to be the most widely used protocols for maintenance of murine ES cells, in vitro hematopoietic differentiation of ES cells, and clonal assays of hematopoietic progenitors.

Key Words: ES cells; embryoid body; hematopoietic progenitors; hematopoietic differentiation; colony assay.

1. Introduction

Murine ES cells are totipotent cells derived from the inner cell mass (ICM) of the day 3.5 blastocyst. These cells possess properties of both the ICM and the

From: *Methods in Molecular Biology, vol. 430: Hematopoietic Stem Cell Protocols*
Edited by: K. D. Bunting © Humana Press, Totowa, NJ

ectoderm-like cells. Under the appropriate culture conditions, ES cells retain the capacity to contribute to all cell lineages when reimplanted back into a blastocyst *(1)*. This potential, combined with their ease of genetic manipulation and selection, has revolutionized many fields by facilitating the ability to generate transgenic, chimeric, "knock-out," and "knock-in" mice for gene function studies in vivo.

ES cells can be maintained in an undifferentiated state by culturing them on feeder cell layers or on gelatinized plates with the addition of leukemia inhibitory factor (LIF) *(2,3)*. When allowed to differentiate in culture, ES cells can differentiate in vitro into complex structures called embryoid bodies (EBs) that contain a number of different cell types. Assay systems have been devised for the detection of various cell types including endothelial, neuronal, muscle, and hematopoietic progenitors *(4–6)*. Various techniques have been used to promote hematopoietic differentiation, including culture on stromal layers *(7–10)*, in chemically defined suspension media in the presence of inducers of hematopoiesis *(11)*, and in methylcellulose-based semisolid media containing cytokines *(12,13)*.

The ES cell-EB model has been particularly useful in elucidating the early events involved in the development of the hematopoietic system and has enabled the identification of a progenitor with characteristics of the hemangioblast, the putative precursor of the hematopoietic and endothelial lineages. These progenitors, known as blast colony-forming cells (BL-CFCs), arise within 2–4 days of EB differentiation and express the tyrosine kinase receptor Flk1 *(14,15)*. When replated in methylcellulose cultures in the presence of Vascular endothelial growth factor (VEGF), the BL-CFCs generate colonies with endothelial and primitive and definitive hematopoietic potential *(16)*. These characteristics suggest that the BL-CFC could represent the in vitro equivalent of the yolk sac hemangioblast and, as such, the earliest commitment step in the differentiation of mesoderm to the hematopoietic and endothelial lineages. Hematopoietic and endothelial progenitors present within EBs can be successfully analyzed by flow cytometry and by direct replating EB cells into methylcellulose cultures to measure the frequency of hematopoietic progenitors. Additionally, EB cells can be sorted for early hematopoietic and endothelial cell markers to further analyze their hematopoietic and/or endothelial cell potential *(17–19)*.

There are several potential advantages to using the ES system as a means to identify and analyze the molecules that regulate early hematopoietic development. First, at all stages of the developmental process, there is accessibility to sufficient numbers of cells for analysis. Second, the effects of genetic manipulations on the cell types of interest can be examined without concern for embryonic lethality. In addition, the relative ease with which ES cells can be genetically manipulated, clones isolated, and hematopoiesis accurately assessed

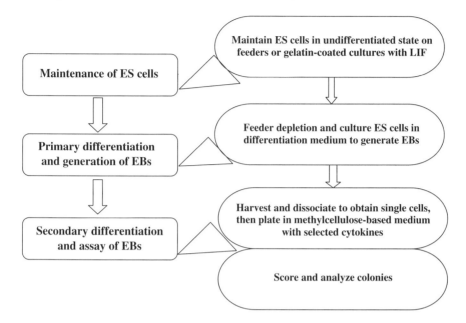

Fig. 1. Scheme for in vitro hematopoietic differentiation of embryonic stem cells.

makes this an exceptionally powerful screening technique for identifying and characterizing genes that may be involved in the process of hematopoiesis.

In this chapter, we will focus on introducing the most widely used protocols for the maintenance of ES cells and their differentiation into two stages of development: the BL-CFC/hemangioblast stage and the primitive/definitive hematopoietic stage. The scheme for this two-step in vitro hematopoietic differentiation of ES cells is depicted in **Fig. 1**.

2. Materials

2.1. Mouse ES Cell Maintenance

1. ES cell growth medium:

   a. Dulbecco's Minimal Essential Medium (DMEM) (high glucose, Invitrogen, Carlsbad, CA, USA, Cat. No. 11960-044).

   b. 15% (v/v) fetal bovine serum (FBS) (Hyclone, heat inactivated, 56°C, 30 min) (*see* **Note 1**).

   c. 2 mM L-glutamine (100× stock, Invitrogen, Cat. No. 25030-149, aliquoted and stored at –20°C, stable in solution for 10 days).

   d. 0.1 mM non-essential amino acids (100× stock, Invitrogen, Cat. No. 11140-050, aliquoted and stored at 4°C).

  e. 1 mM sodium pyruvate (100× stock, Invitrogen, Cat. No. 11360-070, aliquoted and stored at 4°C).
  f. 100 U/µg/ml of Pen/Strep (100× stock, Invitrogen, Cat. No. 15140-122, aliquoted and stored at –20°C).
  g. 0.1 mM 2-mercaptoethanol (Sigma, Atlanta, GA, USA, Cat. No. M7522, aliquoted and stored at –20°C).
  h. 1000 U/ml LIF (Millipore, Billerica, MA, USA, Cat. No. ESG1107, make up 100× stock solution in DMEM with 10% (v/v) or so serum, aliquoted store at –20°C).

2. Mitomycin C 100× stock solution (Sigma, Cat. No. M4287, dissolve mitomycin C at 1 mg/ml in PBS. Store at 4°C in the dark, stable for 1–2 weeks).
3. Dishes containing feeder layers. The STO feeder cell line or primary mouse embryonic fibroblasts are usually used as feeders for ES cell culture. Primary mouse embryonic fibroblasts can be generated following the well-established protocol. STO feeder cells or primary mouse embryo fibroblasts are expanded to confluence in the appropriate number of 100-mm tissue culture grade dishes. One dish yields approximately 1.0×10^7 cells. At this stage, you can choose the following two alternative protocols to treat the feeder cells: 1) The cells are harvested, irradiated at 3000 rad, and frozen at 3×10^6 cells per vial. When thawed, this number of feeders is sufficient to cover the wells of 2×6-well plates; 2) The cells are treated with mitomycin C for 2–3 h, washed, and replenished with fresh culture medium. Change medium with ES cell medium before adding ES cells.
4. 0.25% trypsin–EDTA (Invitrogen, Cat. No. 25200-056)
5. Phosphate-buffered saline (PBS) without Ca^{2+} and Mg^{2+}.
6. Freezing medium containing 25% FBS and 10% DMSO in DMEM, made up freshly and kept on ice.

2.2. In Vitro Differentiation of ES Cells

1. Differentiation medium (the methylcellulose differentiation media contains the same reagents as liquid differentiation media, except that methylcellulose is added to 1% of the final volume):

  a. Iscove's Modified Dulbecco's Medium (IMDM) (Invitrogen, Cat. No. 12440-053).
  b. 15% FBS (Hyclone, regular FBS).
  c. 4.5×10^{-4} M α-Monothioglycerol (MTG, Sigma, Cat. No. M-6145).
  d. 2 mM L-Glutamine (see above).
  e. 50 µg/ml Ascorbic acid (Sigma, Cat. No. A-4544. Make 100× stock solution fresh each time you set up a differentiation. Dissolve ascorbic acid at 5 mg/ml in H_2O and filter through 0.22 µm sterilization filter).
  f. 200 µg/ml Human transferrin (Sigma, Cat. No. T8158).
  g. 1% Methylcellulose (2× stock, Fluka, Cat. No. 64630).

2. ES-IMDM: 15% FBS (ES cell serum, see above), 1000 U/ml LIF, and 1.5×10^{-4} M of MTG in IMDM.
3. Gelatin: Prepare a 0.1% solution of gelatin in H_2O, dissolve, and sterilize by autoclaving.
4. Methylcellulose-based feed medium:

   a) 1/2 (v/v) primary differentiation medium (from above or freshly prepared).
   b) 15% regular FBS.
   c) 1.5×10^{-4}M of MTG; the MTG must be freshly prepared to achieve optimal levels of EB formation. MTG working solution is prepared by diluting MTG 1:100 in IMDM.
   d) 150 ng/ml murine Kit Ligand/Stem Cell Factor (mSCF) (R&D systems, Minneapolis, MN, USA, Cat. No. 455-MC)
   e) 30 ng/ml murine Interleukin-3 (mIL-3) (R&D systems, Cat. No. 403-ML).
   f) 20 ng/ml mouse Interleukin-6 (hIL-6) (R&D systems, Cat. No. 406-ML).
   g) 3 U/ml human Erythropoietin (EPO)(R&D systems, Cat. No. 287-TC-500).
   h) IMDM to the final volume.

5. 2× cellulose. Dissolve cellulose (Sigma, Cat. No. C-1794) in PBS at 2 U/ml. Filter sterilize through 0.45-μm filter.
6. Collagenase. Dissolve 1 g of collagenase (Sigma, Cat. No. C0310) in 320 ml PBS. After filter sterilization, add 80 ml of FBS. Aliquot and keep at −20°C.
7. Hemangioblast colony methylcellulose mixture:

   a) 1% methylcellulose.
   b) 10% FBS.
   c) SCF (100 ng/ml recombinant mSCF or 1% conditioned medium that was derived from medium conditioned by CHO cells transfected with mouse SCF expression vectors).
   d) 25% D4T endothelial cell-conditioned medium (D4T endothelial cells are cultured in IMDM with 10% FBS. Remove media and change to 4% FBS in IMDM when it becomes 80% confluent. Culture additional 72 h and collect the supernatant. Spin down for 5 min at 228 g, Beckman GS-6, GH-3.8 rotor to remove the cell debris and filter sterilize the supernatants by 0.45-μm filter. Make 5- to 10-ml aliquots and keep at −80°C. Once thawed, it is kept at 4°C for about 1 week).
   e) 5 ng/ml mouse VEGF (R&D systems, Cat. No. 494-VE).
   f) 10 ng/ml human IL-6.
   g) IMDM to a final volume.

8. Hemangioblast expansion Medium:

   a) 10% FBS.
   b) 10% horse serum (Invitrogen, Cat. No. 26050-088).
   c) 5 ng/ml mouse VEGF.
   d) 10 ng/ml mouse insulin-like growth factor-1 (IGF-1) (R&D systems, Cat. No. 791-MG).

   e) 2 U/ml human EPO.

   f) 10 ng/ml mouse basic fibroblast growth factor (bFGF) (R&D systems, Cat. No. 3139-FB).

   g) 50 ng/ml mouse IL-11 (R&D systems, Cat. No. 418-ML).

   h) 100 ng/ml SCF.

   i) IL-3 [30 ng/ml recombinant murine IL-3 or 1% conditioned medium obtained from medium conditioned by X63 AG8-653 myeloma cells transfected with a vector expressing IL-3 *(20)*].

   j) 1% L-Glutamine.

   k) 4.5×10^{-4} M of MTG.

   l) IMDM to a final volume.

9. Matrigel-coated wells (the stock bottle of Matrigel should be thawed slowly on ice, diluted 1:1 with IMDM, aliquoted (0.5 ml), and frozen at $-20°C$). Expansion of blast colony cell populations is carried out in microtiter wells pretreated with a thin layer of Matrigel. The wells are coated by first spreading 5 μl of diluted Matrigel over the surface with an Eppendorf pipette tip. The plate should be kept on ice during this procedure. When the required number of wells has been coated, incubate the plate on ice for 10–15 min. Following this incubation, remove excess Matrigel from each well and then incubate at 37°C for an additional 15 min before use.

10. Endothelial Expansion Medium:

   a) 20% FBS.

   b) 50 ng/ml mouse VEGF.

   c) 20 ng/ml mouse bFGF.

   d) 1.5×10^{-4} M of MTG.

   e) IMDM to the final volume.

11. Hematopoietic differentiation medium:

   a) 1% methylcellulose.

   b) 10% plasma-derived serum (Antech, Inc., Colorado, USA, Tyler, TX, USA).

   c) 5% protein-free hybridoma medium (PFHM-II; Invitrogen, Cat. No. 12040-077).

   d) SCF (100 ng/ml mSCF or 1% conditioned medium).

   e) 5 ng/ml mouse thrombopoietin (R& D System, Cat. No. 488-TO).

   f) 2 U/ml human EPO.

   g) 25 ng/ml mouse IL-11.

   h) IL-3 (30 ng/ml recombinant mIL-3 or 1% conditioned medium).

   i) 30 ng/ml mouse granulocyte–macrophage colony-stimulating factor (GM-CSF) (R&D systems, Cat. No. 415-ML).

   j) 30 ng/ml mouse G-CSF (R&D systems, Cat. No. 414-CS).

   k) 5 ng/ml mouse M-CSF (R&D systems, Cat. No. 416-ML).

   l) 5 ng/ml mouse IL-6.

   m) IMDM to the final volume.

3. Methods

3.1. Maintenance of Murine ES Cells

ES cells are traditionally grown at 37°C/5% CO_2/95% humidity in dishes coated with a feeder layer of mitotically inactivated mouse embryonic fibroblasts or gelatin-coated dishes with LIF in culture medium. But, for long-term culture and maintenance, ES cells should be grown on monolayers of mitotically inactivated fibroblasts. Primary embryonic fibroblasts (MEF) or the STO fibroblast cell line are the most commonly used feeder layers.

3.1.1. Thawing ES Cells (Quickly) and Plating-

When thawing ES cells, always have feeder plates prepared.

1. Remove ES cells from freezer/liquid nitrogen and quickly thaw in a 37°C water bath.
2. Transfer cell suspension (cell concentration is not very important) to a sterile tube containing several milliliters of warm medium.
3. Gently mix and pellet the cells by centrifugation at low speed for 5 min.
4. Aspirate off supernatant (removal of DMSO in freezing medium) and resuspend cells into 8 ml (4 ml) of warm ES medium and plate out in a 10-cm (6 cm) feeder plate (*see* **Note 2**).
5. Ideally re-feed cells daily with fresh ES medium (*see* **Note 3**).
6. Upon subconfluence, cells need to be passaged or frozen or used for experiments.

3.1.2. Passaging ES Cells

ES cells are routinely passaged every 2–3 days (except having colonies under selection or so), otherwise cells will spontaneously differentiate (*see* **Note 4**).

1. Check cells under the microscope for subconfluence.
2. Re-feed cells 3 h before passing them (very important), warm up reagents briefly before use (*see* **Note 3**).
3. Aspirate medium off, wash once with PBS, add about 1 ml (2 ml) of trypsin–EDTA to each 6-cm (10 cm) dish and incubate at 37°C until colonies float off when flicking the plate.
4. Carefully transfer trypsin/cell suspension to a sterile falcon tube and trypsinize for a few more minutes at 37°C.
5. Dissociate colonies into single cells by "Gilson pipetting," then add several millilitres of medium to inactivate the trypsin, pellet cells by low-speed centrifugation (*see* **Note 5**).
6. Remove supernatant, resuspend cells in appropriate volume of ES medium depending on plate format and splitting ratio (*see* **Note 2**).

3.1.3. Freezing ES Cells (Slowly)

ES cells can be frozen, like other tissue culture cells. As a general rule, freeze cells slowly and thaw them quickly. For long-term storage, cells should be kept under liquid nitrogen, and for short-term storage, they can be kept in a –80°C freezer. It is important to reduce the time the cells are in culture before freezing and freeze at a density that allows recovery of the culture even if 90% of the cells die during the freezing and thawing process.

1. Check cells under the microscope for subconfluence.
2. Re-feed cells about 3 h before freezing them. Have a pre-cooled styrofoam box as well as freezing vials and freezing medium ready on ice.
3. ES cells are trypsinized and dissociated to single cells following the procedures described above, add several milliliters of medium to inactivate the trypsin, pellet cells by low-speed centrifugation.
4. Remove supernatant, resuspend cells in appropriate volume of pre-cooled freezing medium, and immediately transfer into freezing vials on ice (1 ml per vial). Transfer vials into pre-cooled styrofoam box (inside ideally about 0°C) and then to a –80°C freezer.
5. Next day or later, transfer cells to a liquid nitrogen freezer.

3.2. In Vitro Hematopoietic Differentiation of Murine ES Cells

3.2.1. Primary Differentiation Step, Formation of EBs

Two different culture methods usually have been used to promote hematopoietic differentiation: 1) Methylcellulose-based semisolid media, a highly viscous media that does not encourage cellular migration or aggregation once seeded (12,13,21); 2) Liquid suspension culture, where cells are free to aggregate and move within the culture media (11,21). Although differentiation in semi-solid media such as methylcellulose is the most quantitative method for the formation of EBs from ES cells and generally yields the highest numbers of hematopoietic progenitors per input ES cell, other techniques exist that might be better suited to particular situations. For example, when it is desirable to isolate EBs at early stages of the primary differentiation process, differentiation in suspension culture facilitates the harvest of the small EBs (18).

3.2.1.1. METHYLCELLULOSE-BASED SEMISOLID CULTURE

1. Two days before setting up differentiation (*see* **Note 6**), split ES cells (4×10^5 ES cells per 60-mm dish) into ES-IMDM medium without feeder cells in the dishes. All plates should be gelatinized (*see* **Notes 7–9**).
2. Change the medium the next day.
3. Aspirate the medium from the dishes.
4. Add 1 ml of trypsin–EDTA, swirl, and remove quickly.

5. Add 1 ml trypsin and wait until cells start to come off. It usually takes about 1–2 min. Do not over-trypsinize cells.

6. Stop the reaction by adding 1 ml FBS and 4 ml IMDM and pipette up and down to make single cell suspension. Transfer to a 14-ml tube.

7. Centrifuge for 5–10 min at 228 g.

8. Wash the cell pellet in 10 ml IMDM (without FBS). Spin at 228 g for 5–10 min.

9. Resuspend the cell pellet in 5 ml IMDM (with 10% FBS) and count viable ES cells and check these cells.

10. Set up differentiation as follows: add 6,000–10,000 ES cells per milliliter of methylcellulose differentiation media to obtain day 2.75–3 EBs. Add 4,000–5,000 cells per milliliter to obtain day 4–5 EBs. Add 500–2,000 cells per milliliter to obtain day 6–10 EBs (*see* **Notes 10–13**).

11. Place dishes into a larger covered Petri dish along with an open 35-mm Petri dish containing 3 ml of sterile water and incubate at 37°C in a 5% CO_2 and moisture-saturated incubator until further analysis is performed.

3.2.1.2. SUSPENSION CULTURE

1. Follow **steps 1–10** as described in **Subheading 3.2.1.1** (*Methylcellulose-based semisolid culture*).

2. Plate into low-adherence Petri dishes at 4×10^5 cells per dish. Small aggregates (simple EBs) will be visible in 24 h. These simple EBs can be transferred into methylcellulose between 24 and 48 h.

3. If you are continuing in the liquid culture system, the media must be changed every 3–4 days. The EBs will tend to aggregate into clumps with regions of necrosis. To avoid this, break clumps apart by using a large mouth pipet (25 ml) such that you do not disrupt the EBs themselves. Transfer the EBs to a tube and allow them to sink to the bottom. Carefully aspirate off the old media, replace with fresh medium, and replate into the Petri dish.

3.2.2. Harvest of EBs

1. Harvest EBs

For EBs in liquid: transfer media containing EBs into 50-ml tubes. Wash the plate with IMDM. Let it sit at room temperature for about 10–20 min. EBs will settle down to the bottom of the tube.

For EBs in methylcellulose (*see* **Note 14**): add equal volume of cellulose (2 U/ml, final 1 U/ml) and incubate 20 min at 37°C. Collect EBs in 50-ml tubes (*see* **Note 15**). Wash the plate with IMDM and add this to the tube to ensure all EBs are collected. Let it sit at room temperature for about 10–20 min to allow EBs to settle down to the bottom of the tube (*see* **Note 16**).

2. Aspirate off media, add Trypsin–EDTA, or collagenase depending upon the age of the EBs, as outlined below:

   a) For EBs that are up to 8 days old, add 2–3 ml Trypsin–EDTA and incubate for 2–3 min at 37°C. Add IMDM containing 5% FBS to neutralize trypsin. Disrupt EBs by passing through a 20-G needle on a 3-cc syringe three times (up and down).

   b) For EBs that are 9 or more days old, add 2–3 ml of Collagenase and incubate at 37°C for 1 h, swirling gently following 30 min of incubation. Ensure the EBs stay in solution and are not on walls of tubes. Add IMDM containing 5% FBS to neutralize collagenase. Disrupt EBs by passing through a 20-G needle on a 3-cc syringe three times as above.

3. Transfer to a 14-ml tube and pellet cells by centrifugation at 350 g for 5–8 min.
4. Remove supernatant and resuspend the cells in a minimum volume of IMDM with 2% FBS.
5. Count the viable cells.

3.2.3. Second Differentiation Step, Clonal Assays of EBs

Analysis of early EBs, before the hematopoietic and endothelial commitment stages, revealed the presence of a progenitor with hemangioblast potential *(22)*. In the presence of vascular endothelial growth factor (VEGF) in methylcellulose cultures, these EB-derived precursors generate blast cell colonies that display hematopoietic and endothelial potential *(16,23)*. Kinetic studies demonstrated that these progenitors or BL-CFC represent a transient population that is present within the EBs for approximately 36 h, between day 2.5 and 4 of differentiation, preceding the onset of primitive erythropoiesis. The developmental potential of the BL-CFC strongly suggests that it represents the in vitro equivalent of the hemangioblast and, as such, the earliest stage of hematopoietic and endothelial commitment *(24)*. Beyond day 4.0 of differentiation, the number of BL-CFC declines with the commitment to the hematopoietic program as indicated by the appearance of significant numbers of primitive erythroid progenitors. Hematopoietic stage EBs (day 6–7 of differentiation) can be assayed for hematopoietic progenitors' potential. Hematopoietic and endothelial progenitors present within EBs can be successfully analyzed by flow cytometry and by direct replating EB cells into methylcellulose cultures to measure the frequency of hematopoietic progenitors. Additionally, EB cells can be sorted for early hematopoietic and endothelial cell markers to further analyze their hematopoietic and/or endothelial cell potential *(18,25)*.

3.3. Hemangioblast Stage

Most BL-CFC express Flk1, and a subpopulation of Flk1$^+$ cells also expresses the transcription factor Scl *(26,27)*. Recently, it was reported that Runx1 expression was up-regulated at the hemangioblast stage of EB differentiation and that Runx1 was essential for hematopoietic commitment at the hemangioblast stage of development in vitro *(24,28)*. Although the presence of Flk1 within an EB population does not guarantee the presence of large numbers of BL-CFC, the lack of significant Flk1 expression does indicate that the population has not yet progressed to the hemangioblast stage of development. So, BL-CFC can be initially screened by levels of expression of the receptor Flk1 or other marker gene colonies *(29,30)*. Also, to characterize blast colonies, we need to further identify them by morphological analysis and to analyze the hematopoietic and endothelial potential of the blast colonies. Blast colonies develop within 3–4 days of culture and can be recognized as clusters of cells that are easily distinguished from secondary EBs that develop from residual undifferentiated ES cells. Blast colonies and secondary EBs are the predominant type of colonies present in these cultures.

1. Add 3–6×10^4 EB cells per milliliter of hemangioblast colony methylcellulose mixture. Add 1 ml of the mixture into each of 35-mm Petri dishes. Prepare three replica dishes for each sample (*see* **Notes 17** and **18**).
2. Gently swirl the dishes to disperse the mixture evenly.
3. Place dishes into a larger covered Petri dish along with an open 35-mm Petri dish containing 3 ml of sterile water and incubate at 37°C in a humidified 5% CO_2 incubator for 3–4 days.
4. To analyze blast colonies by fluorescence-activated cell sorting (FACS) or morphological detection (*see* **Note 19**). For FACS analysis, antibody staining is carried out as follows: Cells are collected and resuspended in 100 µl of PBS containing 10% FBS and 0.02% sodium azide. An appropriate amount of antibody is added, and the cells are incubated on ice for 20 min. Following the staining step, the cells are washed two times with the same media and then resuspended in 300 µl of staining buffer, then transferred to a 5-ml polypropylene tube for analysis.
5. To analyze the hematopoietic and endothelial potential of the blast colonies, individual colonies are picked from the methylcellulose and cultured further in hemangioblast expansion medium on matrigel-coated microtiter wells. After 4 days of growth, the non-adherent cells of each well can be harvested by gentle pipetting and assayed for hematopoietic progenitor potential in 1 ml hematopoietic differentiation medium used for the growth of hematopoietic precursors. The remaining adherent population is cultured for an additional 4 days in endothelial expansion medium. At this point, the adherent population can be lysed directly in the well and subjected to reverse transcription-polymerase chain reaction (RT-PCR) for the analysis of expression of genes associated with endothelial development.

3.4. Hematopoietic Stage

Shortly following the peak of the hemangioblast stage of development, committed hematopoietic progenitors can be detected within the EBs. The numbers and types of hematopoietic colonies detected in methylcellulose cultures derived from disaggregated EBs are dependent on the ES cell line, the day of harvest of EBs, and hematopoietic cytokines used in the secondary differentiation. These developing colonies should be identified by morphology or cytological staining.

When EB cells are directly replated, day 5–6 EBs are typically used for a primitive erythroid colony, and day 7–10 EBs for definitive erythroid and myeloid progenitor analysis (*23*). The following is the protocol for direct EB replating.

1. Prepare methylcellulose-based hematopoietic differentiation medium.
2. Add 0.3 ml of cells at $1–5 \times 10^5$ per milliliter to each tube containing the 3-ml hematopoietic differentiation medium and vortex thoroughly. Let stand 3–5 min to allow bubbles to dissipate.
3. Plate 1.1 ml of the cell suspension per 35-mm low-adherence Petri dish.
4. Place dishes into a larger covered Petri dish along with an open 35-mm Petri dish containing 3 ml of sterile water and incubate at 37°C in a humidified 5% CO_2 incubator.
5. Primitive erythroid colonies are scored at day 5–6 of culture, whereas definitive erythroid, macrophage, and multilineage colonies are counted after 7–10 days of culture.

4. Notes

1. Serum quality is very important. Different batches of serum from different manufacturers should be tested. Typically, ES cells are maintained in gelatin-coated dishes in test serum for 5 or 6 passages and scored for morphology. A good lot of serum should keep the ES cells in an undifferentiated state. Although ES cell FBS pre-selected by some vendors, such as Hyclone and Stem Cell Technologies, are available, it is advisable that the serum provided still needs be tested before using for specific ES cell lines.
2. ES cells do not like to be alone, choose appropriate size of dish at thawing; do not routinely split by more than 1/10.
3. It is best to feed ES cells, i.e., change medium, every day; also, re-feed cells about 3 h before passaging or freezing them.
4. Do not overgrow ES cells. Split when subconfluence, otherwise cells will differentiate.
5. To prevent differentiation, always dissociate ES cell cultures into single cells after trypsinization.
6. Do not grow ES cells for too long before starting an experiment. Always try to reduce the time the cells are in culture.

7. Before initiating EB development, ES cells must be separated from feeder layers as they will alter the kinetics of differentiation if present in the EB culture. This can be done by removing the feeder layer and providing LIF from other sources *(31)*.

8. To gelatinize culture dishes, cover the surface of the vessel with the gelatin solution and incubate for 20 min at room temperature. It is possible to prepare dishes in advance and store them with the gelatin solution at 4°C for up to 1 week. Stored plates should be sealed with parafilm. Remove excess gelatin solution before use.

9. ES colonies should show little or no evidence of differentiation. If cultures differ dramatically from this, the efficiency of EB formation will be significantly decreased. Morphologically, undifferentiated ES cells have a larger nucleus, minimal cytoplasm, and one or more prominent dark nucleoli. It should be difficult to identify individual cells within the ES colony. Colonies appear amorphous without a distinct or common shape. Signs of differentiation include the ability to distinguish individual cells within the ES colony by the defined cytoplasmic membrane for the cells. The colony may appear to spread and cells appear flattened. Cells may lift off the dish. A great deal of variability exists among different ES cell lines in their ability to differentiate in vitro. In addition, the ability of ES cells to generate hematopoietic progenitors in vitro is also highly dependent upon the maintenance of the cells before setting up the differentiation cultures. In general, it is best to use low-passage ES cells that have been maintained in vitro for less than 10 days.

10. Methylcellulose is too viscous to be used with regular pipettes. The stock methylcellulose is handled with a 10-ml syringe without any needle. The final mixture can be distributed with a 3- or 5-ml syringe with a 16-G needle.

11. Primary differentiation is set up based on the cell number required for subsequent experiments. The number of starting ES cells to be used should be optimized. Ideally, total EB cell numbers obtained are as follows: day 2.75, $\sim$0.5–1 × 10^6 EB cells/10 ml of differentiation; day 4, $\sim$2–3 × 10^6 EB cells/10 ml of differentiation; day 6, $\sim$3–5 × 10^6 EB cells/10 ml of differentiation. Add higher cell number for ES lines that differentiate poorly.

12. To ensure the viability of the primary differentiation culture over an extended period of time, the cultures are usually fed on day 7 with a dilute methylcellulose medium containing hematopoietic growth factors. Firstly, prepare the methylcellulose-based feed medium as described in **Subheading 2.2**; Secondly, lay 0.5 ml of feed medium onto the surface of each differentiation culture drop-wise using a 3-cc syringe and 16-G blunt-end needle.

13. Differentiation is done in bacterial Petri dishes. Do not use tissue culture dishes. EBs are generated in non-adherent Petri grade dishes. If tissue culture grade dishes are used, the ES cells will differentiate and form a complex adherent cell population that does not generate hematopoietic progeny in a reproducible fashion.

14. EBs will be visible within 2–3 days and will be large enough to quantitate using an inverted microscope by day 5 or 6 of culture. If counted too early, EB estimates may be high because some EBs fail to thrive. Morphologically, an EB appears as a dense mass of cells surrounded by a cellular envelop. Clumps of disorganized or non-viable cells should not be scored as EBs.

15. A pipette with a 1-ml tip works best for mixing and transferring the diluted methylcellulose culture to the tubes. Polystyrene tubes are preferable because EBs are easier to see and they do not stick to the sides.

16. It is important to minimize the number of centrifugation steps when collecting EBs, because EB cells at the early stage are very fragile. So, you can collect cells by centrifugation at 228 g for 1 min or by settling down at room temperature for 10–20 min.

17. The cell numbers and the components of the medium used for the generation of the two stages of EB development are different.

18. The actual number of EB cells plated for differentiation of blast colonies will vary depending on the cell line and conditions used, as well as the age of the EBs. When first establishing optimal plating densities, it is advisable to try two different cell concentrations that differ by two- to three fold.

19. You can experience slight changes in the kinetics of BL-CFC development even when you adhere strictly to the protocol and use identical reagents. Thus, it is advisable to assay EBs from several time points when assessing BL-CFC potential.

References

1. Martin, G. R. (1981) Isolation of a pluripotent cell line from early mouse embryos cultured in medium conditioned by teratocarcinoma stem cells. *Proc Natl Acad Sci USA* **78**, 7634–7638.

2. Williams, R. L., D. J. Hilton, S. Pease, T. A. Willson, C. L. Stewart, D. P. Gearing, E. F. Wagner, D. Metcalf, N. A. Nicola, and N. M. Gough (1988) Myeloid leukaemia inhibitory factor maintains the developmental potential of embryonic stem cells. *Nature* **336**, 684–687.

3. Smith, A. G., J. K. Heath, D. D. Donaldson, G. G. Wong, J. Moreau, M. Stahl, and D. Rogers (1988) Inhibition of pluripotential embryonic stem cell differentiation by purified polypeptides. *Nature* **336**, 688–690.

4. Rohwedel, J., V. Maltsev, E. Bober, H. H. Arnold, J. Hescheler, and A. M. Wobus (1994) Muscle cell differentiation of embryonic stem cells reflects myogenesis in vivo: developmentally regulated expression of myogenic determination genes and functional expression of ionic currents. *Dev Biol* **164**, 87–101.

5. Fraichard, A., O. Chassande, G. Bilbaut, C. Dehay, P. Savatier, and J. Samarut (1995) In vitro differentiation of embryonic stem cells into glial cells and functional neurons. *J Cell Sci* **108** (Pt 10), 3181–3188.

6. Bain, G., D. Kitchens, M. Yao, J. E. Huettner, and D. I. Gottlieb (1995) Embryonic stem cells express neuronal properties in vitro. *Dev Biol* **168**, 342–357.

7. Nakayama, N., I. Fang, and G. Elliott (1998) Natural killer and B-lymphoid potential in CD34+ cells derived from embryonic stem cells differentiated in the presence of vascular endothelial growth factor. *Blood* **91**, 2283–2295.

8. Cho, S. K., T. D. Webber, J. R. Carlyle, T. Nakano, S. M. Lewis, and J. C. Zuniga-Pflucker (1999) Functional characterization of B lymphocytes generated in vitro from embryonic stem cells. *Proc Natl Acad Sci USA* **96**, 9797–9802.

9. Nakano, T. (1996) In vitro development of hematopoietic system from mouse embryonic stem cells: a new approach for embryonic hematopoiesis. *Int J Hematol* **65**, 1–8.

10. Guo, Y., B. Graham-Evans, and H. E. Broxmeyer (2006) Murine embryonic stem cells secrete cytokines/growth modulators that enhance cell survival/anti-apoptosis and stimulate colony formation of murine hematopoietic progenitor cells. *Stem Cells* **24**, 850–856.

11. Johansson, B. M., and M. V. Wiles (1995) Evidence for involvement of activin A and bone morphogenetic protein 4 in mammalian mesoderm and hematopoietic development. *Mol Cell Biol* **15**, 141–151.

12. Keller, G., M. Kennedy, T. Papayannopoulou, and M. V. Wiles (1993) Hematopoietic commitment during embryonic stem cell differentiation in culture. *Mol Cell Biol* **13**, 473–486.

13. Keller, G. M. (1995) In vitro differentiation of embryonic stem cells. *Curr Opin Cell Biol* **7**, 862–869.

14. Huber, T. L., V. Kouskoff, H. J. Fehling, J. Palis, and G. Keller (2004) Haemangioblast commitment is initiated in the primitive streak of the mouse embryo. *Nature* **432**, 625–630.

15. Kyba, M., R. C. Perlingeiro, R. R. Hoover, C. W. Lu, J. Pierce, and G. Q. Daley (2003) Enhanced hematopoietic differentiation of embryonic stem cells conditionally expressing Stat5. *Proc Natl Acad Sci USA* **100** Suppl 1, 11904–11910.

16. Choi, K., M. Kennedy, A. Kazarov, J. C. Papadimitriou, and G. Keller (1998) A common precursor for hematopoietic and endothelial cells. *Development* **125**, 725–732.

17. Carlsson, L., E. Wandzioch, O. P. Pinto do, and A. Kolterud (2003) Establishment of multipotent hematopoietic progenitor cell lines from ES cells differentiated in vitro. *Methods Enzymol* **365**, 202–214.

18. Zhang, W. J., Y. S. Chung, B. Eades, and K. Choi (2003) Gene targeting strategies for the isolation of hematopoietic and endothelial precursors from differentiated ES cells. *Methods Enzymol* **365**, 186–202.

19. Kennedy, M., and G. M. Keller (2003) Hematopoietic commitment of ES cells in culture. *Methods Enzymol* **365**, 39–59.

20. Karasuyama, H., and F. Melchers (1988) Establishment of mouse cell lines which constitutively secrete large quantities of interleukin 2, 3, 4 or 5, using modified cDNA expression vectors. *Eur J Immunol* **18**, 97–104.

21. Burkert, U., T. von Ruden, and E. F. Wagner (1991) Early fetal hematopoietic development from in vitro differentiated embryonic stem cells. *New Biol* **3**, 698–708.

22. Fehling, H. J., G. Lacaud, A. Kubo, M. Kennedy, S. Robertson, G. Keller, and V. Kouskoff (2003) Tracking mesoderm induction and its specification to the hemangioblast during embryonic stem cell differentiation. *Development* **130**, 4217–4227.

23. Kennedy, M., M. Firpo, K. Choi, C. Wall, S. Robertson, N. Kabrun, and G. Keller (1997) A common precursor for primitive erythropoiesis and definitive haematopoiesis. *Nature* **386**, 488–493.

24. Lacaud, G., L. Gore, M. Kennedy, V. Kouskoff, P. Kingsley, C. Hogan, L. Carlsson, N. Speck, J. Palis, and G. Keller (2002) Runx1 is essential for hematopoietic commitment at the hemangioblast stage of development in vitro. *Blood* **100**, 458–466.

25. Zhang, W. J., C. Park, E. Arentson, and K. Choi (2005) Modulation of hematopoietic and endothelial cell differentiation from mouse embryonic stem cells by different culture conditions. *Blood* **105**, 111–114.

26. Chung, Y. S., W. J. Zhang, E. Arentson, P. D. Kingsley, J. Palis, and K. Choi (2002) Lineage analysis of the hemangioblast as defined by FLK1 and SCL expression. *Development* **129**, 5511–5520.

27. D'Souza, S. L., A. G. Elefanty, and G. Keller (2005) SCL/Tal-1 is essential for hematopoietic commitment of the hemangioblast but not for its development. *Blood* **105**, 3862–3870.

28. Lacaud, G., V. Kouskoff, A. Trumble, S. Schwantz, and G. Keller (2004) Haploin-sufficiency of Runx1 results in the acceleration of mesodermal development and hemangioblast specification upon in vitro differentiation of ES cells. *Blood* **103**, 886–889.

29. Faloon, P., E. Arentson, A. Kazarov, C. X. Deng, C. Porcher, S. Orkin, and K. Choi (2000) Basic fibroblast growth factor positively regulates hematopoietic development. *Development* **127**, 1931–1941.

30. Kabrun, N., H. J. Buhring, K. Choi, A. Ullrich, W. Risau, and G. Keller (1997) Flk-1 expression defines a population of early embryonic hematopoietic precursors. *Development* **124**, 2039–2048.

31. Kearney, J. B., and V. L. Bautch (2003) In vitro differentiation of mouse ES cells: hematopoietic and vascular development. *Methods Enzymol* **365**, 83–98.

8

Hematopoietic Development of Human Embryonic Stem Cells in Culture

Xinghui Tian and Dan S. Kaufman

Summary

The successful isolation and characterization of human embryonic stem cells (hESCs) provides a powerful tool to study the cellular and genetic mechanisms that mediate cell-fate decisions toward distinct developmental lineages. hESC-derived cells may also be suitable for novel cellular therapies. Significant progress in hematopoietic development of hESCs has demonstrated production of many types of blood cells from hESCs including myeloid, erythroid and lymphoid lineage cells, and possibly hematopoietic stem cells. Current established approaches to generate specific hematopoietic lineages are based on the initial pre-differentiation of hESCs into a heterogeneous mixture of cell populations. In this chapter, we describe two methods that have been successfully used in our laboratory: *(1)* co-culture with stromal cells derived from hematopoietic microenvironments and *(2)* embryoid body (EB) formation. Subsequent to this early differentiation step, distinct progenitor cell populations can be derived, sorted, and utilized for further lineage-specific developmental studies.

Key Words: hESCs; hematopoiesis; stromal cell; embryoid body; differentiation; hematopoietic precursor.

1. Introduction

Human embryonic stem cells (hESCs) derived from the inner cell mass of pre-implantation blastocysts have the ability to self-renew as undifferentiated cells for prolonged periods in culture, yet retain the potential to differentiate into any cell type within the adult body *(1,2)*. In vitro differentiation of hESCs recapitulates events that occur during normal human embryogenesis, allowing

From: *Methods in Molecular Biology, vol. 430: Hematopoietic Stem Cell Protocols*
Edited by: K. D. Bunting © Humana Press, Totowa, NJ

hESCs to serve as a model system to elucidate cellular and genetic mechanisms that mediate commitment to specific lineages. Moreover, the ability to generate mature cell populations, such as hematopoietic cells, from hESCs offers promising resources for future therapies to repair or replace cells and tissues that have become diseased or damaged. Indeed, the field of hematology has pioneered many aspects of cellular therapies, and hematopoietic cell transplantation (HCT) has been successfully performed in the clinic for over 30 years *(3)*. These transplants use cells from adult bone marrow, mobilized peripheral blood, and umbilical cord blood *(4–6)*. However, this process of HCT remains fraught with problems such as disease relapse, graft-versus-host disease, poor engraftment of transplanted cells, and toxicities associated with the chemotherapy and radiation therapy that accompany this treatment. In many cases, the availability of a source of hematopoietic cells is problematic because of the lack of a suitable histocompatible donor. hESCs may provide an alternative resource for production of hematopoietic cells. These hESC-derived blood cells may be suitable not only for hematopoietic stem cells (HSCs) needed for HCT, but also development of red blood cells and platelets suitable for transfusion medicine, and lymphocytes for improved immune-based therapies against malignancies and infectious diseases.

The first studies of hematopoietic differentiation of hESCs utilized a co-culture system with cell lines derived from hematopoietic microenvironments, such as the murine bone marrow stromal cell line S17, in culture medium containing fetal bovine serum (FBS), but no other additional exogenous cytokines or growth factors *(7)*. This approach provides the basis to elucidate early cellular/molecular events of human hematopoiesis and to further optimize the culture conditions necessary to promote more efficient hematopoietic differentiation from hESCs. Subsequent studies have characterized more specific growth factors required for hematopoietic differentiation of hESCs in stromal cell-based culture system *(8)* and during embryoid body (EB) development *(8–10)*. A combination of cytokines, including bone morphogenetic protein-4 (BMP-4), can strongly promote hematopoietic differentiation in a serum-containing medium by EB formation *(9)* and in serum-free medium by co-culture with S17 *(8)*. In the presence of several cytokines including BMP-4, vascular endothelial growth factor (VEGF-A165) selectively promotes erythropoietic development toward the primitive lineage *(10)*. Differentiation of hESCs induced by EB formation has been found to go through sequential hemato-endothelial, primitive and definitive hematopoietic stages resembling human yolk sac development *(11)*. More mature hematopoietic cells can be derived from a subpopulation of hemogenic precursors during EB development *(12)*. These hemogenic precursors express PECAM-1 (CD31), Flk-1 and VE-cadherin, but not CD45. Their potential to differentiate into both endothelial and hematopoietic lineages has been shown starting with a clonal cell population *(12)*. Another pan-hematopoietic maker, leukosialin (CD43) has been further used to functionally distinguish hematopoietic and endothelial progenitors derived from hESCs *(13)* in a co-culture system using OP9 stromal cells. CD34$^+$ cells derived from hESCs by co-culture with stromal cells exhibit myeloid *(8)* and lymphoid (B, and natural killer cells) lineage developmental potential *(14,15)*. Furthermore, functional dendritic cells *(16,17)* and macrophages *(18)* have also been successfully derived from these differentiated hESCs.

Hemoglobin expression has also been studied from hESC-derived erythroid cells. A developmental switch from embryonic ε-globin to fetal γ-globin has been demonstrated in erythroid cells derived from hESCs co-cultured with S17 or the fetal liver FH-B-hTERT stromal cell, though these cells do not mature to express adult β-globin *(19)*. Another study using hESCs-derived EBs subsequently cultured in adherent conditions also demonstrates embryonic and fetal globin expression with little adult (β) globin expression *(20)*. However, one study did demonstrate a transition from primitive into definitive erythro-poiesis, primarily based on the expression of embryonic, fetal and adult hemoglobins *(11)*.

Although diverse myeloid and some lymphoid lineages can be routinely generated from hESCs using these in vitro assays, putative HSCs derived from hESCs can only be defined by in vivo engraftment experiments. To date, immunodeficient mice and fetal sheep models have been used to define the candidate HSCs derived from hESCs. Stable, but only modest, engraftment of hematopoietic cells derived from hESCs has been achieved in these animal models *(21–23)*.

Taken together, the stromal co-culture method is technically straightforward and offers the advantage to potentially characterize and modify the stromal cells to define specific components they contribute to hematopoiesis. For example, OP9 cells that express the Notch ligand Delta-like 1 have been useful to demonstrate the requirement for Notch signaling to derive T cells from mouse ES cells and hematopoietic precursor cells isolated from human cord blood *(24,25)*. However, differentiation of hESCs through EB formation also offers another suitable method to promote hematopoiesis when investigators want to avoid more complex interactions with stromal cells. In this chapter, we will describe these two methods to promote hematopoietic differentiation of hESCs: *(1)* generation of hematopoietic cells from hESCs by co-culture with mouse stromal cell line S17 in a serum-containing medium; *(2)* promotion of hematopoietic differentiation by EB formation (*see* **Fig. 1**). Phenotypic and function analyses are used to characterize the hematopoietic potential of cells derived from hESCs using these methods.

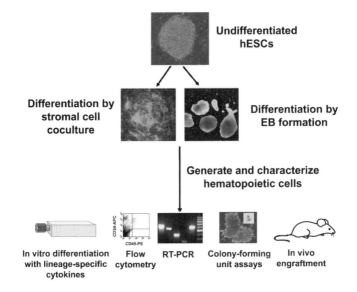

Fig. 1. Diagram of differentiation and analysis of human embryonic stem cell (hESC)-derived hematopoietic cells. Undifferentiated hESCs are induced to differentiate into a heterogeneous cell population by two methods: (1) co-culture with stromal cells derived from bone marrow environment and (2) EB formation. Hematopoietic cells can be characterized by flow cyotmetric analysis for specific surface markers, reverse transcription-polymerase chain reaction (RT-PCR) for hematopoietic genes and colony-forming unit (CFU) assays for hematopoietic progenitors. Hematopoietic progenitors can also be isolated and cultured with specific cytokines to promote development of desired mature blood cell lineages. In vivo assays can be used to examine the ability of hematopoietic cells derived from hESCs to mediate long-term multi-lineage engraftment when transplanted into immunodeficient mice.

2. Materials

2.1. Co-Culture of hESCs and S17 Cells

2.1.1. Cell Culture Media

1. Dulbecco's phosphate-buffered saline (DPBS), Ca^{2+}- and Mg^{2+}-free (Cellgro/Mediatech, Herndon, VA, USA; Cat. No. 21-031-CV).
2. For culture of undifferentiated hESCs, L-glutamine is routinely prepared fresh from powder by mixing 0.146 g of L-glutamine (Invitrogen Corporation/Gibco, Grand Island, NY, USA; Cat. No. 21051-024) and 7 μl of β-mercaptoethanol (Sigma, St. Louis, Mo, USA; Cat. No. M7522) in 10 ml of DPBS. L-Glutamine-DPBS solution (2.5 ml) is added into 250 ml of hESC medium, for a final concentration of 2 mM L-glutamine and 0.1 mM β-mercaptoethanol.
3. To prepare basic fibroblast growth factor (bFGF; Invitrogen; Cat. No. 13256-029) working solution, 10 μg of bFGF powder is reconstituted in 5 ml of 0.1%

Fraction V bovine serum albumin (BSA; Roche, Indianapolis, IN, USA; Cat. No. 03117332001, prepared in sterile DPBS). Aliquot 0.5 ml into sterile tubes and store at –80ºC. Use 0.5 ml of the reconstituted bFGF in 250 ml of hESC medium.

4. hESC medium: DMEM/F12 (Invitrogen Corporation/Gibco; Cat. No. 11330-032) supplemented with 15% knockout serum replacement (Invitrogen Corporation/Gibco; Cat. No. 10828028), 0.1 mM β-mercaptoethanol, 2 mM L-glutamine, 1% MEM non-essential amino acid solution (Invitrogen Corporation/Gibco; Cat. No. 11140-050), and 4 ng/mL bFGF.

5. S17 cells (*see* **Note 1**) culture medium: RPMI1640 (Cellgro/Mediatech; Cat. No. 10-040-CV) medium containing 10% FBS certified (Invitrogen Corporation/Gibco; Cat. No. 16000-044), 0.1 mM β-mercaptoethanol (Invitrogen Corporation/Gibco; Cat. No. 21985-023), 1% MEM non-essential amino acids, 1% penicillin–streptomycin (P/S; Invitrogen Corporation/Gibco; Cat. No. 15140-122), 2 mM L-glutamine (Cellgro/ Mediatech; Cat. No. 25-005-CI). S17 cells are courtesy of Dr. Ken Dorshkind (UCLA; *see* **Note 1**) *(26)*.

6. R-15 differentiation medium (*see* **Notes 2** and **3**): RPMI1640 supplemented with 15% defined FBS (Hyclone, Logan, UT, USA; Cat. No. SH30070.03), 2 mM L-glutamine, 0.1 mM β-mercaptoethanol, 1% MEM non-essential amino acids solution, and 1% P/S.

7. R-10 Medium (used for washing): RPMI1640 supplemented with 10% FBS and 1% P/S.

8. Collagenase split medium: DMEM/F12 medium containing 1 mg/ml collagenase type IV (Invitrogen Corporation/Gibco; Cat. No. 17104-019). Collagenase medium is filter sterilized with a 50-ml, 0.22-μm membrane Steriflip (Millipore, Billerica, MA, USA; Cat. No. SCGP00525).

9. Trypsin–ethylene diamine tetra-acetic acid (EDTA) + 2% chick serum: 0.05% trypsin–0.53 mM EDTA solution (Cellgro/Mediatech; Cat. No. 25-052-CI) with 2% chicken serum (Sigma; Cat. No. C5405; *see* **Note 4**).

10. Mitomycin C (American Pharmaceutical Partners, Los Angeles, CA, USA, Product No. 109020).

2.1.2. Cell Culture Supplies

1. Six-well tissue culture plates (NUNC™ Brand Products, Nalgene Nunc, Rochester, NY, USA; Cat. No. 152795).

2. Gelatin (Sigma; Cat. No. G-1890): 0.1% (w/v) in water. Autoclave for sterility.

3. Disposable serological pipets (all from VWR Scientific Products, West Chester, PA ,USA): 10 ml (Cat. No. 53283-740); 5 ml (Cat. No. 53283-738); and 1 ml (Cat. No. 53283-734; *see* **Note 5**).

4. 70-μm Cell strainer filter (Becton Dickinson/Falcon, Bedford, MA, USA; Ref. No. 352350).

5. 0.4% Trypan Blue Stain (Invitrogen Corporation/Gibco; Cat. No. 15250).

6. Blue polypropylene 15-ml conical tubes (Becton Dickinson/Falcon; Cat. No. 352097)

2.2. EB Formation from hESCs

2.2.1. Cell Culture Media

1. R-15 differentiation medium: RPMI1640 supplemented with 15% defined FBS (Hyclone; Cat. No. SH30070.03), 2 mM L-glutamine, 0.1 mM β-mercaptoethanol, 1% MEM non-essential amino acid solution, and 1% P/S (*see* **Note 6**).
2. Dispase split medium: 0.25 g of dispase powder (Invitrogen Corporation/Gibco; Cat. No. 17105-041) in 50 ml of DMEM/F-12, then filter sterilized with a 50-ml, 0.22-μm membrane Steriflip (Millipore; Cat. No. SCGP00525); Dispase final concentration 5 mg/ml.

2.2.2. Cell Culture Supplies

1. Blue Max polypropylene 50-ml conical tubes (Becton Dickinson/Falcon; Cat. No. 352098)
2. Blue polypropylene 15-ml conical tubes (Becton Dickinson/Falcon; Cat. No. 352097)
3. Costar® 6 Well Clear Flat Bottom Ultra Low Attachment Microplates (Corning, Inc., Corning, NY, USA; Cat. No. 3471)

2.3. Flow Cytometric Analysis (see Note 7)

1. Fluorescence-activated cell-sorting (FACS) wash medium: PBS containing 2% FBS and 0.1% sodium azide (Fisher chemicals, Pittsburgh, PA, USA; Cat. No. S227I)
2. 12 × 75-mm Polystyrene round-bottom tube (Becton Dickinson/Falcon; Cat. No. 352054)
3. 7-Amino-actinomycin D (7-AAD; Sigma; Cat. No. A9400-5MG; *see* **Note 8**)

2.4. Hematopoietic Colony-Forming Unit Assays

1. MethoCult GF+ H4435 (StemCell Technologies, Vancouver, BC, Canada; Cat. No. 04435) consisting of 1% methylcellulose, 30% FBS, 1% BSA, 50 ng/ml stem cell factor, 20 ng/ml granulocyte–macrophage colony-stimulating factor, 20 ng/ml interleukin (IL)-3, 20 ng/ml, IL-6, 20 ng/ml granulocyte colony stimulating factor, and 3 U/ml erythropoietin. This medium is optimized for detection of most primitive colony-forming cells (CFCs). A 100-ml bottle of Methocult can be aliquoted into 2.5-ml samples. Alternatively, Methocult can also be purchased pre-aliquoted into 3-ml samples.
2. I-2 Medium: Iscove's modified Dulbecco's medium (IMDM, Invitrogen Corporation, Gibco; Cat. No. 12440-053) containing 2% FBS.
3. Non-tissue culture-treated 35-mm Petri dish (Greiner Bio-One, Kaysville, UT, USA; Cat. No. 627102; *see* **Note 9**).
4. Stripette disposable serological pipet (2 ml; Corning Costar, Corning, NY, USA; Cat. No. 53283-915).

2.5. RNA Isolation (see Note 10)

1. Rneasy® Micro Kit (QIAGEN; Cat. No. 74004).
2. QIA shredder™ (QIAGEN, Valencia, CA, USA; Cat. No. 79645).
3. TURBO DNA-free™ Kit (Ambion, Austin, TX, USA; Cat. No. 1907).

3. Methods

3.1. Culture of Undifferentiated hESCs

Undifferentiated hESCs were cultured as previously described *(1,27)*. Briefly, hESCs were maintained in hESC medium by co-culture with MEF cells (inactivated through irradiation or treatment with mitomycin C) or in MEF-conditioned medium on Matrigel-coated plates. hESCs are fed daily with fresh hESC medium and are passed onto fresh feeder plates or Matrigel-coated plates at weekly intervals to maintain undifferentiated growth.

3.2. Preparation of S17 Feeder Layer

Mouse bone marrow S17 cells *(26)* are maintained in S17 culture medium. To prepare feeder layers, the S17 cells are inactivated by incubating cells with S17 culture medium containing 10 µg/ml of mitomycin C for 3 h at 37°C, 5% CO_2 (*see* **Note 11**). The cells are then washed twice in DPBS and dissociated with trypsin–EDTA. Inactivated S17 cells are plated onto 0.1% gelatin-coated six-well plates at a density of 2.5×10^5 cells per well. Feeder layers should be prepared at least 1 day before co-culture with hESCs and remain suitable for use up to 1–2 weeks when kept in a 37°C, 5% CO_2 incubator.

Other stromal cell lines can also be used in a similar manner, although irradiation dose and cell density may vary (*see* **Fig. 2** and **Note 12**).

3.3. Co-Culture of hESCs with S17 Stromal Cells

To improve the viability of hESCs, small colonies or clusters of ES cells, rather than a single cell suspension, should be plated onto S17 stromal cells. Before co-culture with S17, hESCs from one well grown under the same conditions can be used to obtain a single cell suspension for counting. hESCs can be added into S17 cultures at a density of $2–3 \times 10^5$ cells per well of a 6-well plate.

1. Warm collagenase split medium to 37°C in a water bath.
2. Aspirate medium off of hESC culture and add 1.5 ml/well collagenase split medium. Place the plate in 37°C incubator for 5–10 min, observing at about 5-min intervals. Cells are ready to be harvested when the edges of the colony are rounded up and curled away from the MEFs or from the matrigel plate. Note that

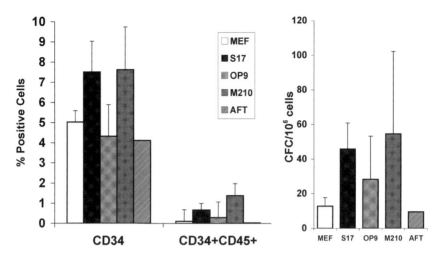

Fig. 2. Comparison of induction of hematopoietic differentiation from human embryonic stem cells (hESCs) by co-culture with different stromal cell lines. hESCs were co-cultured with irradiated MEF, S17, OP9, M210B4, and AFT024 stromal cells for 17 days. Flow cytometry analysis and colony-forming unit (CFU) assays are performed to characterize the hematopoietic cells derived from hESCs. Significantly higher percentages of CD34$^+$, CD34$^+$CD45$^+$, and CFUs can be derived from hESCs by co-culture with S17 and M210B4 cells compared with other cell lines. Two to five experiments are done with each stromal cell line.

  colonies will not detach from plate during collagenase treatment, but edges will become more angular and distinct.

3. Using a 5-ml pipet, scrape and gently pipet to wash the colonies off of the plate, transfer cell suspension to a 15-ml conical tube, and add another 3–6 ml of R-15 differentiation medium. Centrifuge at $400 \times g$ for 5 min. Aspirate medium and wash cells with additional 3–6 ml R-15 differentiation medium by centrifugation again at $400 \times g$ for 5 min.

4. During the last centrifugation step, prepare the S17 feeder layers by aspirating off the S17 culture medium and washing once with DPBS. Add 2 ml R-15 medium into each well.

5. After centrifugation of the hESCs, aspirate medium, resuspend cells in an appropriate volume with R-15 medium, and add 1 ml of cell suspension per well onto S17 plate. To evenly distribute cells, gently shake the plate from side to side while placing them in 37°C/5% CO_2 incubator. Do not disturb plates for several hours or preferably overnight.

6. During differentiation, culture medium is changed every 2–3 days. For the first few days, colonies maintain an undifferentiated appearance. Subsequently, they will show obvious evidence of differentiation, forming three-dimensional cystic and other loosely adherent structures (*see* **Fig. 1**).

3.4. Dissociation and Harvesting of Differentiated hESCs

The optimal time required for differentiation into hematopoietic populations, such as $CD34^+$, $CD45^+$ cells, and CFCs varies somewhat depending on the hESC line and stromal cells used. In general, a culture period of 14–21 days results in the best differentiation for the S17 stromal cells. However, another group demonstrated that the appearance of $CD34^+$ and $CD45^+$ cells peaked at 7–9 days in the OP9 co-culture system *(14)*. A time course experiment in which cells are sampled every 2–3 days is recommended to find the optimal time for formation of specific lineages.

For flow cytometry and colony-forming unit (CFU) assays, it is necessary to produce a single cell suspension of hESCs that have differentiated on S17 or other stromal cells. Because stromal cells are inactivated before co-culture with ES cells (*see* **Note 11**), >90% of cells harvested will be derived from hESCs.

To prepare single cell suspension:

1. Dissociate the differentiated hESCs by incubation with collagenase split medium for 5–10 min until stromal cell layer becomes move spindle-shaped and begins to break up. Scrape with a 5-ml pipet and transfer hES/S17 cell suspension into 15-ml conical tube. Add another 6 ml of Ca^{2+}- and Mg^{2+}-free DPBS and break up the colonies by pipetting up and down (vigorously) against the bottom of the tube until a fine suspension of cells is produced. Centrifuge cell suspension at $400 \times g$ for 5 min.
2. Remove the supernatant, add 1.5 ml of trypsin–EDTA + 2% chick serum solution per well and incubate for 5–15 min in a 37°C water bath. Vigorously vortex and observe samples at 3–5 min intervals until there are few, if any, clumps of undispersed cells.
3. Add 6 ml of R-10 medium to neutralize the trypsin–EDTA and pipet up and down to further disperse the cells. Centrifuge at $400 \times g$ for 5 min. Resuspend the cell pellet with 5–10 ml of R-10 medium and filter the cell suspension through a 70-μm cell strainer filter to remove any remaining clumps of cells. Count viable cells after staining with 0.4% Trypan Blue, using a hemocytometer. From a nearly confluent well, $1–2 \times 10^6$ single cells can be obtained.
4. Aliquot cells as needed for flow cytometry analysis, RNA, protein isolation, and CFU assays. Performing multiple assays from the same collection of differentiated hESCs will ensure uniformity of results. Depending on the density of cells, two to three wells can be harvested at a single time point to collect enough cells for flow cytometry analysis, RNA, protein isolation, and CFU assays.

3.5. EB Formation

1. Preparation of hESCs: Use hESC colonies that contain (by morphological criteria) only undifferentiated cells. EB formation works best when colonies are neither very large nor very small. With experience, one develops a feeling for those

colony sizes that work well. If they are too small, the colonies will dissociate within 2–4 days. If the colonies are too large, they will not dissociate very easily, and in an attempt to break them apart, they may be damaged. Also, with larger colonies, EB formation may not occur efficiently. An ideal time for hESCs to form EBs is typically 6–7 days after their last passage date. One six-well plate can generate up to 1–2 Costar® 6 Well Clear Flat Bottom Ultra Low Attachment Microplates.

2. Forming EBs: Aspirate the medium from each well, leaving the colonies adherent. Treat hESCs with 1.5 ml/well of 5 mg/ml dispase solution. Incubate at 37°C/5% CO_2 until 50% of the colonies are detached. This usually takes 5–10 min with freshly made dispase and can take up to 15 min with an older dispase solution. Gently shake the plate until the remaining colonies detached. If they do not detach, use a 5-ml pipet to wash them off.

3. Add 2 ml of R-15 to each well and gently separate the colonies by pipetting up and down. Pool the cell suspension into a 50-ml conical tube and let the colonies settle to the bottom. Remove the top supernatant as much as possible, leave the colonies with about 3 ml medium. Add fresh medium to the colonies for two total washes.

4. Resuspend the colonies with appropriate volume of R-15 medium and aliquot the cell suspension into untreated Costar® 6 Well Clear Flat Bottom Ultra Low Attachment Microplates and then add R-15 to a final volume of 4 mL. Incubate at 37°C/5% CO_2 over night.

5. EB resuspension: The day after EB formation, the cells must be "cleaned-up" to remove remaining stromal and dead cells from the suspension. Pool all the EBs from one plate to a 50-mL conical tube and let the EBs settle to the bottom. If smaller EBs are desired, or the cells have "clumped" together overnight, pipet up and down before the EBs settle to the bottom. Gently aspirate the supernatant and try to remove the single cells that float in the supernatant. Resuspend the colonies in R-15 medium and add to a fresh Costar® 6 Well low-attachment plate. Incubate at 37°C/5% CO_2.

3.6. Dissociation of EBs

1. Add EBs to a 15-ml conical tube, let them settle by gravity for approximately 1 min, and gently aspirate medium and floating cells that have not yet settled out.

2. Wash with 5 ml of Ca^{2+}- and Mg^{2+}-free DPBS, then centrifuge at $400 \times g$ for 3 min.

3. Aspirate supernatant, add 1.5–2 ml/well of trypsin–EDTA with 2% chick serum, vigorously pipet up and down several times, and vortex to break up EBs.

4. Incubate in 37°C water bath for 5 min, vortex, and pipet vigorously to further dissociate EBs. Return tube to water bath for another 5 min and again remove to vortex and pipet vigorously. Repeat incubation, pipetting, and vortexing at 5-min intervals until EBs seem maximally dissociated (about 10–20 min total). Some clumps may still remain, but longer incubation usually does not improve this digestion.

5. After the EBs have been maximally digested, add 4 mL of R-10 medium and centrifuge at $400 \times g$ for 3 min. Aspirate supernatant and wash twice with

additional 5 ml of R-10 medium and centrifuge at $400 \times g$ for 3 min for each wash step.

6. Resuspend cells in desired medium and filter the cell suspension with 70-μm cell strainer filter to remove any remaining clumps of cells. Count viable cells using a hemocytometer after staining with 0.4% Trypan Blue.

3.7. Flow Cytometric Analysis

1. Transfer single cell suspension prepared from hESC/S17 or EB differentiation at approximately 2×10^5 cells per tube for staining with different antibodies. Wash one or two times with FACS medium before starting staining.
2. Stain with either antigen-specific antibodies or isotype control for at least 15 min on ice. If the primary antibodies are unconjugated, a conjugated secondary antibody will be used to incubate for another 15–30 min after washing with FACS medium between these two steps.
3. Wash one or two times with FACS medium and resuspend the cell pellet in 200–300 μl of FACS medium. Perform flow cytometric analysis by standard methods. Importantly, to increase specificity, dead cells should be excluded by 7-AAD staining. Fixation of cells is not performed, as this precludes the use of 7-AAD staining.

3.8. Hematopoietic CFC Assay

1. hESC/S17 cells are cultured for the desired number of days before harvesting. A single cell suspension is prepared, as above. Aliquot 6×10^5 cells into a sterile 1.5-ml tube. Centrifuge at $400 \times g$ for 5 min. Resuspend the cell pellet in 100 μl of I-2 medium after washing once with I-2 medium.
2. Thaw the MethoCult GF+ medium to room temperature before starting the CFU assay. Add cells into 2.5 ml of MethoCult™ GF+ and vortex until the cells distribute evenly within the medium. Keep the tube of cells in methylcellulose upright at room temperature for approximately 15 min to let the bubbles rise and dissipate.
3. Transfer the cells in Methocult GF+ medium into sterile Petri dishes. The medium (2.5 ml) should be divided into two 35-mm non-tissue culture Petri dishes using a wide blunt 2-ml stripette (1.1 ml cells = 2.5×10^5 cells per dish). Place these two dishes and a third, open dish containing water into a 100-mm culture dish. The third dish helps maintain humidity and thus prevents drying of the methylcellulose-based medium.
4. Incubate at 37°C, 5% CO_2 for 2 weeks and score for colony-forming units according to standard criteria.

3.9. RNA Isolation

The single cell suspension used for RNA isolation is prepared as above. Total RNA is extracted using provided protocol for isolation of total RNA from animal cells in the RNeasy® Micro Kit and QIA shredder™, according to the manufacturer's instructions. DNA is removed as described in the TURBO DNA-free™ Kit.

4. Notes

1. S17 stromal cells were kindly provided by Dr Kenneth Dorshkind, University of California, Los Angeles, CA. S17 cells are best passed at a low ratio (~1:3) to ensure hematopoietic support is not lost. Lower passages of S17 have better ability to support hematopoietic differentiation of hESCs than higher passages. S17 cells typically have less ability to support hematopoietic development after passage 20.

2. R-20 medium has equivalent or even better differentiation efficiency than the D-20 medium (*see* **Fig. 3**). R-20 can be used as the differentiation medium instead of D-20 medium. IMDM medium supplemented with 20% FBS was also tested in the S17 co-culture system. But the results show that IMDM cannot promote efficient hematopoietic differentiation of hESCs.

3. Supplementing 15% FBS in the differentiation medium yields similar results as using 20% FBS in hESC/S17 differentiation medium. There is no statistical difference regarding the promotion of CD34$^+$, CD34$^+$CD45$^+$, and CFCs using R-15 or D-15 medium versus R-20 or D-20 medium (data not shown). In this chapter, we use R-15 as a representative differentiation medium.

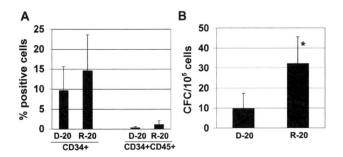

Fig. 3. Hematopoietic differentiation of human embryonic stem cells (hESCs) in R-20 medium and comparison with D-20 medium. Flow cytometric analysis and colony-forming unit (CFU) assays were performed to characterize hematopoietic cells derived from H1 and H9 hESCs after co-culture with S17 for 17–21 days. (A) Similar development of phenotypic CD34$^+$ and CD34$^+$CD45$^+$ cells is demonstrated in the two different media. (B) However, a higher number of CFCs are derived from hESCs differentiated in the R-20 medium (*$p < 0.01$). Results are cumulative from six different experiments. Identical hESCs were used in each differentiation experiment.

4. Chick serum is added to trypsin–EDTA solution to improve cell viability, but unlike FBS, chick serum does not contain trypsin inhibitors. Trypsin–EDTA + 2% chick serum should be warmed to 37°C before use.

5. Disposable glass pipets are used for culture of undifferentiated hESCs. Some researchers feel these pipets help maintain the ES cells in an undifferentiated state by minimizing exposure to plastics, which may vary between lots, or to detergents used to clean reusable glass pipets.

6. Other serum-free media can be used for culture of embryoid bodies. Additional cytokines can be added to serum-free media to promote hematopoietic development.

7. We present protocols for assays that are not specific to analysis of hESC-derived blood cells. Many variations are possible for flow cytometric analysis, CFU assays, and RNA isolation. We offer these methods as one example.

8. 1 mg/ml Propidium iodide (PI; Sigma; Cat. No. P4170) dissolved in DPBS can also be used instead of 7-AAD.

9. Other non-tissue culture-treated dishes have been tested. Only Greiner dishes had no adherent cells when this complex mixture of cells was plated in this CFC assay. If cells do adhere and grow, these proliferating cells will likely interfere with results.

10. Other means of RNA and/or protein isolation are also available. For samples with less than 500,000 cells, RNeasy Mini kit (QIAGEN; Cat. No. 74106) with a similar protocol is suitable. If both RNA and protein are desirable from a sample, TRIzol (Invitrogen; Cat. No. 15596-026) is another alternative.

11. Reports using S17 cells to support hematopoietic differentiation of rhesus monkey ES cells did not irradiate or otherwise mitotically inactive the S17 stromal cells *(28)*. These authors felt that growth inhibition of S17 cells when confluent was sufficient to prevent overgrowth when co-cultured with ES cells. We prefer stromal cells inactivated with Mitomycin C or irradiated (20 Gy) to prevent growth and proliferation that may complicate interpretation of subsequent assays (such as CFU assay as below).

12. Five stromal cell lines have been tested for hematopoietic differentiation function in our laboratory. S17 and M210 cells result in the best differentiation efficiency (*see* **Fig. 2**). We do not see markedly improved hematopoietic promotion with co-culture with OP9 cells, as described by other groups. Notably, OP9 cells are very sensitive to variations in confluence of the cells during culture, medium source, and serum lot *(14)*. And, also, the inactivation of OP9 and the ratio of hESC:OP9 input for the differentiation may affect the hematopoietic differentiation.

Acknowledgments

We thank Julie Morris for assistance with EB protocols; Petter S. Woll for the effort of comparing different stromal cell lines and Dr Colin H. Martin for editing this manuscript.

References

1. Thomson, J. A., Itskovitz-Eldor, J., Shapiro, S. S., Waknitz, M. A., Swiergiel, J. J., Marshall, V. S., and Jones, J. M. (1998) Embryonic stem cell lines derived from human blastocysts. *Science* **282**, 1145–1147.
2. Odorico, J. A., Kaufman, D. S., and Thomson, J. A. (2001) Multilineage differentiation from human embryonic stem cell lines. *Stem Cells* **19**, 193–204.
3. Thomas, E. D. (1999) Bone marrow transplantation: a review. *Semin Hematol* **36**, 95–103.
4. Korbling, M., and Anderlini, P. (2001) Peripheral blood stem cell versus bone marrow allotransplantation: Does the source of hematopoietic stem cells matter? *Blood* **98**, 2900–2908.
5. Grewal, S. S., Barker, J. N., Davies, S. M., and Wagner, J. E. (2003) Unrelated donor hematopoietic cell transplantation: Marrow or umbilical cord blood? *Blood* **101**, 4233–4244.
6. Brunstein, C. G., and Wagner, J. E. (2006) Umbilical cord blood transplantation and banking. *Annu Rev Med* **57**, 403–417.
7. Kaufman, D. S., Hanson, E. T., Lewis, R. L., Auerbach, R., and Thomson, J. A. (2001) Hematopoietic colony-forming cells derived from human embryonic stem cells. *Proc Natl Acad Sci USA* **98**, 10716–10721.
8. Tian, X., Morris, J. K., Linehan, J. L., and Kaufman, D. S. (2004) Cytokine requirements differ for stroma and embryoid body-mediated hematopoiesis from human embryonic stem cells. *Exp Hematol* **32**, 1000–1009.
9. Chadwick, K., Wang, L., Li, L., Menendez, P., Murdoch, B., Rouleau, A., and Bhatia, M. (2003) Cytokines and BMP-4 promote hematopoietic differentiation of human embryonic stem cells. *Blood* **102**, 906–915.
10. Cerdan, C., Rouleau, A., and Bhatia, M. (2004) VEGF-A165 augments erythropoietic development from human embryonic stem cells. *Blood* **103**, 2504–2512.
11. Zambidis, E. T., Peault, B., Park, T. S., Bunz, F., and Civin, C. I. (2005) Hematopoietic differentiation of human embryonic stem cells progresses through sequential hematoendothelial, primitive, and definitive stages resembling human yolk sac development. *Blood* **106**, 860–870.
12. Wang, L., Li, L., Shojaei, F., Levac, K., Cerdan, C., Menendez, P., Martin, T., Rouleau, A., and Bhatia, M. (2004) Endothelial and hematopoietic cell fate of human embryonic stem cells originates from primitive endothelium with hemangioblastic properties. *Immunity* **21**, 31–41.
13. Vodyanik, M. A., Thomson, J. A., and Slukvin, II (2006) Leukosialin (CD43) defines hematopoietic progenitors in human embryonic stem cell differentiation cultures. *Blood* **108**, 2095–2105.
14. Vodyanik, M. A., Bork, J. A., Thomson, J. A., and Slukvin, II (2005) Human embryonic stem cell-derived CD34+ cells: efficient production in the coculture with OP9 stromal cells and analysis of lymphohematopoietic potential. *Blood* **105**, 617–626.
15. Woll, P. S., Martin, C. H., Miller, J. S., and Kaufman, D. S. (2005) Human embryonic stem cell-derived NK cells acquire functional receptors and cytolytic activity. *J Immunol* **175**, 5095–5103.

16. Slukvin, II, Vodyanik, M. A., Thomson, J. A., Gumenyuk, M. E., and Choi, K. D. (2006) Directed differentiation of human embryonic stem cells into functional dendritic cells through the myeloid pathway. *J Immunol* **176**, 2924–2932.

17. Zhan, X., Dravid, G., Ye, Z., Hammond, H., Shamblott, M., Gearhart, J., and Cheng, L. (2004) Functional antigen-presenting leucocytes derived from human embryonic stem cells in vitro. *Lancet* **364**, 163–171.

18. Anderson, J. S., Bandi, S., Kaufman, D. S., and Akkina, R. (2006) Derivation of normal macrophages from human embryonic stem (hES) cells for applications in HIV gene therapy. *Retrovirology* **3**, 24.

19. Qiu, C., Hanson, E., Olivier, E., Inada, M., Kaufman, D. S., Gupta, S., and Bouhassira, E. E. (2005) Differentiation of human embryonic stem cells into hematopoietic cells by coculture with human fetal liver cells recapitulates the globin switch that occurs early in development. *Exp Hematol* **33**, 1450–1458.

20. Chang, K. H., Nelson, A. M., Cao, H., Wang, L., Nakamoto, B., Ware, C. B., and Papayannopoulou, T. (2006) Definitive-like erythroid cells derived from human embryonic stem cells coexpress high levels of embryonic and fetal globins with little or no adult globin. *Blood* **108**, 1515–1523

21. Wang, L., Menendez, P., Shojaei, F., Li, L., Mazurier, F., Dick, J. E., Cerdan, C., Levac, K., and Bhatia, M. (2005) Generation of hematopoietic repopulating cells from human embryonic stem cells independent of ectopic HOXB4 expression. *J Exp Med* **201**, 1603–1614.

22. Tian, X., Woll, P. S., Morris, J. K., Linehan, J. L., and Kaufman, D. S. (2006) Hematopoietic engraftment of human embryonic stem cell-derived cells is regulated by recipient innate immunity. *Stem Cells* **24**, 1370–1380.

23. Narayan, A. D., Chase, J. L., Lewis, R. L., Tian, X., Kaufman, D. S., Thomson, J. A., and Zanjani, E. D. (2006) Human embryonic stem cell-derived hematopoietic cells are capable of engrafting primary as well as secondary fetal sheep recipients. *Blood* **107**, 2180–2183.

24. de Pooter, R. F., Cho, S. K., Carlyle, J. R., and Zuniga-Pflucker, J. C. (2003) In vitro generation of T lymphocytes from embryonic stem cell-derived prehematopoietic progenitors. *Blood* **102**, 1649–1653.

25. La Motte-Mohs, R. N., Herer, E., and Zuniga-Pflucker, J. C. (2005) Induction of T-cell development from human cord blood hematopoietic stem cells by Delta-like 1 in vitro. *Blood* **105**, 1431–1439.

26. Collins, L. S., and Dorshkind, K. (1987) A stromal cell line from myeloid long-term bone marrow cultures can support myelopoiesis and B lymphopoiesis. *J Immunol* **138**, 1082–1087.

27. Xu, C., Inokuma, M. S., Denham, J., Golds, K., Kundu, P., Gold, J. D., and Carpenter, M. K. (2001) Feeder-free growth of undifferentiated human embryonic stem cells. *Nat Biotechnol* **19**, 971–974.

28. Li, F., Lu, S., Vida, L., Thomson J. A., Honig, G. R. (2001) Bone morphogenetic protein 4 induces efficient hematopoietic differentiation of rhesus monkey embryonic stem cells in vitro. *Blood* **98**, 335–342.

9

In Vitro Human T Cell Development Directed by Notch–Ligand Interactions

Génève Awong, Ross N. La Motte-Mohs, and Juan Carlos Zúñiga-Pflücker

Summary

Traditionally, the study of human T cell development has relied on the availability of human and mouse thymic tissue. In this chapter, we outline a simple in vitro protocol for generating large numbers of human T-lineage cells from umbilical cord blood (CB)-derived hematopoietic stem cells (HSCs) using a bone marrow stromal cell line. This protocol is broken into three major steps: (1) the maintenance of a working stock of OP9 bone marrow stromal cells expressing the Notch receptor ligand Delta-like 1 (OP9-DL1), (2) the purification of human HSCs from umbilical CB, and (3) the initiation and maintenance/expansion of OP9-DL1 cocultures over time (*see* **Fig. 1**). The use of this system opens avenues for basic research as it equips us with a simple in vitro method for studying human T cell development.

Key Words: T cell development; lymphopoiesis; umbilical cord blood; Notch; Delta-like 1; CD34; hematopoietic stem cells; stromal cells; IL-7.

1. Introduction

Over 40 years ago, Miller established the critical role of the thymus for supporting T lymphopoiesis in the mouse *(1)*. Similarly, the stringent requirement of the thymus for human T cell development is demonstrated in patients lacking a thymus, such as those afflicted with DiGeorge's syndrome *(2)*. Because of this requirement, the initial study of human T lymphopoiesis proved more challenging than their mouse counterparts. However, the SCIDhu(thy/liv) in vivo model *(3)* and the in vitro hybrid human/mouse fetal thymic organ

From: *Methods in Molecular Biology, vol. 430: Hematopoietic Stem Cell Protocols*
Edited by: K. D. Bunting © Humana Press, Totowa, NJ

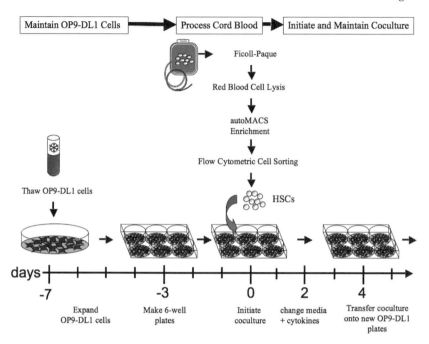

Fig. 1. Schematic overview and temporal guideline for initiating and maintaining human HSC/OP9-DL1 cocultures.

culture *(4)* were two ground-breaking models that have greatly impacted our knowledge on human T-lymphocyte development. Whereas these models were seminal to the field, the requirement for human or mouse thymic tissue remained as limitations.

Recent insights into the key molecular players responsible for the thymic dependency of T cell development *(5,6)* have permitted the differentiation of expanded hematopoietic progenitor cells into T cells by the use of a simple coculture system *(7)*. This coculture system employs the OP9 bone marrow stromal cell line that has been genetically modified to express high levels of the Notch receptor ligand Delta-like 1 (OP9-DL1 cells) *(8)*. The OP9-DL1 cells were derived from the OP9 parental cells, which were obtained from the bone marrow of osteopetrotic (op/op) mice *(9)*. These cells are deficient for the cytokine macrophage-colony stimulating factor (M-CSF) *(10)* and thus support the generation of lymphocytes over myeloid cells. Specifically, OP9 cells support the differentiation of many hematopoietic lineages such as B and NK cells, but not T cells. In striking contrast, OP9-DL1 cells support the development of T cells at the expense of B cells *(8,11)*. Our laboratory has demonstrated that this system is capable of supporting the initiation and differentiation of large numbers of human progenitor T cells in vitro from HSCs obtained from umbilical CB *(12)*. Here, we describe a protocol for the generation of human T-lineage cells from umbilical cord blood HSCs by coculture with OP9-DL1 stromal cells as an alternative approach for the study of human T lymphopoiesis.

2. Materials

It is important to note that the first step in successful cell culture is to practice sterile culture technique. All reagents should be prepared and maintained under sterile conditions. Extreme care should be taken with both the cells and the reagents needed to culture them. This protocol describes the use of primary human cells obtained from whole blood and should be handled with caution and in accordance to local Institutional Ethical Board policies. It should also be noted that this protocol requires access to an autoMACS and a flow cytometric cell sorter.

2.1. Maintenance of Cellular Components and Coculture

1. OP9-DL1 cells: OP9 cells (Riken repository, Tsukuba, Japan, http://www.rtc.riken.go.jp) retrovirally transduced to express the gene *Delta-like 1 (Dll-1)*, as previously reported *(8)*.
2. α-Modified Eagle's Medium (α-MEM) (Gibco, Carlsbad, California, USA, 12561-056). Store at 4°C.
3. Fetal bovine serum (FBS) (*see* **Note 1**). Heat-inactivate (hi) at 56°C for 30 min. Store at 4°C.
4. Penicillin/Streptomycin: 100× or 10,000 U/ml penicillin and 10,000 U/ml Streptomycin (Hyclone, Logan, Utah, USA, SV30010). Use at 1×. Store at 4°C once opened.
5. 1× Phosphate-buffered saline (PBS) without Ca^{2+}/Mg^{2+} (Gibco, 14190-144).
6. Trypsin 2.5% (Gibco, 15090). Dilute with PBS to 0.25% solution. Store at 4°C.
7. OP9 medium: α-MEM supplemented with 20% hiFBS and 1× Penicillin/Streptomycin.
8. 40 µm cell strainers (BD Falcon, Mississauga, ON, Canada, 352340).
9. 70 µm nylon mesh filters (N70R; Biodesign, Inc., Carmel, NY, USA).
10. Human IL-7 (Peprotech, Rocky Hill, NJ, USA, 200-07). Reconstitute at 5 µg/ml (1000×) in OP9 media. Aliquot and store at –80°C.
11. Human Flt-3L (R&D, Minneapolis, Minnesota, 308-FK). Reconstitute at 5 µg/ml (1000×) in OP9 media. Aliquot and store at –80°C.
12. Freezing media: 90% hiFBS, 10% dimethyl sulfoxide (DMSO). Sterile filtered.
13. Tissue culture ware (10-cm dishes, 6-well plates, cryovials), tissue culture treated (Sarstedt, Newton, North Carolina, USA).

2.2. Mononuclear Cell Isolation and Enrichment to Obtain HSCs

1. Human umbilical CB: Obtained in accordance with Institutional Ethical Review Board approval and upon parental consent following healthy deliveries in blood collection bags containing anticoagulant (Baxter, Deerfield, Illinois, USA, 4R3610nm) (*see* **Note 2**).

2. Ficoll-Paque PLUS (GE healthcare, Uppsala, Sweden, 17-1440-03).
3. 10× Lysing buffer (red blood cell lysis) (BD biosciences, 555899).
4. 1× Hank's balanced salt solution (HBSS) without phenol red, Ca^{2+} and Mg^{2+} (Hyclone, SH30268.01).
5. DNAse I (Stem Cell Technologies, Vancouver, British Columbia, Canada, 07900).
6. StemSep negative selection human progenitor enrichment kit (Stem Cell Technologies, 14056A) or StemSep $CD34^+$ cell selection kit (Stem Cell Technologies, 14756A).
7. autoMACS Running buffer: 1× HBSS, 2 mM ethylenediaminetetraacetic acid (EDTA), 0.5% bovine serum albumin. Filter sterilize.
8. autoMACS Rinsing buffer: 1× HBSS, 2 mM EDTA (keep sterile).

3. Methods

3.1. OP9-DL1 Stromal Cells

It should be noted that all incubations are performed in a standard, humidified, cell culture incubator at 37°C in 5% CO_2. In addition, cells are pelleted by centrifugation at $450 \times g$ for 5 min, unless otherwise indicated.

1. A vial of OP9-DL1 cells should be thawed in a 37°C water bath using a gentle swirling motion and then transferred slowly into a 15-ml conical tube containing OP9 media.
2. Centrifuge the cells to obtain pellet and then seed cells in a 10-cm dish containing 9–10 ml of fresh OP9 media. Change media the following day and be sure to split them when no more than 80% confluent. Appropriate confluency can generally be maintained by splitting 1:4 every 2 days (*see* **Note 3**).
3. To passage OP9-DL1 stromal cells from a 10-cm plate, remove the media then add 5 ml PBS to wash off any remaining media. Remove PBS and incubate with 5 ml 0.25% trypsin for 5 min at 37°C.
4. Following trypsinization, vigorously pipette the cells to remove them from the surface of the plate and add them to a conical tube containing 5 ml OP9 media. Rinse the plate with PBS and add to the contents of the first wash. Pellet the cells, resuspend in media, and divide among 10 cm and/or 6-well plates (*see* **Note 4**). Gently rock the plate back and forth for even cell distribution.

3.2. Isolation of Mononuclear Cells from Umbilical Cord Blood

1. Under sterile conditions, dilute whole cord blood with an equal volume of HBSS containing 2 mM EDTA (helps prevent clotting during centrifugation). Carefully layer 30 ml of the diluted blood into a 50-ml conical tube already containing 15 ml Ficoll-paque solution (an approximate 2:1 ratio is used). Avoid mixing the blood–Ficoll layer.

2. Centrifuge at $750 \times g$ for 30 min at 18°C with the brake "off." Once the spin is complete, carefully remove the cells at the "cloudy" plasma/Ficoll interface (*see* **Note 5**).

3. Transfer the mononuclear cell fraction to a clean 50-ml conical tube, resuspend in HBSS, and centrifuge at $515–585 \times g$ for 5 min. Carefully remove the supernatant and lyse any contaminating red blood cells by resuspending the pellet in 1× lysis buffer for 10 min at room temperature (RT). Wash the cells by adding HBSS, centrifuge, and carefully remove the supernatant.

4. Cells can be resuspended in PBS and an aliquot used for cell count determination.

5. Pellet the cells once more. It is at this point that the pellet can either be frozen down using freezing media or one can proceed to the next step of lineage-depletion. If cells are to be frozen, resuspend the mononuclear cells in 1.5 ml ice cold freezing media and aliquot into 2 ml cryovial(s).

6. Transfer the vials on ice to a –80°C freezer overnight, then to a liquid nitrogen tank the next day for long-term storage. These vials can be thawed at a later time to proceed with lineage-depletion. Alternatively, the cells can be resuspended in the appropriate buffer for magnetic labeling and selection of human hematopoietic progenitors assuming the use of the autoMACS system (*see* **Note 6**).

3.3. Enrichment of Human Hematopoietic Progenitors by Either Positive or Negative Selection

Before isolating human hematopoietic progenitor cells, one should decide which subfraction of the stem cell compartment (e.g., $CD133^+CD34^-$; $CD34^+CD38^-$) is desired as this may affect which enrichment protocol is performed. The following steps assume the use of Stem Cell Technologies progenitor enrichment kit for either $CD34^+$ cell enrichment or lineage depletion using magnetic labeling and separation on the autoMACS system. Instructions are provided with the manufacturer's product inserts and should be followed according to their guidelines. However, it will be described in brief here.

1. Thawed (*see* **Note 7**) or fresh red blood cell-lysed lymphocytes from human CB should be resuspended in filter-sterilized running buffer at the specified concentration suggested by manufacturer. Save a small aliquot ($1–3 \times 10^4$ cells) for enrichment determination (*see* **Note 8**).

2. Add selection cocktail at 100 µl/ml, mix well and incubate for specified time at 4°C.

3. Add magnetic colloid at 60 µl/ml, mix well, and incubate for 10 min at 4°C. Wash the cells by adding sterile autoMACS-running buffer, centrifuge, remove the supernatant, and resuspend the pellet in 1 ml running buffer.

4. Perform positive or negative selection on the autoMACS following manufacturer's instructions. Save a small aliquot to determine enrichment between pre- and post-magnetic separation (*see* **Note 9**).

3.4. Initiation and Maintenance of Coculture

The autoMACS fraction enriched for human HSCs can now be further purified and sorted based on the surface expression of stem cell markers with a flow cytometric cell sorter. Purified HSCs are seeded onto OP9-DL1 stromal cells in the presence of IL-7 and Flt-3L. Human HSC/OP9-DL1 cocultures are maintained in 3 ml/well of OP9 media in a 6-well plate.

1. Sorted human hematopoietic stem cells (CD34$^+$CD38$^{-/\text{low}}$ cells isolated from human umbilical CB) are seeded onto an 80% confluent well of 6-well plate of OP9-DL1 cells (*see* **Note 10**).
2. The coculture is maintained in OP9-DL1 media. The human cytokines Flt3-L and IL-7 are added to the wells from a 1000× stock solution (1 μl of stock/1 ml of media for a final concentration of 5 ng/ml) starting at day 0 and maintained throughout the coculture (*see* **Note 11**).
3. Depending on overall cellularity (*see* **Note 10**), the media is changed every 2–4 days (*see* **Note 12**).
4. Every 4 days, the entire coculture is disaggregated through vigorous pipetting (5 ml pipette) and passaged through a 70-μm sterile nylon mesh or a 40-μm cell strainer into a 50-ml conical tube.
5. The filtered cells are washed with 5 ml PBS and centrifuged. Remove supernatant and resuspend the pellet in 1 ml of OP9 media.
6. Transfer resuspended cells onto a new 90% confluent 6-well plate of OP9-DL1 cells containing 2 ml media with cytokines and continue the coculture (*see* **Note 12**).

4. Notes

1. New lots of FBS serum should be batch tested against a standard lot of FBS known to support human T lymphopoiesis.
2. Umbilical cord blood can also be collected in heparinized vacutainers; however, blood collection bags aid in the ease of both collecting and processing of the blood. In addition, larger volumes can be obtained when collected in bags.
3. To preserve early passage stocks of OP9 stromal cells, allow OP9 cells to grow to 80% confluency in a 10-cm dish. Split the 10-cm dish into four more dishes and continue the subculturing procedure until at least 16 plates are 80% confluent. Freeze one confluent plate per cryovial in freezing media.
4. As mentioned earlier, one 10-cm dish can be split to obtain four 10-cm dishes that will be confluent in 2 days. In addition, one 10-cm dish is equivalent to one 6-well plate so that four 6-well plates can be made from one 10-cm dish.
5. If the plasma fraction is desired, it can be saved at this step by sterile filtering followed by storage at –20°C. Remember that this plasma has been diluted.
6. If an autoMACS is not available, a benchtop magnet and columns can be used for enrichment.

7. Previously frozen mononuclear cells should be incubated with DNAse I (1 mg/ml) to prevent clumping of cryopreserved cells and ultimately cell loss.

8. Enrichment of hematopoietic progenitors can be assessed, by saving small aliquots of cells pre- and post-selection on the autoMACS. This can be done with antibody staining for CD34 and CD38 and analysis with a flow cytometer.

9. Following autoMACS enrichment, progenitor cells can be frozen at this step (if not previously frozen at **step 5**, **Subheading 3.2**). Cells should only be frozen down once.

10. At least $5 \times 10^3 - 1 \times 10^4$ human HSCs are seeded per well in order to observe T cell differentiation markers at 4-day intervals. Should human HSCs be limiting, the coculture can be initiated with lower cell numbers of human HSCs, but differences in the temporal expression of T cell markers, expansion of T cells, and efficiency of T cell generation may be observed. Thus, it is recommend that cocultures be seeded with greater numbers of input human HSCs.

11. These cocultures are long term: 40–60 days for mature T cells. Developmental kinetics published by La Motte-Mohs, R.N., et al. *(12)* is with 1×10^4 cells. Fewer input cells require longer kinetics.

12. During media changes or OP9-DL1 transfers, each well of a 6-well plate contains a final volume of 3 ml of OP9 media. When changing media in large numbers of wells with identical conditions, we pool the conditioned media into a 50-ml conical tube and centrifuge to recover non-adherent cells. The adherent cells then receive 1 ml of fresh OP9 media per well. The pellet containing the non-adherent cells is suspended in 1 ml of OP9 media and filled to the *n* ml (where *n* equals number of total wells). These cells are plated in 1 ml aliquots either onto new OP9-DL1 cells or existing human HSC/OP9-DL1 coculture cells. A stock solution containing cytokines is also added in 1 ml aliquots.

References

1. Miller, J. F. (1961) Immunological function of the thymus. *Lancet* **2,** 748–9.

2. Thomas, R. A., Landing, B. H., and Wells, T. R. (1987) Embryologic and other developmental considerations of thirty-eight possible variants of the DiGeorge anomaly. *Am J Med Genet Suppl* **3,** 43–66.

3. McCune, J. M., Namikawa, R., Kaneshima, H., Shultz, L. D., Lieberman, M., and Weissman, I. L. (1988) The SCID-hu mouse: murine model for the analysis of human hematolymphoid differentiation and function. *Science* **241,** 1632–9.

4. Fisher, A. G., Larsson, L., Goff, L. K., Restall, D. E., Happerfield, L., and Merkenschlager, M. (1990) Human thymocyte development in mouse organ cultures. *Int Immunol* **2,** 571–8.

5. Pui, J. C., Allman, D., Xu, L., DeRocco, S., Karnell, F. G., Bakkour, S., Lee, J. Y., Kadesch, T., Hardy, R. R., Aster, J. C., and Pear, W. S. (1999) Notch1 expression in early lymphopoiesis influences B versus T lineage determination. *Immunity* **11,** 299–308.

6. Radtke, F., Wilson, A., Stark, G., Bauer, M., van Meerwijk, J., MacDonald, H. R., and Aguet, M. (1999) Deficient T cell fate specification in mice with an induced inactivation of Notch1. *Immunity* **10,** 547–58.
7. Zúñiga-Pflücker, J. C. (2004) T-cell development made simple. *Nat Rev Immunol* **4,** 67–72.
8. Schmitt, T. M., and Zúñiga-Pflücker, J. C. (2002) Induction of T cell development from hematopoietic progenitor cells by delta-like-1 in vitro. *Immunity* **17,** 749–56.
9. Kodama, H., Nose, M., Niida, S., and Nishikawa, S. (1994) Involvement of the c-kit receptor in the adhesion of hematopoietic stem cells to stromal cells. *Exp Hematol* **22,** 979–84.
10. Yoshida, H., Hayashi, S., Kunisada, T., Ogawa, M., Nishikawa, S., Okamura, H., Sudo, T., Shultz, L. D., and Nishikawa, S. (1990) The murine mutation osteopetrosis is in the coding region of the macrophage colony stimulating factor gene. *Nature* **345,** 442–4.
11. Schmitt, T. M., de Pooter, R. F., Gronski, M. A., Cho, S. K., Ohashi, P. S., and Zúñiga-Pflücker, J. C. (2004) Induction of T cell development and establishment of T cell competence from embryonic stem cells differentiated in vitro. *Nat Immunol* **5,** 410–7.
12. La Motte-Mohs, R. N., Herer, E., and Zúñiga-Pflücker, J. C. (2005) Induction of T-cell development from human cord blood hematopoietic stem cells by Delta-like 1 in vitro. *Blood* **105,** 1431–9.

10

In Vitro Assays for Cobblestone Area-Forming Cells, LTC-IC, and CFU-C

Ronald P. van Os, Bertien Dethmers-Ausema, and Gerald de Haan

Summary

Various assays exist that measure the function of hematopoietic stem cells (HSCs). In this chapter, in vitro assays are described that measure the frequency of progenitors (colony-forming unit in culture; CFU-C), stem cells (long-term culture-initiating cell; LTC-IC), or both (cobblestone area-forming cell assay; CAFC). These assays measure the potential of a test cell population retrospectively, i.e., at the time its activity is evident when the stem cell itself is often not detectable anymore. Although the in vitro LTC-IC and CAFC assays have been shown to correlate with in vivo activity, in vivo transplantation assays, where it can be shown that cells possess the ability to indefinitely repopulate all blood lineages, are the ultimate proof for HSC activity. Nevertheless, these in vitro assays provide an excellent method to screen for stem cell activity of a putative stem cell population or for screening the effect of a certain treatment on HSCs.

Key Words: Hematopoietic stem cell; CAFC assay; cobblestone area; LTC-IC assay; CFU-C assay; colony-forming unit granulocyte/macrophage (CFU-GM); long-term bone marrow culture.

1. Introduction

Blood contains multiple distinct cell types, such as erythrocytes, granulocytes, lymphocytes, monocytes, and platelets. The finite lifetime of these mature cells requires a well-organized system of replenishment and cell renewal, which in mammals is predominantly confined to the bone marrow. Numerous assays exist that measure the function of stem cells. In this chapter, we will describe in vitro assays commonly used to measure hematopoietic stem and/or progenitor

From: *Methods in Molecular Biology, vol. 430: Hematopoietic Stem Cell Protocols*
Edited by: K. D. Bunting © Humana Press, Totowa, NJ

cell activity. It should be realized that most assays assess stem and progenitor content retrospectively, and true stem cell function can only be determined by in vivo transplantation assays where stem cells can be shown to indefinitely repopulate all blood lineages, myeloid and lymphoid, of an irradiated recipient after bone marrow transplantation (BMT). The various assays that exist to measure the frequency of hematopoietic stem cells (HSC) or progenitors make use of the capacity of the cells to rapidly generate many cells. These assays include in vitro clonogenic assays in stroma cell-supported long-term bone marrow cultures or in semi-solid medium.

Long-term bone marrow cultures showed that HSC could be cultured on a pre-established stromal layer allowing growth of stem cells in close association with supporting stroma *(1)*. This method was modified to measure the frequencies of different hematopoietic cell subsets growing as cobblestone area-forming cells (CAFC) underneath the stromal layer *(2–4)*. Instead of scoring cobblestone areas, the long-term culture-initiating cell (LTC-IC) determines the presence of committed progenitors by replacing the culture medium with a semi-solid medium and scoring for colonies 7–14 days later *(5,6)*. The CAFC and LTC-IC assay are both miniaturized long-term bone marrow cultures set up in 96-well plates in which a stromal cell layer is first allowed to grow to confluency. Because of reproducibility and other practical considerations, it is easiest to use a defined stromal cell line for this purpose, but the assay works similarly well when fresh bone marrow is used as a source of stromal cells. Stromal layers derived from fresh bone marrow have to be irradiated to eradicate endogenous hematopoiesis. Stromal cell lines are highly variable in the extent to which they sustain such long-term bone marrow cultures, and therefore, it is of crucial importance to select one that has been demonstrated to be a good "supporter." Although not many single-laboratory data have been published on comparison of the various cell lines that have been described, two excellent papers that may be useful for this purpose are from Muller-Sieburg et al. *(7,8)*. Some lines with reasonable to good supportive activity include S17 *(9)* (originating from the Dorshkind laboratory), CFC034, 2012, AFT024 *(10,11)* (from the Lemischka laboratory), and flask bone marrow dexter-clone 1 (FBMD-1), the cell line we routinely use *(12)* (from Steve Neben). Recently, cell lines from aorta and mesenchyme (AM) and urogenital ridges (UG) have been generated from midgestational embryos and were shown to have excellent supportive capacity *(13)*. Some cell lines need to be irradiated before use, others are sufficiently contact inhibited to prevent excessive growth of the stromal cells.

When the stromal cells have become confluent, the layer is seeded by the hematopoietic cell suspension to be tested in a limiting dilution fashion. For unknown reasons, primitive cells present in the inoculated cell sample will migrate through the stromal layer to inhabit a putative hematopoietic stem cell (HSC) niche microenvironment beneath the stromal cells and begin to proliferate at time points defined by their primitiveness. At sequential time points after initiation of the assay, individual wells are microscopically screened for the presence or absence of "cobblestone areas," which we define as colonies of at least five small, non-refractile cells that grow underneath the stromal layer. It was demonstrated that CAFC frequencies determined at various culture times showed good correlation with the different hematopoietic subsets as tested with other assays [colony-forming units in culture (CFU-C), colony-forming units in the spleen on day 12 (CFU-S-12), and marrow-repopulating ability (MRA)]. An assay using a similar approach but a different end-point is the LTC-IC assay. The stem cells are also grown on stromal layers but the end-point is the presence of progenitor cells, with colony-forming ability, at later time-points that demonstrate the persistence of hematopoiesis with time. To this end, usually after 4–5 weeks, the medium is replaced by semi-solid medium containing hematopoietic growth factors that stimulate the outgrowth of progenitors present in the cultures. The progenitors will develop into colonies that can be counted. This assay was initially developed for studying human stem cell growth in vitro *(5)* but was shown to be also useful for measuring murine stem cell frequencies, and culture conditions can even be altered to support growth of lymphomyeloid progenitors and allow their quantitation *(6)*. Until now, the CAFC and LTC-IC assay are the only testing systems that have the ability to reliably measure mouse primitive stem cell frequencies other than in vivo long-term engraftment studies and have provided a basis for useful comparisons with, for example, long-term repopulating activity (LTRA) radiosensitivity *(14)*, or cytotoxicity *(15)*.

Whereas the CAFC and (to a lesser extent) LTC-IC assay can measure both stem cell-like and progenitor cell-like activity, semi-solid medium-based clonogenic cell assays are only able to measure progenitor cell activity and can serve as a screening method. The semi-solid medium is usually provided by methylcellulose that provides viscosity, which supports the three-dimensional growth of the hematopoietic colonies and prevents migration of the cells so that they remain within a colony. A colony is derived from a single cell and can contain mature cells of different lineages depending on the growth factors added to the cultures and the multipotentiality of the progenitor cell.

Finally, as the most strict definition of a pluripotent HSC states that such a cell should be able to fully reconstitute all the blood cell lineages of a properly conditioned recipient, the most important weakness of the CAFC (and any in vitro) assay is that it will always remain a surrogate method to quantify stem cells. It is possible that conditions exist in which stem cells will flourish upon in vivo transplantation but will fail to thrive in a CAFC assay, but more often, stem cells may be detected in the CAFC assay whereas their existence cannot be confirmed in an in vivo competitive repopulation setting as we have shown with serially transplanted stem cells *(16,17)*. In such situations, one again has to realize that a stem cell may reveal its presence only in one environment (in vitro or in vivo) and not in the other.

2. Materials

1. Medium: Iscove's modified DMEM medium (IMDM) (e.g., Invitrogen, Breda, The Netherlands, No. 041-90898; contains Glutamax)
2. Phosphate-buffered saline (e.g., Invitrogen)
3. Flat-bottomed 96-well plates (e.g., Corning Inc. Life Sciences, Lowell, MA, USA, No. 3595)
4. Antibiotics: Penicillin/streptomycin solution (e.g., Invitrogen, No. 15140-114), β-mercaptoethanol (e.g., Sigma M7522, St. Louis, MI, USA); Make a stock (0.1 M) solution of β-mercaptoethanol: 70 μl in 10 ml HBSS, sterilize over a filter. Add this to 100 ml Pen/strep. Make aliquots of 5.5 ml and store in a –20°C freezer. Each aliquot is sufficient to supplement 500 ml IMDM. The final concentration of β-mercaptoethanol is 10^{-4} M Hydrocortisone (HC; Sigma, H-4881); Dissolve 24 mg HC 21-hemisuccinate in 50 ml HBSS. Sterilize over a 0.2-μm filter (Whatman Schleicher & Schuell, Dassel, Germany). Make 5 ml aliquots and store at –20°C. The final concentration of HC is 10^{-5} M.
5. Tryspin/EDTA Invitrogen (No. 15090-046). Dilute 2.0 ml of 2.5% Tryspin and 0.5 ml 0.1 M EDTA [37.32 g of EDTA (Merck, No. 108418) in 10 ml of PBS] in 97.5 ml PBS.
6. Horse serum: Tested for optimal growth of hematopoietic cells.
7. Fetal bovine serum: Tested for optimal growth of hematopoietic and stromal cells.
8. Ready-to-use FBMD-1 culture medium: Add 25 ml horse serum (5%), 50 ml of FBS (10%), 5.5 ml Pen/strep/β-mercaptoethanol solution, and 5 ml HC to a 500-ml bottle of IMDM medium. Mix well.
9. Ready-to-use CAFC culture medium: Add 100 ml of horse serum (20%), 5.5 ml Pen/strep/β-mercaptoethanol solution, and 5 ml HC to a 500-ml bottle of IMDM medium. Mix well.
10. Multichannel pipette (or Eppendorf repeat pipetter).
11. Reagent reservoirs.
12. 14-ml polystyrene tubes for making dilutions (e.g., Greiner Bio-One, Frickenhausen, Germany, No. 191180).
13. Inverted (phase-contrast) microscope (100×).

3. Methods

3.1. Stromal Cell Lines

Several investigators have generated stromal cell lines that can be used in the CAFC assay. Some of these cell lines, such as AFT024 or MS-5, require irradiation (>20 Gy) of the confluent layer before the assay can be initiated, to prevent excessive growth of the stromal cells. FBMD-1 or MS-5 (stromal cell lines) should be carefully grown by passaging them 2–3 times a week preventing the cells from reaching confluency. Stromal cell lines are usually maintained by passaging cells from a 60–90% confluent flask into a new flask, thereby diluting the cells 3–5 times. If cells are too confluent before passaging, trypsinization will release cells as sheets rather than as individual cells and cell clumps will be evident. Clumps will produce uneven monolayers in the new culture flask and should be avoided (filter if clumps are evident). Obtaining a stromal cell suspension requires washing of the culture flask with PBS to remove serum-containing medium followed by trypsinization with 0.05% trypsin/EDTA. Cells can be viewed microscopically and observed to round-up and detach from the culture flask. Gentle tapping will release loosely adherent cells. Medium containing serum should then be added to neutralize the trypsin and the cell suspension gently pipetted up and down and washed over the growth surface to further detach remaining loosely attached cells. Cells are ready to be further cultured or seeded for the CAFC assay.

Prepare 96-well plates for the CAFC assay by seeding FBMD-1 cells into each of the 60 inner wells of the plate [leave the outer wells of the plate (rows A and H and rows 1 and 12) empty or fill with sterile water or 0.1 N NaOH]. In this way, evaporation of the inner wells is prevented. FBMD-1 cells from a single T75 flask can be used to seed 10–15 plates. Plates should be prepared 7–14 days before an experiment to allow the stromal cells to grow to confluency in each well. FBMD-1 cells can be used in the CAFC assay up to passage 20. After passage 20, you may notice morphological changes such as the appearance of more fat cells. FBMD-1 cells seeded for the CAFC assay are cultured at 34°C and 5% CO_2. After becoming confluent, the culture medium can be replaced by CAFC medium to reduce the chance of differentiation of the cells into fat cells.

In the mouse system, we have used the CAFC assay to measure progenitor and stem cell activity in bone marrow, spleen, peripheral blood, and fetal liver. Great care should be taken to harvest the test cells as sterile as possible, because the risk of contamination in these long-term culture systems is substantially higher than with short-term CFU assays.

3.2. Performing a CAFC Assay

3.2.1. Set-Up and Serial Dilutions

1. In the case of normal mouse bone marrow, the highest concentration of cells is usually 81,000 cells per well. Cell concentrations can be reduced in threefold dilution steps all the way down to 333 cells per well. For good statistical practice, use 10–20 wells per dilution each with 200 μl culture volume per well. Use

higher cell numbers for test cell populations with lower frequencies of stem cells such as spleen, peripheral blood, or liver. When a wide range of frequencies is expected, make more dilutions to cover a wide range of dilutions. When enriched stem cell populations will be used, the cell numbers should be adjusted. See **Table 1** for some guidelines.

Table 1
Cobblestone Area-Forming Cell Assay (CAFC) Frequencies and Recommended Dilutions for Several Sources of Hematopoietic Cells

Source	CAFC-7 frequency[a]	CAFC-35 frequency[a]	Recommended dilutions
Bone marrow (BM)	500–1500	10–30	81,000–27,000–9,000–3,000–1,000–333
Spleen	3–12	0.5–5	972,000[c]–243,000–81,000–27,000–(9,000–3,000)
Blood	0.5–4	0.1–1.2	972,000[c]–243,000–81,000–27,000–9,000–3,000
G-CSF Mobilized blood	500–2000	1–5	243,000–81,000–27,000–9,000–3,000–1,000
Fetal liver	25,000–60,000	2–10	243,000–81,000–27,000–9,000–3,000–1,000
BM Lin⁻Sca-1⁺kit⁺	120,000–200,000	10,000–20,000	100–30–10–3–1
BM Lin⁻Sca-1⁺kit⁺Rhˡᵒ	10,000–35,000	60,000–120,000	30–10–3–1
Human BM	20–200	3–80	243,000[c]–81,000–27,000–9,000–3,000–1,000
Human mobilized blood[b]	0.8–100	0.4–50	972000[c]–243,000–81,000–27,000–9,000–3,000

[a] Frequency ranges are given per 10^6 cells. Mouse cells are from C57Bl/6 mice. Numbers may differ for other mouse strains. Human CAFC were grown with IL-3 and G-CSF (both at 10 ng/ml).

[b] Mobilized by G-CSF after chemotherapy.

[c] High cell numbers may increase detachment of stroma (*see* **Notes**).

2. For 20 wells per dilution 20 × 0.2 ml = 4 ml cell suspension needed. Prepare 5 ml to ensure sufficient suspension is available.
3. We routinely set up 6 × 14 ml pop-top tubes containing 5 ml CAFC medium.
4. In tube No. 1, prepare 7.5 ml of your highest concentration.
5. A limiting dilution is then performed by diluting threefold in every step. In the first 14-ml tube, 7.5 ml CAFC medium is pipetted. Volume includes the volume needed for—in this case—81.000 × 5 × 7.5 = 3.0375 × 10^6 cells).
6. Add cells to the first tube.
7. Then mix by inverting the tube or pipetting up and down using a 5-ml pipet. Thorough mixing is essential!
8. Pipet 2.5 ml into the next tube, mix, and continue on to all the required dilutions. Mixing is crucial in this assay, so do it well!
9. Remove the medium in the 96-well plates, for instance, by using a suction device with a vacuum pump.
10. Add the cells, starting with the lowest concentration in the lowest 20 wells, using a repetitive pipet with an Eppendorf combitip or use a multichannel and reagent reservoirs. Add 200 μl per well.
11. Important: Culture at 33–34°C in a 5% CO_2 in air fully humidified atmosphere.
12. Refresh medium every week. Some people remove 100 μl and then add 100 μl, and some take off all medium and add 200 μl fresh medium (multichannel pipet).
13. For plating and refreshing the medium, it is best to start with the most dilute wells and move to the most concentrated wells.
14. Unfractionated cell populations usually have higher frequencies of more mature hematopoietic progenitors (early appearing CAFC, e.g., days 7–14) than more primitive stem cells (late appearing CAFC, e.g., days 28–35). Therefore, cell dilutions with low cell numbers are scored at early time points and the higher cell numbers are scored later.

3.2.2. Scoring CAFC

A well can be scored as either "positive" (containing one or more cobblestone areas) or "negative" (containing no cobblestone areas). The power of the CAFC assay comes from its use of limiting dilution analysis (LDA). LDA uses a Poisson-based probability statistic to calculate frequencies through the use of serial dilutions. In effect, known dilutions are performed until the feature being assayed is diluted away. This strategy can be used in many fields.

A well is considered positive when at least 6 cells (in proximity of each other) are growing underneath the stroma (*see* **Fig. 1**). These cells are non-refractile, and although most pictures show cobblestone-like cells with a phase dark appearance, this is usually not the case in 96-well plates because of the deflection of light. Only dilutions with both negative and positive wells are informative for frequency analysis. Thus, most of the time, only three dilutions need to be scored at a given time. When there is output (see below), this does

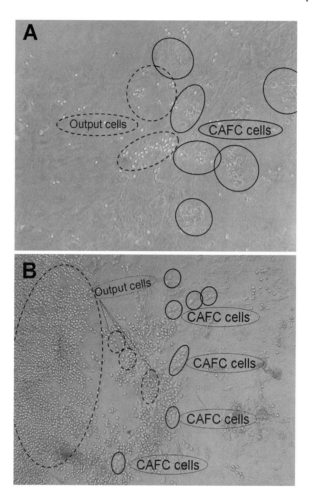

Fig. 1. Cobblestone areas CAFC cells in real life. Representations of CAFC cells in the CAFC assay. In the top panel, an almost ideal situation is depicted (a CAFC day 6 colony). Some cobblestone areas (CA) are shown (straight-lined circles) as well as two areas with output (phase bright) cells shown in dashed circles. However, most of the time colonies are less "ideal" as shown in the bottom panel. This is more representative for later time points within the CAFC assay. Many more output cells (dashed circles) can be seen, but also some (straight circles). These pictures were taken with a 20 × 10 magnification.

not necessarily mean that there is a CAFC colony, and when there is no output, this does not necessarily mean there is no colony. CAFC cells are "masters of disguise." It is not easy to learn to see them, especially in very crowded wells.

In the pictures shown in **Fig. 1** the CAFC colonies can be observed in the straight-lined circles. They are non-granulated, almost invisible under the microscope, are of the same size, and a bit flattened at the sides. It is best to continuously microfocus with small diaphragm or with phase contrast. Most of the times, the CAFC have already produced output cells, either progenitors or mature cells. These cells are much smaller, lay over each other on top of the stroma, and can fill the entire well. The top panel of **Fig. 1** shows an almost "ideal" cobblestone colony, the lower panel shows a more representative example. The cells in the straight-lined circles are CAFC cells, and in the circles (dotted lines) output cells can be seen.

3.3. LTC-IC Assay

The LTC-IC assay is very similar to the CAFC assay; the endpoint is not the presence of cobblestone areas, but the ability of the cells to form a progenitor cell colony *(5,6)*. Usually, this is evaluated after 5 weeks of culture of the test cells set up in limiting dilution as described in the CAFC assay. At the time of assay, the culture medium is removed and replaced by semi-solid methylcellulose culture medium supplemented with hematopoietic growth factors. Stem Cell Technologies (www.stemcell.com) produce a range of Methocult reagents available for the assay of hematopoietic progenitor cells. Hematopoietic progenitor cells that are present in the culture and sensitive to the hematopoietic growth factors are stimulated to form colonies. Although in theory one can check for the presence of hematopoietic progenitor cells at all times throughout culture, the addition of methylcellulose is usually only performed when cultures have grown for 5 weeks. The persistence of hematopoiesis in the microwell culture for this period provides a measure of stem cell activity in the initial test population. Once the medium has been replaced by methylcellulose, one cannot revert back to liquid medium anymore.

3.3.1. Set-Up of LTC-IC Assay

1. Grow test cells on pre-established stromal layer (FBMD-1, MS-5, AFT1024) in limiting dilution (as for the CAFC assay).
2. Refeed cultures on a weekly basis by removal and replacement of medium.
3. After 5 weeks, aspirate off *all* the medium and add 100 µl of methylcellulose medium for your progenitor cell assay of choice (*see* **Subheading 3.6.**).
4. Leave for an additional 7–14 days until colonies can be observed.

3.3.2. Scoring of an LTC-IC Assay

As with the CAFC assay, a well is scored either "positive" or "negative" at assay. A positive well is defined as one containing one or more colonies (indicative of the presence of hematopoietic progenitor cells and the evidence for the persistence of hematopoiesis and therefore of stem cell activity), and a negative well is defined as one that contains no colonies (no evidence of hematopoietic progenitor cells). As with the CAFC analysis, LDA allows the measurement of the frequency of stem cell activity as determined by the persistence of hematopoietic activity for 5–6 weeks in culture.

3.4. Assays for Human Cells (CAFC and LTC-IC)

The CAFC assay has been adapted for use with human cells *(18)*. The only difference is the extended culture time (up to 6 weeks) and the addition of low doses of human-specific growth factors, G-CSF and IL-3, which are added to the CAFC culture medium at 20 and 10 ng/ml, respectively.

The LTC-IC assay for human stem cell quantification does not require the addition of growth factors during the initial culture. However, a methylcellulose medium specific for human progenitor cells, containing several growth factors, should be used. After a period of 5 or 6 weeks, the progenitor cell medium containing methylcellulose is added instead of LTC-IC medium and is left for 14 days until colonies are visible.

3.5. The CFU-C Assay

Hematopoietic progenitor cell assays utilize the ability of progenitor cells to rapidly produce a large number of progeny. To prevent their migration through the culture dish, a semi-solid medium composed of either 1% methylcellulose or 0.3% agar is used. Progenitor cell assays are clonal assays, i.e., a single progenitor cell clonally expands into a colony. The addition of optimal levels of hematopoietic growth factors provides a convenient, reproducible system for quantification of colony-forming cells. The choice of growth factor(s) determines whether erythrocyte, granulocyte/macrophage, megakaryocyte, or a combination of these precursors is measured. (*See* **Table 2** for details) Most of these output cells can be distinguished based on morphological criteria, but for mouse erythroid cells or megakaryocytic cells, specific staining can be helpful.

StemCell Technologies, Inc., offers many ready-to-use culture systems (e.g., MethoCult) and also has a lot of background information and instructional manuals on their Web site (http://www.stemcell.com/). In both the human and the mouse systems, several colony types can be distinguished.

Homemade methylcellulose can also be considered. A 1–1.2% methylcellulose (e.g., Sigma, M0512) solution should be made in medium [e.g., minimal essential medium (MEM)-alpha] containing 30% serum (FBS). Usually, concentrated medium is used or made from powder (e.g., MEM-alpha, Invitrogen, 12000-

Table 2
Progenitor Cell Types That Can Be Measured in Semi-Solid Cultures

Progenitor cell type	Essential cytokine	Time of scoring (mouse/human)
CFU-E (colony-forming unit erythroid)	EPO	2–3/5–7
BFU-E (burst-forming unit-erythroid)	EPO	6–7/14–16
CFU-GM (colony-forming unit granulocyte/macrophage)	GM-CSF (or G-CSF and M-CSF)	6–7/14–16
CFU-G (colony-forming unit granulocyte)	G(M)-CSF	6–7/14–16
CFU-M (colony-forming unit macrophage)	(G)M-CSF	6–7/14–16
CFU-Mk (colony-forming unit megakaryocyte)		6–7/14–16
CFU-GEMM (colony-forming unit granulocyte/erythroid/macrophage megakaryocyte)	GM-CSF, EPO, TPO	6-7/14-16

50 ng/ml rSCF is often added to CFU-GM cultures. Instead of rGM-CSF (1.25 ng/ml), 50 ng/ml rSCF, 10 ng/ml rIL-3, and 10 ng/ml rIL-6 might be added.

3 U/ml rEPO is added to assays for erythroid precursors.

SCF, GM-CSF, IL-3 are species specific.

To visualize megakaryocytes, immunohistochemical staining with antibodies against CD41 (GPIIb/IIIa) can be performed. Murine CFU-Mk can be identified by the detection of acetyl-cholinesterase activity of megakaryocytes.

063). Growth factors can then be added at own choice. This is a laborious but cheap method. Very few laboratories still use homemade methylcellulose.

3.5.1. Setting up a Progenitor Cell (CFU-C) Assay

1. Use a single cell suspension of your tissue of interest and add a recommended number of cells (or a range of dilutions when progenitor content is unknown) to a thawed and preferably pre-warmed (waterbath or incubator) vial or tube with methylcellulose containing the appropriate growth factor.
2. Mix thoroughly and allow bubbles to disappear (approximately 5 min).
3. Using an 18-G needle and syringe, add 1 ml of suspension to a pre-labeled 30-mm Petri dish, and be sure that the entire bottom of the dish is covered with methylcellulose medium.
4. Place several dishes with one dish containing water for humidity in a larger dish and incubate for the appropriate time at 37°C, 5% CO_2 in incubator.
5. Prevent dispersion of colonies (methylcellulose remains liquid) while taking the dishes out of the incubator to count.

3.5.2. Scoring Progenitor Cell Colonies

Starting with one end of the dish, count all the colonies of interest under high power (e.g., 4–5×) objective of an inverted microscope, or else using a dissecting microscope set up with dark field illumination. A finger-operated counter is also recommended to assist with scoring. Typically a "colony" (derived from a single colony-forming cell) is defined as an aggregate of more than 40 cells (>7 divisions). Depending on the combination of growth factors present in the semi-solid culture medium, a degree of hemoglobinization may be observed within individual colonies apparent as a red coloration. Colonies can also be defined as of "loose," "tight," or "mixed" morphology. Loose colonies are derived from macrophage progenitors, tight colonies from granulocyte progenitors and mixed colonies from bipotential granulocyte-macrophage progenitors. Other rarer morphologies include colonies derived from a single multipotential progenitor that can give rise to granulocyte, erythroid, macrophage, and megakaryocyte cells. (These are reviewed photographically by StemCell Technologies at www.stemcells.com.)

Furthermore, CFU-E (colony-forming unit erythroid) colonies are typically grown with lower (~0.8%) methylcellulose concentrations to allow the small colonies to descend to the bottom of the dish allowing easier scoring, which for CFU-E is best done under higher power magnification.

It is necessary to continually focus up and down to identify all colonies present in the three-dimensional culture, those at the edges, and to distinguish individual colonies that are close together but present in different planes. Move the dish up and down using the stage control knob to count all colonies in each column, rather than across. This will minimize the sensation of motion sickness common to individuals new to scoring. Once the entire dish has been viewed under high power, switch to a lower magnification (2.5× objective, total magnification 25× with a 10× ocular eyepiece). Switch to a higher power if necessary to help with colony identification. Burst-forming unit-erythroid (BFU-E), colony-forming unit granulocyte/erythroid/macrophage megakaryocyte (CFU-GEMM), and CFU-GM are colonies that can typically be scored under low power.

3.6. Quantification of Stem Cell Frequencies

Data obtained with the CAFC assay should be converted into a frequency by using LDA. Formulas for the estimation of the frequency of stem or progenitor cells within a test population are derived by the statistical method of likelihood maximization (*19*). For each dilution the cell number per well, the number of negative wells, and total number of wells are used in the calculation. Whereas Fazekas de St.Groth provides a method that can be used for incorporation into an Excel sheet and subsequent calculation (available upon request from the authors), Stem Cell Technologies offers a simple, free software program for researchers (L-Calc™ Software for LDA, available at http://www.stemcell.com/). Finally, the CAFC, LTC-IC, and CFU assays are techniques that reveal the presence of primitive and mature hematopoietic progenitors in a given tissue of interest, e.g., bone marrow, blood, spleen, or fetal liver. Data from such analytical techniques allow the calculation of the frequency and the absolute number of progenitors or stem cells in a given tissue. This may reveal changes within various compartments of the hemato30poietic system.

4. Notes

1. Over time, many cells will accumulate in wells that at one time contain a stem or progenitor cell. For the CAFC assay, counting cells on top of the stroma (cells that appear phase-bright) will result in a gross overestimation of the number of stem cells present. Experiments in which the entire content of a well was replated in a methylcellulose assay or transplanted into lethally irradiated recipient mice, have revealed that only wells containing true cobblestone areas, growing underneath the stroma contain CFU-GM or LTRA (varying from a few to several dozens), indicating that primitive hematopoietic cells can only be found in wells containing cobblestone areas. Also, this shows that LTC-IC and CAFC really measure the same kind of cell. It should be remembered though that the CAFC and LTC-IC assays are based simply on the presence or absence of cobblestone areas or CFU, respectively, in a well at a given concentration, rather than the actual numbers of cobblestone areas or CFU. These features make these assays relatively robust.

2. One of the major problems with the long-term culture assays is detachment of the stromal layer from the surface of the well. This is particularly evident when many cells are being produced at early time points after seeding or when longer culture times are used such as with human cells. When the stromal layer has completely disappeared, no reliable quantification is possible and that well should be excluded from the analysis. To minimize the probability of detachment, several options can be considered. Firstly, be sure to use tissue culture treated, flat bottom 96-well plates. Secondly, reduce the cell number when possible or introduce an extra medium change at day 10 (when cell production is at its optimum). Thirdly, the 96-well plates may be pre-coated with gelatin (0.1–0.3%; add 0.1–0.3 g to 100 ml distilled water and autoclave). Add 200 µl of solution to each well, incubate the plates for 2 h at 37°C, or overnight at room temperature. After coating, remove the entire contents of each well, e.g., by aspirating it off with a suction device. Leave them to dry in air in the laminar flow hood (15–30 min will usually do, but overnight is also good). Dry with lid on but lifted at one edge to avoid the risk of contamination. After drying, the plates can be used immediately or stored at 4°C for several weeks when packed in aluminum foil.

3. Because CAFC and LTC-IC cultures are maintained for extended periods of time with regular medium changes, contamination can be a major problem. Most contaminations are caused by fungi and should be treated as soon as they are spotted. The best solution is to add 1 M NaOH to the contaminated well. Also, treat surrounding wells when fungi are dry. The best chance of controlling contamination is when they are spotted when still completely immersed in the medium. After addition of 1 M NaOH, remove the entire contents of the well and replace with 0.1 M NaOH. Check for spreading of contamination during the next few days and treat suspicious wells if necessary.

Acknowledgments

The authors thank Dr. Rob Ploemacher for his support and for introducing the CAFC assay in hematopoietic stem cell research. We also thank Dr. Simon Robinson, Department of Blood and Marrow Transplantation, University of Texas, MD, Anderson Cancer Center, Houston, USA, for critically reading the manuscript.

Reference

1. Dexter, T. M., Allen, T. D., and Lajtha, L. G. (1977) Conditions controlling the proliferation of haemopoietic stem cells in vitro. *J Cell Physiol* **91,** 335–344.
2. Ploemacher, R. E., van der Sluijs, J. P., Voerman, J. S. A., and Brons, N. H. C. (1989) An in vitro limiting-dilution assay of long-term repopulating hematopoietic stem cells in the mouse. *Blood* **74,** 2755–2763.
3. Ploemacher, R. E., van der Sluijs, J. P., Van Beurden, C. A. J., Baert, M. R. M., and Chan, P. L. (1991) Use of limiting-dilution type long-term marrow cultures in frequency analysis of marrow-repopulating and spleen colony-forming hematopoietic stem cells in the mouse. *Blood* **10,** 2527–2533.
4. Ploemacher, R. E., van der Loo, J. C. M., Van Beurden, C. A. J., and Baert, M. R. M. (1993) Wheat germ agglutinin affinity of murine hemopoietic stem cell subpopulations is an inverse function of their long-term repopulating ability in vitro and in vivo. *Leukemia* **7,** 120–130.
5. Sutherland, H. J., Eaves, C. J., Eaves, A. C., Dragowska, W., and Lansdorp, P. M. (1989) Characterization and partial purification of human marrow cells capable of initiating long-term hematopoiesis in vitro. *Blood* **74,** 1563–1570.
6. Lemieux, M. E., Rebel, V. I., Lansdorp, P. M., and Eaves, C. J. (1995) Characterization and purification of a primitive hematopoietic cell type in adult mouse marrow capable of lymphomyeloid differentiation in long-term marrow "switch" cultures. *Blood* **86,** 1339–1347.
7. Deryugina, E. I., Muller-Sieburg, C. E. (1993) Stromal cells in long-term cultures: keys to the elucidation of hematopoietic development? *Crit Rev. Immunol.* **13,** 115–150.

8. Muller-Sieburg, C. E., Deryugina, E. (1995) The stromal cells' guide to the stem cell universe. *Stem Cells* **13,** 477–486.
9. Collins, L. S., Dorshkind, K. (1987) A stromal cell line from myeloid long-term bone marrow cultures can support myelopoiesis and B lymphopoiesis. *J. Immunol.* **138,** 1082–1087.
10. Moore, K. A., Ema, H., and Lemischka, I. R. (1997) In vitro maintenance of highly purified, transplantable hematopoietic stem cells. *Blood* **89,** 4337–4347.
11. Wineman, J., Moore, K., Lemischka, I., and Muller-Sieburg, C. (1996) Functional heterogeneity of the hematopoietic microenvironment: rare stromal elements maintain long-term repopulating stem cells. *Blood* **87,** 4082–4090.
12. Neben, S., Anklesaria, P., Greenberger, J., and Mauch, P. (1993) Quantitation of murine hematopoietic stem cells in vitro by limiting dilution analysis of cobblestone area formation on a cloned stromal cell line. *Exp Hematol* **21,** 438–443.
13. Oostendorp, R. A., Harvey, K. N., Kusadasi, N., de Bruijn, M. F., Saris, C., Ploemacher, R. E., Medvinsky, A. L., and Dzierzak, E. A. (2002) Stromal cell lines from mouse aorta-gonads-mesonephros subregions are potent supporters of hematopoietic stem cell activity. *Blood* **99,** 1183–1189.
14. Down, J. D., Boudewijn, A., van Os, R., Thames, H. D., and Ploemacher, R. E. (1995) Variations in radiation sensitivity and repair among different hematopoietic stem cell subsets following fractionated irradiation. *Blood* **86,** 122–127.
15. Wierenga, P. K., Setroikromo, R., Kamps, G., Kampinga, H. H., and Vellenga, E. (2002) Peripheral blood stem cells differ from bone marrow stem cells in cell cycle status, repopulating potential, and sensitivity toward hyperthermic purging in mice mobilized with cyclophosphamide and granulocyte colony-stimulating factor. *J. Hematother. Stem Cell Res.* **11,** 523–532.
16. Kamminga, L. M., van Os, R., Ausema, A., Noach, E. J., Dontje, B., Vellenga, E., and de Haan, G. (2005) Impaired hematopoietic stem cell functioning after serial transplantation and during normal aging. *Stem Cells* **23,** 82–92.
17. van Os, R., Kamminga, L. M., and de Haan, G. (2004) Stem cell assays: something old, something new, something borrowed. *Stem Cells* **22,** 1181–1190.
18. Breems, D. A., Blokland, E. A., Neben, S., and Ploemacher, R. E. (1994) Frequency analysis of human primitive haematopoietic stem cell subsets using a cobblestone area-forming cell assay. *Leukemia* **8,** 1095–1104.
19. Fazekas de St.Groth, S. (1982) The evaluation of limiting dilution assays. *J Immunol Meth* **49,** R11–R23.

IV

TRANSPLANTATION ASSAYS FOR MOUSE AND HUMAN HSC

11

Hematopoietic Stem Cell Transplant in Mice by Intra-Femoral Injection

Yuxia Zhan and Yi Zhao

Summary

In traditional bone marrow transplantation, the majority of peripherally introduced stem cells are trapped in peripheral organs, such as the lung and liver. The frequency of cells homed in bone marrow by such method is extremely low. This circumstance adds difficulty to the research of hematopoietic stem cell (HSC), a rare population to begin with. By introducing HSC directly into bone marrow cavity, the peripheral loss of HSC can be minimized. Thus, intra-femoral injection of HSC is a useful method for HSC study.

Key Words: Hematopoietic stem cell (HSC); bone marrow transplantation (BMT); intra-femoral injection.

1. Introduction

Bone marrow transplantation (BMT) has been commonly used in hematopoietic stem cell (HSC) studies. Although many in vitro assays have been developed to characterize HSC/progenitor cells, the golden-standard test of HSC activity is BMT of the testing cells with other competitor cells or supporting cells *(1,2)*. The conventional BMT introduces testing cells by tail vein or retro-orbital injection. After injection of HSC into peripheral blood, the stem cells need to reach bone marrow, find their "niches," and seed, a process called homing. However, the recovery of testing cells from bone marrow and spleen is extremely low with such methods, as the majority of the cells are trapped in other organs, such as lung and liver *(3)*. As the frequency of HSC is extremely low (one HSC per 10^5 bone marrow cells in mice) *(4)*, such low homing efficiency adds difficulties to HSC study.

From: *Methods in Molecular Biology, vol. 430: Hematopoietic Stem Cell Protocols*
Edited by: K. D. Bunting © Humana Press, Totowa, NJ

Recently, the method to introduce HSC by direct intra-femoral injection has been developed *(5,6)*. By directly introducing HSC into bone marrow, the loss of HSC in peripheral organs is minimized in comparison with the method of tail vein or retro-orbital injection. The sensitivity of assay is thus increased. Furthermore, intra-femoral injection also allows researchers to study questions that are difficult by using traditional tail vein or retro-orbital injection. By directly introducing HSC into bone marrow cavity, donor cell homing and proliferation under normal physiological conditions (non-ablated) can be studied. For example, it was long believed that ablation was essential to open the niches in bone marrow for the transplanted stem cells. However, using intra-femoral injection, it is shown that the homing efficiency in ablated and non-ablated animals is similar, but the stem cells in non-ablated animals remain as quiescent while the cells in ablated animals start to proliferate shortly after transplantation. So the traditional observed "low engraftment" in non-ablated animals is from the lack of proliferation, whereas the high "engraftment" in ablated animals is from the up-regulation of donor cell proliferation (*see* **Fig. 1**).

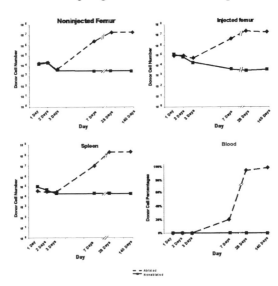

Fig. 1. Engraftment of donor cells in ablated and non-ablated mice. After injection of 5×10^5 Lin$^-$ Ly5.1 cells into the right femur, cells were collected from the injected femur, non-injected femur, spleen, and peripheral blood and were subjected to FACS analysis to determine the donor cell percentages. Two experiments were performed, with 5–8 mice per group. From 1 to 3 days, donor cells were identified by PKH26 labeling and antibody staining. After 3 days, donor cells were identified by PE-conjugated anti-Ly5.1 antibody. Donor cells rapidly proliferated in bone marrow (BM) and spleen of ablated mice but did not proliferate in non-ablated mice. The peripheral blood donor cell percentages also reflected this rapid proliferation of donor cells in ablated mice (reproduced from **ref. *6***).

Besides HSC transplantation, intra-femoral injection is also a very useful method for introducing gene therapy vectors into HSC in situ, thus bypassing the in vitro manipulation of HSC, which usually causes the change of HSC physiological characteristic *(7)*.

2. Materials

2.1. Animals

1. Mice aged 8 weeks or older can be used as recipients. The recipient mice strain varies for different testing cells and different experimental purposes. In general, C57BL/Ly5.1 or C57BL/Ly5.2 mice are the first choice to test mouse stem cells, as the donor cells can be identified from recipient's cells by cell surface markers *(8)*. To test human stem/progenitor cells, non-SCID mice are commonly used *(9)*. Different genetic defect mice strains can be used for different purposes to test gene therapy vectors or stem cell therapy.
2. Depending on the goal of the experiments, the recipient mice would receive "conditioning treatment" with lethal/sublethal dose irradiation (ablated recipients) or no irradiation (non-ablated recipients).

2.2. Anesthesia Equipment

1. Isoflurane inhalation provides safe general anesthesia for surgical procedures. Isoflurane system (Cat. No. 30–301 Summit medical equipment company. Bend, OR, USA, 97701) including oxygen flow meter and anesthetic vaporizer, can be purchased from Summit Medical Equipment Company.
2. Isoflurane can be purchased from Halocarbon Products Corporation, River Edge, NJ.
3. Oxygen tanks can be purchased from Gilmore liquid air company (South El Monte, CA, USA).

2.3. Surgical Tools (from ROBOZ, Gaithersburg, MD, USA)

1. One straight sharp scissors (Cat. No. RS-5840).
2. One micro-dissecting scissors (Cat. No. RS-5850).
3. One micro-dissecting forceps (Cat. No. RS-5150).
4. One micro-embryonic capsule curved forceps (Cat. No. RS-5163).
5. Reflex wound closure staple and clip (Cat. No. RS-9260, RS-9262).

2.4. Other Tools

1. Hamilton syringe (50 μl, Cat. No. 705; Hamiton, Reno, NV, USA).
2. Animal Shaving Clipper (Cat. No. 78015-010 VWR, Bristol, CT, USA).
3. 9 mm mouse nose cone (Item number 921609, VetEquip, Pleasanton, CA, USA).
4. Induction chamber (Item number 941443, VetEquip).

5. Trypan Blue staining buffer from BioWhittaker (Cat. No. 17-942E, Walkersville, MD, USA).
6. Bone wax (Cat. No. MED-dynjbw25 Devine Medical Supplies, Whittier, CA, USA).
7. Sterile cotton tips, wipes, 25-G needle, and disposable scalpel.
8. Spray bottle with 70% ethanol.
9. Table-top centrifuge (Beckmen Coulter GS-6 centrifuge, CH-3.8/CH3.8A bucket).
10. FALCON 5-ml polystyrene round-bottom tube (part number 352054) and FALCON 15-ml polypropylene conical tube (part number 352097).

2.5. Reagents

1. Red blood cell-lysing buffer (Cat. No. R7757, Sigma, St. Louis, MO, USA).
2. Phosphate-buffered saline (P3813, Sigma).
3. Staining buffer: 1% fetal bovine serum (Hyclone, Logon, UT, USA) in PBS.
4. All antibodies are from BD Pharmingen, San Jose, CA, USA: PE-anti-CD45.1(Cat. No. 553776); FITC-anti-mouse CD3e (Cat. No. 553062); FITC-anti-mouse CD45R/B220 (Cat. No. 553088); FITC-anti-mouse Ly-6G (Gr-1, Cat. No. 553126); FITC-anti-mouse CD11bc (Cat. No. 557396).

3. Methods
3.1. Anesthetizing the Animals
3.1.1. Preparation

1. Check all the supplies: (isoflurane, oxygen tank, wipes, etc.)
2. Assemble and check all the equipment (anesthesia machine, induction chamber, nosecone, etc)
3. Check to ensure that the control dial of the vaporizer is in the off position.
4. Check isoflurane level in the sight glass and fill if needed.
5. Check F-air cannister and replace the cannister if needed.
6. Check all the hoses and connections.
7. Check the oxygen tank. If the pressure gauge register is less than 200 PSI, the tank needs to be replaced.
8. Turn the oxygen flow meter on then off to determine whether the gas supply is operational.
9. Depress and release the oxygen flush valve(s).

3.1.2. Induction

1. Place the mouse in the induction chamber.
2. Turn on the oxygen flow meter to read 1 l/min.
3. Turn on the isoflurane vaporizer to 4–5%, continue until the mouse's breathing slows down.
4. Monitor oxygen flow and isoflurane flow rates during the surgical procedure.

3.1.3. Maintenance

1. Take the mouse out from the induction chamber and place it on nose cone (*see* **Note 1**).
2. Reduce the isoflurane flow on the vaporizer dial to 2–2.5%.
3. Reduce the oxygen flow rate to 0.8 l/min.

3.1.4. Recovery (After Surgery Procedure)

1. Turn off the vaporizer and flush the animal's circulatory system with oxygen.
2. Remove the mouse from nose cone, and place it in a warmed recovery area/cage.

3.2. Surgical Procedures

1. Injection material should be ready and kept on ice before the surgical procedures (*see* **Note 2**).
2. Place the induced mouse (on nose cone) on top of a clean napkin.
3. With a razor, shave the fur in the knee joint area.
4. Locally spray 70% ethanol in the leg/knee joint area.
5. Flex the knee to 90°, then open the skin just above the patella with the scalpel.
6. Move the patella to side by cutting off the patellar ligament.
7. Place a 25-G needle at the end of femur and twist the needle gently to make an intra-femoral tunnel into the bone marrow cavity. Stop the twisting when the resistance is gone.
8. Insert the needle tip of Hamilton microsyringe (filled with the testing cells at desired concentration) into the top of the bone marrow cavity (*see* **Note 3**).
9. For testing, inject Trypan Blue staining buffer into the bone marrow cavity.
10. The undamaged femur cavity can accommodate 5 µl of injection volume without leaking.
11. As an example of the injected HSC (5 µl) in bone marrow cavity, we stained lineage negative cells (Lin⁻) with carboxyfluorescein diacetate succinimidyl ester (CFDA) and examined if there is leakage from the injection (*see* **Fig. 2**). However, Trypan Blue staining can be used for the same purpose. For studies other than homing regulation, researchers can try higher volume in the injection (*see* **Note 4**).
12. Seal the needle insertion hole with bone wax.
13. Close the wound-by-wound staples (*see* **Note 5**)

3.3. Immunostaining to Detect Donor Cells in Peripheral Blood

1. Preparing two antibody mixes with staining buffer: (1) to detect donor cell-derived lymphoid cells: PE-anti-CD45.1, FITC-anti-mouse CD3e, and FITC-anti-mouse CD45R/B220; (2) to detect donor cell-derived myeloid cells: PE-anti-CD45.1, FITC-anti-mouse Ly-6G(Gr-1), FITC-anti-mouse CD11bc (Mac-1). Each 100-µl antibody mix contains 0.5 µl of each antibody. Keep the antibody mix on ice at dark.

Fig. 2. Fluorescence microscopic picture of a femur longitudinal section showing the location of injected cells in the femur cavity. Lin⁻ cells were labeled with carboxyfluorescein diacetate succinimidyl ester (CFDA) and injected into a femur cavity with a Hamilton syringe. The needle insertion hole was sealed with bone wax. The mouse was killed 5 min after injection, and the injected femur was removed for fluorescence microscopic analysis. Original magnification, 2×20 (reproduced from **ref. 6**).

2. Prepare 15-ml Falcon tubes with 10 ml ice-cold PBS buffer, keep the tubes on ice.
3. Collect 70–100 µl blood through tail clipping or orbital sinus, and mix the blood with PBS buffer immediately to prevent coagulation.
4. Centrifuge at $622 \times g$ in table-top centrifuge (1650 rpm for Beckmen Coulter GS-6 centrifuge, CH-3.8/CH3.8A bucket) at 4°C for 5 min, discard the supernatant.
5. Add 1 ml of red blood cell-lysing solution to the tube; immediately resuspend the blood cell pellet with polyethylene transfer pipette gently (*see* **Note 6**).
6. When the cell suspend becomes "clear" red solution (about 1 min), add 14 ml ice-cold PBS buffer to the tube (*see* **Note 7**).
7. Centrifuge at $367 \times g$ for 5 min at 4°C (1250 rpm for Beckmen Coulter GS-6 centrifuge, CH-3.8/CH3.8A bucket), discard the supernatant, the cell pellet should be pale gray (*see* **Note 8**).

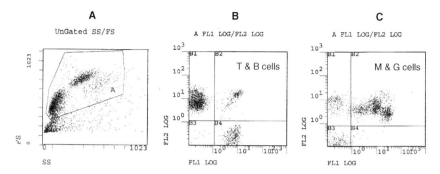

Fig. 3. Fluorescent-activated cell sorter analysis of the multilineage cell repopulation by donor cells. Donor-originated cells are stained with anti-Ly5.1 (*Y*-axis, PE-conjugated, FL-2) and lineage-specific antibodies (FITC-conjugated, *X*-axis, FL-1). (**A**) shows the sample gating in FACS analysis; (**B**) shows donor-derived T and B cells stained with anti-CD3 and anti-B220; (**C**) shows donor-derived M and G cells (macrophage and granulocyte) stained with anti-CD11b and anti-Gr1. PE: phycoerythrin; FITC: fluorescein isothiocyanate (Reproduced from **ref. 10**).

8. Add 1 ml ice-cold staining buffer; transfer 500 µl to each of two FALCON 5-ml polystyrene round-bottom tubes.
9. Centrifuge at $622 \times g$ for 5 min at 4°C, discard the supernatant.
10. Add 100 µl of antibody mix to the cell pellet, resuspend the cells. For each sample, half of the cells with antibody mix to detect donor lymphoid cells and the other half with antibody mix for donor myeloid cells.
11. Incubate the cells on ice in the dark for 30 min.
12. Add 1 ml of ice-cold staining buffer to the cells, centrifuge at $622 \times g$ for 5 min at 4°C.
13. Discard the supernatant; add 500 µl of PBS buffer for fluorescence-activated cell-sorting (FACS) analysis.
14. A typical FACS analysis result is shown in **Fig. 3** by using Coulter EPICS XL (*see* **Note 9**).

4. Notes

1. It is important to keep the animal at the position to the nose cone. Make sure the animal is sleeping during the entire time of the surgical procedure.
2. Keep the prepared stem cells on ice all the time. When preparing the cells, the concentration should not be too high; the increased viscosity may cause difficulty during injection. In our experience, 3×10^5/µl in Hanks tissue culture medium is the maximal cell concentration.
3. Align the Hamilton syringe needle along the femurs. Change direction carefully if resistance is felt during the injection. Insert the needle into the hole on top of the bone cavity. After injection, hold the syringe for 5–10 s, then pull out the needle, and seal the hole with bone wax.

4. In our experience, the maximum injection volume is 5 µl without leaking and the damage to the bone marrow is minimal. Our purpose was to study the homing regulation; it was important to keep the bone marrow structure intact. However, for the purpose of introducing HSC or gene therapy vectors into bone marrow cavity, higher volume can be considered, as other researchers injected 25–50 µl of cell/vector to the bone cavity *(7)*.

5. The surgical operation time is about 15 min for each animal, depending on the experience. The animals should wake up 10 min after removal from isoflurane inhalation. There should be very little bleeding during the operation.

6. It is important to resuspend the cell pellet in lysing buffer immediately after adding the buffer. Delaying may result in difficulty to separate the cells.

7. The timing of red blood cell lysing is important; overtime may cause white blood cell damage.

8. If lysis is incomplete, **step 5** may be repeated.

9. If available, three-color FACS analysis can be used to replace two-color analysis. In that case, one antibody mix contains five antibodies in three different colors: *(1)* anti-Ly5.1 (for donor cell marker); *(2)* anti-CD3e and anti-B220 (lymphoid cells); *(3)* anti-Gr-1 and anti-CD-11b (myeloid cells).

Acknowledgment

The authors appreciate the careful reading and editing of the manuscript by Ms. Yumin He.

References

1. Harrison, D.E. (1980) Competitive repopulation: a new assay for long-term stem cell functional capacity. *Blood* **55**, 77–81.

2. Osawa M., Hanada K.I., Hamada H., and Nakauchi H. (1996) Long-term lympho-hematopoietic reconstitution by a single CD34-low/negative hematopoietic stem cell. *Science* **273**, 242–245.

3. Szilvassy S.J., Bass M.J., Zant G.V., and Grimes B. (1999) Organ-selective homing defines engraftment kinetics of murine hematopoietic stem cells and is compromised by ex vivo expansion. *Blood* **93**, 1557–1566.

4. Harrison D.E., Jordan C.T., Zhong R.K., and Astle C.M. (1993) Primitive hemopoietic stem cells: direct assay of most productive populations by competitive repopulation with simple binomial, correlation and covariance calculations. *Exp Hematol* **21**, 206–219.

5. Kushida T., Inaba M., Hisha H., Ichioka N., Esumi T., Ogawa R., Lida H., and Ikehara S. (2001) Intra-bone marrow injection of allogeneic bone marrow cells: a powerful new strategy for treatment of intractable autoimmue diseases in MRL/lpr mice. *Blood* **97**, 3292–3299.

6. Zhong J.F., Zhan Y., Anderson W.F., and Zhao Y. (2002) Murine hematopoietic stem cell distribution and proliferation in ablated and nonablated bone marrow transplantation. *Blood* **100**, 3521–3526.

7. McCauslin C.S., Wine J., Cheng L., Klamann K.D., Candotti F., Clausen P.A., Spence S.E., and Keller J.R. (2003) In vivo retroviral gene transfer by direct intrafemoral injection results in correction of the SCID phenotype in Jack knock-out animals. *Blood* **102**, 843–848.

8. Zhao Y., Lin Y., Zhan Y., Yang J., Louie J., Harrison D.E., and Anderson W.F. (2000) Murine hematopoietic stem cell characterization and its regulation in BM transplantation. *Blood* **96**, 3016–3022.

9. Mazurier F., Doedens M., Gan O.I., and Dick J.E. (2003) Rapid myeloerythroid repopulation after intrafemoral transplantation of NON-SCID mice reveals a new class of human stem cells. *Nat Med* **9**, 953–963.

10. Zhao, Y., Zhan, Y., Burke, K.A., and Anderson, W.F. (2005) Soluble factor(s) from bone marrow cells can rescue lethally irradiated mice by protecting endogenous hematopoietic stem cells. *Exp Hematol* 33, 428–434.

12

Hematopoietic Stem Cell Transplant into Non-Myeloablated *W/Wᵛ* Mice to Detect Steady-State Engraftment Defects

Zhengqi Wang and Kevin D. Bunting

Summary

Hematopoietic stem cells (HSC) are capable of self-renewal and reconstitution of the lymphoid and myeloid lineages of transplant recipients. Classical assays for HSC function rely on lethal irradiation to prepare the host for donor engraftment. This assay destroys most of the hematopoietic tissue and the vasculature of the bone marrow space, leading to regeneration of the niche in which HSC are intimately dependent for their survival, self-renewal, and lineage differentiation. The non-ablated transplant setting provides a more physiological background for measuring HSC function during steady-state hematopoiesis. In this chapter, we describe methods for assaying HSC function during the steady-state using *W/Wᵛ* c-Kit mutant mice as recipients. Our previous studies have found that the competition from *W/Wᵛ* allows an additional level of stringency that is not observed in limiting dilution assays of HSC number based on fully ablated recipient competition. The ease of this approach is an advantage, and this method may be particularly useful for teasing apart HSC engraftment phenotypes that are especially dependent on functions related to homing, adhesion, or migration into the niche.

Key Words: HSC transplant; myeloablation; engraftment; flow cytometry; competitive repopulation; limiting dilution.

1. Introduction

Hematopoietic stem cells (HSC) are defined as cells capable of both self-renewal and multi-lineage differentiation *(1–3)*. HSC migrate in a regulated fashion during development to seed the fetal liver, spleen, and eventually

From: *Methods in Molecular Biology, vol. 430: Hematopoietic Stem Cell Protocols*
Edited by: K. D. Bunting © Humana Press, Totowa, NJ

bone marrow, and migrate under certain conditions such as cytokine-induced mobilization later in life. The fine balance among the activity of self-renewal, differentiation, migration, and apoptosis determinates the number of stem cells present in the body *(4)*. The gold standard for HSC assays in vivo is the long-term reconstitution of hematopoiesis in lethally irradiated recipient mice in which donor HSC will regenerate the entire hematopoietic system of host mice for the lifetime of the animal *(5)*. Experiments with genetically marked parabiotic mice indicate that migration of HSC and progenitors from bone marrow to blood and back to functional niches in the bone marrow is a physiological process existing in the unmanipulated mice *(6)*. This provides strong rationale for the feasibility of bone marrow transplantation in the non-myeloablated setting. Recipients that received a lethal dose of irradiation will create a non-competitive host HSC pool that can be easily replaced by the donor HSC. However, irradiation can disrupt the bone marrow and endothelial barriers that separate the extravascular compartment in the bone marrow from the blood circulation *(7,8)* but also can cause an increase in cytokines such as granulocyte-macrophage colony-stimulating factor (GM-CSF) and stem cell factor (SCF) *(9,10)*. The net effects of those environmental changes on the donor engraftment are not fully understood.

Long-term reconstitution of hematopoiesis in lethally irradiated recipients to detect the HSC engraftment defects is well documented and widely used *(5,11)*. Reconstitution with a cell fraction that contains functional HSC can be direct or competitive. In the competitive setting, the HSC-containing donor cells mixed with a radioprotective dose of host marrow cells are injected into lethally irradiated recipients *(12–14)*. Y-chromosome markers can be used to detect male cells in a female host by Southern blot analysis. The allelic CD45 marker, which is expressed on all hematopoietic cells except mature red blood cells, can be used to distinguish host and donor cells in multiple lineages detected by monoclonal antibodies and flow cytometry. With the limiting dilution of donor cells, the competitive repopulating assay can be used to quantitate the frequency of stem cells, termed competitive repopulating units (CRU), in the original test population. The proportion of recipients whose regenerated hematopoietic system is determined to have more than 5% donor lymphoid and myeloid cells is used to calculate the frequency of CRU in the original test cells by Poisson statistics *(14–16)*.

Although myeloablative conditioning is generally required for donor engraftment of bone marrow HSCs, some genetically altered mouse models are very receptive to donor engraftment under non-ablative conditions. These include PU.1 and STAT5 knockout mice *(17,18)*. However, because of the decreased survival of these mice, requirement for neonatal transplantation, and other severe immunologic defects, these mice have not been widely applied for HSC assays.

The classic model has been the stem cell defective WBB6F1-*W/W^v* (*W/W^v*) mice with the mutations at the *white spotting locus (W)*, which are allelic to *c-kit* proto-oncogene, a gene that encodes the receptor for SCF *(19–21)*. The *W* allele encodes a truncated c-Kit protein that lacks in vitro kinase activity, and *W/W* homozygotes die perinatally. The *W^v* allele contains a point mutation at position 2007 (C→T) of the known *c-kit* sequence, which results in the change of the threonine at position 660 to methionine and leads to the partial impairment of the c-Kit kinase activity. *W^v* is less severe than the original *W* mutation, and W^v/W^v

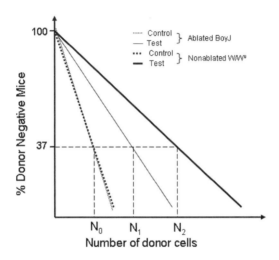

Fig. 1. Schematic representation of the murine competitive repopulation assay in lethally ablated and non-ablated recipients. Limiting numbers of C57BL/6 test or control cells (CD45.2$^+$, Hb$_s$/Hb$_s$) are injected into ablated B6.SJL (Boy J) mice along with 0.1 million B6.SJL normal bone marrow cells or injected into non-ablated *W/W^v* (Hb$_s$/Hb$_d$) recipients. The proportion of donor cells in *W/W^v* recipients 12 weeks after transplantation is determined by peripheral blood hemoglobin (Hb) gel electrophoresis for erythroid engraftment and by Southern blot analysis in the other hematopoietic tissues. The proportion of donor cells in ablated B6.SJL recipients 12 weeks after transplantation is determined by flow cytometry using CD45.2 and lymphoid- or myeloid-specific antibodies. Positive mice are defined as those with greater than 5% donor cells in both lymphoid and myeloid lineages. The variation in the proportion of positive mice at each test cell dose is analyzed by Poisson statistics. This test provides an absolute measure of competitive repopulating units (CRU) frequency in the test graft. The CRU frequency with control donor cells is very similar using either assay (1 per N_0 control cells). However, the CRU frequency with test donor cells is 1 per N_1 test cells when transplanted into ablated recipients but is 1 per N_2 test cells when transplanted into non-ablated recipients. The difference ($N_2 - N_1$) detected by the comparison of two repopulating assays is likely associated with the steady-state engraftment defects.

homozygotes are viable and sterile whereas the heterozygous W^v/+ animals are fertile *(20)*. W/W^v mice have a severe macrocytic anemia characterized by an underlying stem cell defect especially with the major defects in erythroid progenitors and less-severe deficiencies of granulocyte precursors and megakaryocytes. Because of mutant c-Kit receptor in W/W^v mice, the HSC in these mice are less competitive and allow implantation of wild-type C57BL/6 donor grafts without prior irradiation *(22–25)*. The hemoglobin (Hb) of the recipient W/W^v is Hb_d/Hb_s (50%:50%) whereas the Hb of the donor C57BL/6 is Hb_s/Hb_s. Therefore, blood can be sampled at multiple times after transplantation and Hb electrophoresis can be used to distinguish donor from host and to monitor erythroid engraftment. Because the donor erythrocyte repopulation always precedes leukocyte repopulation in this model *(22,23,25)*, Southern blot analysis can be used to verify the donor engraftment in the other hematopoietic tissues.

Our prior studies showed that HSC transplant into non-myeloablated W/W^v mice has the potential to detect steady-state engraftment defects that otherwise may be undetectable in lethally irradiated recipients. The concept for this is illustrated in **Fig. 1** *(17)*. The major differences for these two reconstitution assays are listed in **Table 1**. W/W^v hosts have very poor HSC competitive ability, and in the absence of irradiation, donor HSC can out compete the endogenous HSC for the HSC niche. This type of engraftment mirrors the normal process that occurs during physiologic migration of HSC from the BM microenvironment to the circulation

Table 1
Comparison Between Repopulating Assays in Lethally Ablated and Non-Ablated Recipients

Condition	Advantages	Disadvantages	References
Ablated B6.SJL	CD45 marker useful for analysis of multi-lineage donor engraftment	Bone marrow sinus endothelium is altered and cytokine production is increased to alter the hematopoietic stem cell (HSC) niche	*(7–10)*
Non-ablated W/W^v	Moderate cost	Limited to hemoglobin markers and Southern blots for detecting donor engraftment	*(6)*
	Engraftment is mirrored by a normal physiological process	Recipients are relatively expensive	

and return *(6)*. Wild-type donor HSC will have significant advantage to compete with the host HSC with no irradiation. Meanwhile, the host still has a weak HSC competitive ability to compete with a defective donor HSC challenge, leading to a potentially lower seeding efficiency for the defective donor in the *W/W^v* mice than that in the lethally irradiated recipients. Therefore, it can be used to assay the functional engraftment ability especially for defective donor HSC.

2. Materials

2.1. Isolation of Bone Marrow Cells

1. Source of test as well as control bone marrow cells derived from C57BL/6 (Ly5.2) strain (*see* **Notes 1–3**).
2. Phosphate-buffered saline (PBS, without calcium and magnesium, Hyclone, Logan, Utah, USA, Cat. No. SH30256.01) with 2% heat-inactivated fetal bovine serum or calf serum.
3. 3-ml syringes with 25- or 21-G needles to flush marrow out of femurs and tibias.
4. Mouse tail buffer: 0.6 mg/ml proteinase K in 50 mM Tris pH 8.0, 1% sodium dodecyl sulfate, 100 mM NaCl, and 10 mM ethylenediaminetetraacetic acid (EDTA) pH 8.0.

2.2. Reconstitution Assays in the Non-Ablative W/W^v Mouse

1. WBB6F1/J-*Kit^W/Kit^{Wv}* (*W/W^v*) mice (6- to 12-week old) are an F1 of a WB-*W/+* and C57BL/6-*W^v/+* cross that can accept C57BL/6 grafts. Commercial supply: Jackson Laboratories, Bar Harbor, ME (*see* **Notes 4–6**).
2. STATSPIN Micro-hematocrit heparinized glass tubes (STATSPIN, Inc., Norwood, Massachusetts, USA).
3. 10× Cystamine solution: For making 15 ml 10× cystamine solution, 1.13 g cystamine, 1.0 ml 1 M DTT, 0.5 ml 100% ammonium hydroxide, and 13.5 ml distilled water. Store in aliquots at −20 °C.
4. 10× TBV buffer: For making 100 ml 10× TBV buffer, 21.8 g Tris-Base, 6.18 g boric acid, and 0.58 g EDTA-versene to the final volume 100 ml, Autoclave.
5. Ponceau S solution: 0.1% (w/v) ponceau S in 5% acetic acid.
6. SepraClear II solution containing 40.0% (v/v) aqueous *N*-methyl pyrrolldone (Diasys Europe Ltd, Workingham, Berkshine, England, Product code: 51283). Avoid contact with skin and eyes. May be irritating to eyes.

3. Methods

3.1. Isolation of Bone Marrow

1. Kill the mouse using protocols approved by the host institution and dissect the femurs and tibias.
2. Cut the ends off the bones with sharp surgical scissors; flush the marrow plug into 3 ml cold PBS with 2% heat-inactivated fetal bovine serum, using a sterile

21-G needle attached to a 3-ml syringe for the femurs and 25-G needle for the tibias.

3. Prepare a single-cell suspension by drawing the marrow and medium through the needle a few times. Count viable nucleated bone marrow cells. Keep the cells on the ice.

3.2. Reconstitution Assays in the Non-Ablated W/W^v Mouse

3.2.1. Transplantation

Prepare limiting dilution of wild-type C57BL/6 control or test bone marrow cells with 0.5 ml containing the various desired number of control or test cells, and inject 0.5 ml per recipient through the lateral tail vein into otherwise untreated W/W^v mice with five animals per dose while ideally the more animals with the lower dose. As the frequency of CRU in the unseparated bone marrow is around 1 of $1–2 \times 10^4$, the ideal limiting dose for wild type is around 5×10^5 to 10^4. For test cells, the limiting dilution should be adjusted accordingly (*see* **Note 7**).

3.2.2. Hb Electrophoresis for Erythroid Engraftment

Erythroid engraftment is measured on the basis of the pattern of distinct major Hb bands by electrophoresis on the cellulose acetate gel at any time at least 6 weeks after the transplantation. C57BL/6 donor-derived red cells contain a single band of hemoglobin (Hb$_s$) whereas the host W/W^v red cells contain 50:50 ratio of Hb$_d$ and Hb$_s$. The gels are stained with Ponceau S and the patterns are scanned on a scanning densitometer, and the relative percent of each band determines the percent of erythroid engraftment. Engraftment levels are calculated by subtracting 50% from the percent of donor type Hb$_s$ and then dividing by 50. For example, 80% Hb$_s$ corresponds to 60% donor engraftment in this model (*see* **Note 8**).

1. Peripheral blood is obtained from the retro-orbital sinus following puncture using a microcapillary tube (StatSpin microhematocrit tubes, 40 mm heparinized glass).
2. Spin tubes in the StatSpin microcentrifuge and record the hematocrit values.
3. Use a mark on a piece of paper to score a 1.0-cm tube containing the packed red blood cells. Place into a 1.5-ml microcentrifuge tube containing 30 μl of 1× Cystamine solution and shake vigorously until all blood cells are suspended. The tubes can be stored here at 4 °C for a day or two.
4. Soak the appropriate number of Gelman cellulose acetate gels in 1× TBV buffer.

5. Fill each buffer chamber in the gel box with 100 ml of 1× TBV buffer.
6. Load 8 μl of sample into the wells of the sample plate.
7. Blot each gel with a paper towel and place on the applicator tray so that both ends of gel are below the buffer level. The buffer will create tension in the gel.
8. Press the sample applicator several times into the wells to pick up samples. Hold for 10 s and then transfer to the applicator tray. Position the applicator over the gel and press down for 10 s. Repeat this four times.
9. Check orientation of gels as the Hb will run from negative to positive.
10. Place the lid on the gel box and run at 300 V for 25 min.
11. Remove gels and soak in Ponceau S solution for 7–10 min with shaking. Rinse gels gently with distilled water.
12. Destain gels in 7% acetic acid for 5 min.
13. Soak gels in SepraClear II solution for 5 min.
14. Transfer gels to a clean glass plate with the acetate side up, cut the edges, and place in oven at 50–60 °C for 15 min or until dry.
15. The patterns are scanned on a scanning densitometer, and the relative percent of each band determined the percent of erythroid engraftment.

3.2.3. Southern Blot Analysis for Hematopoietic Reconstitution

In the *W/W^v* mouse model, the severe macrocytic anemia of the stem cell-deficient *W/W^v* mouse can be alleviated by intravenous injection of normal bone marrow under non-ablative conditions, which is reflected by donor erythroid engraftment. Furthermore, it is well known that the donor erythrocyte repopulation always precedes leukocyte repopulation in this model *(22,23,25)*. Therefore, for true HSC engraftment after transplantation, it is necessary to assess the donor engraftment of more than just erythrocytes, and this can be done by Southern blot analysis with genomic DNA prepared from peripheral blood, bone marrow, and spleen cells from the mice that received donor grafts 12 weeks after transplantation.

1. Genomic DNA is prepared from peripheral blood, bone marrow, and spleen cells from the mice that received transplantation by digestion with tail buffer.
2. DNA is then extracted with an equal volume of phenol/chloroform/isoamyl alcohol (25:24:1) and precipitated with 2.5 vol ice-cold ethanol and one-tenth vol 3 M sodium acetate.
3. 5–10 μg DNA is digested overnight with EcoR I and separated on a 0.8% agarose gel by electrophoresis. The gel is blotted overnight onto Hybond N+ nylon membrane (Invitrogen, Carlsbad, California, USA), UV-cross-linked, and probed with a [^{32}P]-labeled fragment of mouse ß-globin intervening sequence 2.
4. Blots are washed at a final stringency of 0.5 × SSC (3M sodium chloride, 0.3M sodium citrate) /0.5% sodium dodecyl sulfate at 65°C, and autoradiographic images are obtained using a Molecular Dynamics Storm phosphorimager and X-ray film.
5. The donor engraftment is determined as described in the **Subheading 3.2.2**.

4. Notes

1. In cases where the test donor bone marrow is not feasible, unfractionated E14.5 fetal liver cells can be used as donor cells. It should be noted that the fetal liver cells repopulate 3.7–4.9 times better than adult marrow cells *(26)*.

2. To avoid minor immunological responses, donor cells derived from B6.SJL (CD45.1+) should not be used in the reconstitution assay into W/W^v recipients.

3. W/W^v mice can be ordered from the Jackson Laboratory, and the cost is considerable. As W/W^v mice are sterile because of male germ cell self-renewal defects, they can only be generated through crosses of WB-W/+ and C57BL/6-W^v/+ mice. Our calculations of the expense and time effort required for maintaining individual colonies of the WB-W/+ and C57BL/6-W^v/+, and their crossing and genotyping, suggests that purchase of these mice is practical.

4. It is worthy to note that W/W^v mice are F1 hybrids of WB × C57BL/6. Although C57BL/6 marrow repopulates W/W^v far more effective than WB marrow, it does not repopulate as well as F1 marrow, and the difference is most marked when the young recipient is used. This indicates that complicating hybrid effects exist *(27)*.

5. For the stem cell transplantation experiment under non-ablated conditions, it is best to use stem cell-deficient recipients that are otherwise genetically identical with the donor. W^{41J}/W^{41J} on the C57BL/6 background is an alternative model for marrow transplantation experiments as these homozygotes not only survive but are fertile *(28)*. The mice have a less-severe anemia than W/W^v mice but the lymphoid and erythroid repopulation is very similar *(29)*. When C57BL/6-W^{41J}/W^{41J} mice are chosen as recipients, congenic B6.SJL-derived cells could be used as donors and multi-lineage repopulation could be quantitated by CD45 markers. However, see **Note 2** before using this approach.

6. To avoid recipient immune responses to the donor Y-chromosome, it is best to use females as donors. From our trial and error, we have learned that no engraftment will be observed if the donor is from males and the recipient is female. However, male mice can serve as universal recipients in this approach and will accept either male or female donor grafts.

7. This assay might be more sensitive to pick up more severe steady-state engraftment defects associated with defective test donors. When the repopulating ability of the test cells is initially not known, it is best to use a single dose such as one-fifth of a donor equivalent dose for testing. The desired dose range for limiting dilution can then be adjusted afterwards based on the results of this initial test. For the unseparated E14.5 fetal liver, the frequency of CRU is around 1 of 15,000 *(30)*.

8. Besides using the Hb marker to distinguish donor from host, glucose-phosphate isomerase (*Gpi-1*) can also be used as a marker as donor cells derived from C57BL/6 have *Gpi-1ᵃ/Gpi-1ᵃ* whereas the F1 hybrid of W/W^v could carry the marker *Gpi-1ᵇ/Gpi-1ᵇ*. The cell lysates from each lineage can be analyzed by GPI-1 assay *(31)*.

References

1. Abramson, S., Miller, R.G., and Phillips, R.A. (1977) The identification in adult bone marrow of pluripotent and restricted stem cells of the myeloid and lymphoid systems. *J Exp Med* **145**,1567–1579.
2. Dick, J.E., Magli, M.C., Huszar, D., Phillips, R.A., and Bernstein, A. (1985) Introduction of a selectable gene into primitive stem cells capable of long-term reconstitution of the hemopoietic system of W/Wv mice. *Cell* **42**,71–79.
3. Lemischka, I.R., Raulet, D.H., and Mulligan, R.C. (1986) Developmental potential and dynamic behavior of hematopoietic stem cells. *Cell* **45**,917–927.
4. Domen, J., and Weissman, I.L. (1999) Self-renewal, differentiation or death: regulation and manipulation of hematopoietic stem cell fate. *Mol Med Today* **5**,201–208.
5. Morrison, S.J., Uchida, N., and Weissman, I.L. (1995) The biology of hematopoietic stem cells. *Annu Rev Cell Dev Biol* **11**,35–71.
6. Wright, D.E., Wagers, A.J., Gulati, A.P., Johnson, F.L., and Weissman, I.L. (2001) Physiological migration of hematopoietic stem and progenitor cells. *Science* **294**,1933–1936.
7. Daldrup-Link, H.E., Link, T.M., Rummeny, E.J., August, C., Konemann, S., Jurgens, H., and Heindel, W. (2000) Assessing permeability alterations of the blood-bone marrow barrier due to total body irradiation: in vivo quantification with contrast enhanced magnetic resonance imaging. *Bone Marrow Transplant* **25**,71–78.
8. Shirota, T., and Tavassoli, M. (1992) Alterations of bone marrow sinus endothelium induced by ionizing irradiation: implications in the homing of intra-venously transplanted marrow cells. *Blood Cells* **18**,197–214.
9. Gaugler, M.H., Squiban, C., Mouthon, M.A., Gourmelon, P., and van der Meeren, A. (2001) Irradiation enhances the support of haemopoietic cell transmi-gration, proliferation and differentiation by endothelial cells. *Br J Haematol* **113**, 940–950.
10. Shi, P.A., Pomper, G.J., Metzger, M.E., Donahue, R.E., Leitman, S.F., and Dunbar, C.E. (2001) Assessment of rapid remobilization intervals with G-CSF and SCF in murine and rhesus macaque models. *Transfusion* **41**,1438–1444.
11. Bunting, K.D., Bradley, H.L., Hawley, T.S., Moriggl, R., Sorrentino, B.P., and Ihle, J.N. (2002) Reduced lymphomyeloid repopulating activity from adult bone marrow and fetal liver of mice lacking expression of STAT5. *Blood* **99**,479–487.
12. Harrison, D.E. (1980) Competitive repopulation: a new assay for long-term stem cell functional capacity. *Blood* **55**,77–81.
13. Harrison, D.E., Jordan, C.T., Zhong, R.K., and Astle, C.M. (1993) Primitive hemopoietic stem cells: direct assay of most productive populations by competitive repopulation with simple binomial, correlation and covariance calculations. *Exp Hematol* **21**,206–219.
14. Szilvassy, S.J., Humphries, R.K., Lansdorp, P.M., Eaves, A.C., and Eaves, C.J. (1990) Quantitative assay for totipotent reconstituting hematopoietic stem cells by a competitive repopulation strategy. *Proc Natl Acad Sci USA* **87**,8736–8740.

15. Taswell, C. (1981) Limiting dilution assays for the determination of immunocompetent cell frequencies. I. Data analysis. *J Immunol* **126**,1614–1619.
16. Szilvassy, S.J., Nicolini, F. E., Eaves, C. J., and Miller, C. L. (2001) Quantitation of murine and human hematopoietic stem cells by limiting-dilution analysis in competitive repopulated hosts, in *Hematopoietic Stem Cell Protocols* (Klug, C. A. and Jordan, C. T., ed.), Humana, Totowa, NJ, pp. 167–188.
17. Couldrey, C., Bradley, H.L., and Bunting, K.D. (2005) A STAT5 modifier locus on murine chromosome 7 modulates engraftment of hematopoietic stem cells during steady-state hematopoiesis. *Blood* **105**,1476–1483.
18. Mezey, E., Chandross, K.J., Harta, G., Maki, R.A., and McKercher, S.R. (2000) Turning blood into brain: cells bearing neuronal antigens generated in vivo from bone marrow. *Science* **290**,1779–1782.
19. Huang, E., Nocka, K., Beier, D.R., Chu, T.Y., Buck, J., Lahm, H.W., Wellner, D., Leder, P., and Besmer, P. (1990) The hematopoietic growth factor KL is encoded by the Sl locus and is the ligand of the c-kit receptor, the gene product of the W locus. *Cell* **63**,225–233.
20. Nocka, K., Tan, J.C., Chiu, E., Chu, T.Y., Ray, P., Traktman, P., and Besmer, P. (1990) Molecular bases of dominant negative and loss of function mutations at the murine c-kit/white spotting locus: W37, Wv, W41 and W. *EMBO J* **9**,1805–1813.
21. Reith, A.D., Ellis, C., Lyman, S.D., Anderson, D.M., Williams, D.E., Bernstein, A., and Pawson, T. (1991) Signal transduction by normal isoforms and W mutant variants of the Kit receptor tyrosine kinase. *EMBO J* **10**,2451–2459.
22. Barker, J.E., Braun, J., and McFarland-Starr, E.C. (1988) Erythrocyte replacement precedes leukocyte replacement during repopulation of W/Wv mice with limiting dilutions of +/+ donor marrow cells. *Proc Natl Acad Sci USA* **85**,7332–7335.
23. Barker, J.E., Greer, J., Bacon, S., and Compton, S.T. (1991) Temporal replacement of donor erythrocytes and leukocytes in nonanemic W44J/W44J and severely anemic W/Wv mice. *Blood* **78**,1432–1437.
24. Boggs, D.R., Boggs, S.S., Saxe, D.F., Gress, L.A., and Canfield, D.R. (1982) Hematopoietic stem cells with high proliferative potential. Assay of their concentration in marrow by the frequency and duration of cure of W/Wv mice. *J Clin Invest* **70**,242–253.
25. Nakano, T., Waki, N., Asai, H., and Kitamura, Y. (1989) Different repopulation profile between erythroid and nonerythroid progenitor cells in genetically anemic W/Wv mice after bone marrow transplantation. *Blood* **74**,1552–1556.
26. Jordan, C.T., Astle, C.M., Zawadzki, J., Mackarehtschian, K., Lemischka, I.R., and Harrison, D.E. (1995) Long-term repopulating abilities of enriched fetal liver stem cells measured by competitive repopulation. *Exp Hematol* **23**,1011–1015.
27. Harrison, D.E. (1981) F1 hybrid resistance: long-term systemic effects sensitive to irradiation and age. *Immunogenetics* **13**,177–187.
28. Geissler, E.N., McFarland, E.C., and Russell, E.S. (1981) Analysis of pleiotropism at the dominant white-spotting (W) locus of the house mouse: a description of ten new W alleles. *Genetics* **97**,337–361.

29. Harrison, D.E., and Astle, C.M. (1991) Lymphoid and erythroid repopulation in B6 W-anemic mice: a new unirradiated recipient. *Exp Hematol* **19**,374–377.

30. Morrison, S.J., Hemmati, H.D., Wandycz, A.M., and Weissman, I.L. (1995) The purification and characterization of fetal liver hematopoietic stem cells. *Proc Natl Acad Sci USA* **92**,10302–10306.

31. Eppig, J.J., Kozak, L.P., Eicher, E.M., and Stevens, L.C. (1977) Ovarian teratomas in mice are derived from oocytes that have completed the first meiotic division. *Nature* **269**,517–518.

13

Isolation and Functional Characterization of Side Population Stem Cells

Jonathan B. Johnnidis and Fernando D. Camargo

Summary

The "side population" (SP) phenotype is a manifestation of primitive cells' ability to efficiently efflux the fluorescent DNA-staining dye Hoechst 33342 and can be used as the basis by which to isolate these cells using flow cytometry. In the bone marrow (BM), the SP defines a cell subset with a highly homogeneous content of hematopoietic stem cells (HSCs). In this chapter, we describe a protocol to reproducibly isolate murine BM SP cells, as well as analytic measures, such as single cell transplantation, that can be used to assess the functionality of SP-derived stem cells.

Key Words: Side population; Hoechst 33342; stem cell; phenotype; dye efflux; purification.

1. Introduction

In the early 1990s, intravital dyes such as Rhodamine 123 and various Hoechst compounds were observed to differentially stain cell populations in the bone marrow (BM) *(1)*. Extending upon these reports, investigators later perfected a technique that employs dual-wavelength flow-cytometric visualization of fluorescence from the DNA-binding dye Hoechst 33342 to reliably isolate a specific population of very primitive cells that have the capacity to efficiently efflux the dye following a period of incubation *(2)*. This side population (SP), so called because of its peripheral location on a two-dimensional fluorescence-activated cell sorting (FACS) plot of Hoechst-red versus Hoechst-blue fluorescence, consists mostly of highly homogenous hematopoietic stem cells (HSC) *(3)* and represents 0.04–0.07% of the entire

From: *Methods in Molecular Biology, vol. 430: Hematopoietic Stem Cell Protocols*
Edited by: K. D. Bunting © Humana Press, Totowa, NJ

marrow. SPs of primitive cells have been observed in the BM of every mammalian species examined *(4)*, and as well in a number of solid organs and tumors *(5)*.

Hoechst 33342 penetrates living cells and binds to the minor groove of the DNA molecule. The SP phenotype is due to cells' ability to extrude the Hoechst dye at a higher rate than non-side population cells, which are part of the "main population." The phenotype was initially observed to be absent after treatment with verapamil, an inhibitor of ATP-binding cassette (ABC)-class efflux proteins, suggesting a possible molecular mechanism of action *(2)*. Later, the ABC transporter protein Bcrp1/ABCG2 was determined to be a major molecular determinant of the SP phenotype in BM *(6)*, and the knockout (KO) mouse was demonstrated to have a severe reduction in the SP *(7)*. However, Brcp1 was expressed in some cell types residing in the main population, indicating that Bcrp1 expression alone is necessary but not sufficient to confer the SP phenotype.

Bcrp1 expression is highest in the most primitive HSCs and is downregu-lated upon differentiation *(6)*. However, although Bcrp1 KO mice lack an SP, steady-state hematopoiesis is otherwise normal, including HSCs with normal numbers and repopulating capacity *(7)*. If the SP phenotype is dispensable, what then is its function and that of the underlying efflux transporters? Many ABC proteins, most famously the multi-drug resistance gene MDR1, have been known for decades to efflux various compounds from cells, and Bcrp1 KO cells show greatly enhanced sensitivity to the cytotoxic drug mitoxantrone *(7)*. Additionally, Bcrp1 KO HSCs show reduced survival under hypoxic condi-tions, and Bcrp1 is upregulated in hematopoietic cells during hypoxia, in order to reduce detrimental intracellular accumulation of porphyrins *(8)*. Therefore, the SP phenotype may be a manifestation of primitive cells' need for and capacity to protect themselves from harmful compounds, either endogenous or exogenous, during adverse conditions. Despite these findings, additional physiological functions for the SP phenotype may remain to be uncovered.

1.1. Relationship Between SP Cells and HSCs Isolated by Alternative Methods

Regardless of mechanism and function, the SP method represents an extremely useful tool by which to prospectively isolate HSCs and other primitive populations without reliance upon cell-surface antigens. Other vital dyes may also be used in a similar fashion *(9,10)*, although Hoechst 33342 staining and dual-wavelength visualization thereof are the most established methods. Within the context of the hematopoietic system, the literature is scattered with different HSC purification protocols, and it is often unclear whether one is more advantageous over another; in contrast, enrichment of stem cells by SP has been shown to be as efficient as traditional methods using surface markers *(4)*, and furthermore, these stem cells have been shown to exhibit surface marker phenotypes that overlap with those in reports published by many other groups *(3)*. Isolation of SP cells with the greatest Hoechst efflux ("tip" SP cells) with the addition of another stem cell marker such as Sca-1, can be as, or even more, efficient than isolating HSCs through complex antibody-based protocols that usually rely on more than 10 different markers *(3)*.

1.2. Functional Characterization of Single SP Cells

Clonal assays are crucial for analysis of HSCs because HSC differentiation and self-renewal potentials need to be demonstrated at the single cell level. In the hematopoietic system, the most stringent way to prove function at the individual cell level is through in vivo single-cell reconstitution assays. In this assay, single HSCs are isolated using FACS and individually injected into lethally irradiated mice. To guarantee the survival of the hosts during the lag phase between the injection of the donor HSCs and the development of single cell–derived mature blood cells, a *carrier* population is co-injected with the single cell. To identify the hematopoietic progeny of the single HSC, we routinely use C57BL/6 (B6) mice congenic for the CD45 locus: B6-CD45.2 mice as single cell donors and B6-CD45.1 mice as hosts. As CD45 is expressed in all cells of the hematopoietic lineage with the exception of mature erythrocytes, CD45.1 and CD45.2 epitopes expressed in circulating leukocytes can be specifically recognized by anti-CD45.1 and anti-CD45.2 monoclonal antibodies.

We and others have performed single-cell reconstitution assays with SP cells and found that approximately 1 in 3 SP HSCs are able to engraft long term and produce multi-lineage hematopoietic progeny in recipient animals *(3,11,12)*. Given that the probability of any individual HSC to reach and "seed" the BM, when injected intravenously, has previously been calculated to be 20% *(13,14)*, these results suggest that the SP population is functionally homogeneous. Similar results have been obtained utilizing single HSCs isolated by conventional cell surface–staining approaches *(15,16)*. Whether additional surface markers or dye-stains allow for further fractionation of the SP population is still a matter of debate.

In the following sections, we describe an SP protocol that can be used to reliably isolate stem cells from murine BM; variations on the technique for other organ systems and species can be made, and have been published elsewhere *(4,5)*. We also describe a protocol for the transplantation of single SP cells and the methodology used to track single cell–derived engraftment. This method can also be used to test the reconstituting potential of any putative HSC or progenitor cell.

2. Materials

2.1. Isolation and Analysis of BM SP Cells

1. C57Bl/6 mice (*see* **Note 1**).
2. 10-cm tissue culture dishes.
3. Two 10-ml syringes.
4. 18-G and 27-G needles.
5. 70-μm nylon mesh cell strainers (Becton Dickinson, San Jose, CA, "Cell Strainers").
6. Hemocytometer.
7. DME+: Dulbecco's modified Eagle's medium, high glucose (Gibco, Rockville, MD, USA, Cat. No. 11965-092) supplemented with 2% fetal bovine serum (FBS) and 10 mM HEPES buffer.
8. Hoechst 33342 at 1 mg/ml in water. The dye can be obtained from Sigma (bis-benzamide, Cat. No. B2261) as a powder and resuspended, filter-sterilized, and frozen in 250-μl aliquots.
9. Verapamil, dissolved in 95% ethanol as a 5-mM 100× stock, and stored at –20°C in 100-μl aliquots.
10. Propidium iodide (PI) in PBS at 200 μg/ml. PI powder is obtained from Sigma, and a stock solution (10 mg/ml) is dissolved in water and stored at –20°C in 100-μl aliquots. The 100× working stock, covered in aluminum foil and kept at 4°C, is at 200 μg/ml in PBS. The final concentration of PI in your sample should be 2 μg/ml.
11. Circulating water bath at exactly 37°C.
12. Flow cytometer with an ultraviolet (UV) laser.

2.2. Stem Cell Transplantation

1. C57Bl/6 (B6) CD45.2 and B6-CD45.1 congenic mice (8–12 weeks old).
2. Neomycin solution (2 mg/ml final concentration in drinking water).
3. Autoclaved acidified water, pH 2.0.
4. Gamma irradiator.
5. Insulin syringes with attached 29-G needles.
6. StemPro medium (Invitrogen, Carlsbad, CA, USA).
7. 96-well, flat-bottomed microtiter plate.

2.3. FACS Analysis of Peripheral Blood

1. Heparinized microhematocrit capillary tubes for blood collection (Fisher Scientific, Pittsburgh, PA, USA, Cat. No. 22-362-566).

2. Red blood cell (RBC) lysis buffer stock solutions: 0.16 M NH_4Cl (8.3 g/l) and 0.17 M Tris-HCl, pH 7.6. Mix 9 parts NH_4Cl to 1 part Tris-solution right before use.
3. Refrigerated table-top centrifuge.
4. Monoclonal antibody anti-CD16/CD32 for blocking of Fc-mediated interactions.
5. Monoclonal antibodies against CD45.1 (PE-Cy5.5 conjugated), CD45.2 (APC), B220 (FITC), CD3 (PE), Gr1 (FITC) and Mac1 (PE).
6. PI solution (*see* **Subheading 2.1., step 10**).
7. Four-color FACS analyzer.

3. Methods

3.1. Isolation of SP Cells

The SP phenotype is a manifestation of an active biological process, namely the activity of ATP-dependent cell-surface efflux proteins. Because of this, it is critical to apply the protocol with utmost precision and rigid attention to detail; even minor deviations from such critical parameters as cell or dye concentrations or temperatures may adversely impact detection of the SP phenotype. If followed carefully (*see* **Note 2**), this method will yield a SP containing highly homogenous HSCs.

3.1.1. Extraction of BM Cells

1. Euthanize C57Bl/6 mice 8–10 weeks of age, and excise femur and tibia bones, taking care to dissect away as much as muscle tissue as possible. Place bones into a tissue-culture dish containing chilled DME+; if sterility is required (e.g., the cells will be used for culture or transplant), the procedure may be executed in a biological safety cabinet, and the medium may be supplemented with antibiotics.
2. Draw up chilled DME+ into a 10-ml syringe tipped with a 27-G needle, and flush BM into a fresh culture dish, on ice. Flushing from both ends of each bone will ensure maximum marrow recovery; bones emptied of marrow will appear very pale.
3. Change the needle to 18-G and draw the medium-marrow mixture up and down several times to homogenize marrow and ensure a single-cell suspension. This is done gently to avoid high sheer forces, which can be detrimental to cell viability.
4. Filter the cell suspension through a 70-μm cell strainer into a 50-ml polypropylene tube.
5. Carefully count nucleated cells using the hemocytometer (*see* **Note 3**); we find an average of 5×10^7 nucleated cells per 8- to 12-week-old mouse from the marrow of two femurs and two tibias.
6. Pellet cells in a centrifuge at $800 \times g$ for 7 min at 4°C.
7. Resuspend cells at 10^6 cells/ml in prewarmed (to 37°C) DME+.

3.1.2. Hoechst 33342 Staining of Extracted Cells

1. Ensure the water bath is at precisely 37°C by checking with a thermometer.
2. Add Hoechst 33342 to a final concentration of 5 μg/ml (a 200× dilution of the stock). Mix the cells gently and place in the water bath to incubate for exactly 90 min (*see* **Note 4**).
3. Pellet cells in a centrifuge at $800 \times g$ for 7 min at 4°C.
4. Resuspend in ice-cold DME+. If samples will be used for FACS analysis, supplement medium with 2 μg/ml PI to discriminate dead cells.
5. Note, any and all further cell manipulations must be done at 4°C to prohibit further efflux of the dye.

3.1.3. Antibody Staining of Hoechst-Stained Cells

To confirm identification of HSCs, SP cells can be co-stained with antibodies against canonical markers such as Sca-1, c-Kit, and lineage antigens (*see* **Note 5**).

1. Aliquot Hoechst-stained BM cells into staining tubes at 10^7 cells per tube in 100 μl of DME+. To avoid perturbing the SP profile, we emphasize this step should be executed on ice.
2. Add appropriately titered antibodies (usually a 1/100 dilution of rat anti-mouse antibodies).
3. Incubate the cells with antibodies for 15 min on ice.
4. Wash cells once by centrifugation with a 10-fold excess of DME+.
5. For flow cytometric analysis, resuspend cells in PI solution as described in **Subheading 3.1.2.**, **step 4**.

3.1.4. Flow Cytometry Analysis for SP Cells

The SP is analyzed on flow cytometers equipped with an UV laser (*see* **Note 6**), which excites both the Hoechst dye and the PI. Additional lasers (such as those at 488 nm) may be used to excite additional fluorochromes, such as those conjugated to antibodies against HSC-surface markers. The key component of SP analysis, dual-wavelength visualization of Hoechst fluorescence, comes from UV excitation of the Hoechst dye, followed by measurement of the fluorescence in both "blue" and "red," using a 450/20 band pass (BP) filter and a 675-edge filter long pass (EFLP), respectively. A 610 dichroic mirror short pass (DMSP) is used to separate the emission wavelengths. The fluorescence of PI, having been excited by the UV laser, is also measured through the 675 EFLP; note that PI-staining dead cells are significantly brighter red than the Hoechst red signal.

1. Load Hoechst-stained cells on the cytometer, preferably keeping them cold by means of a chilling apparatus.

2. Visualize a plot of Hoechst blue versus red fluorescence, with the former on the *y*-axis, and the latter on the *x*-axis.

3. Ensure that the detectors are in linear mode, and adjust voltages such that red blood cells are located in the very bottom-left corner of the plot, and dead cells are against the far-right side of the plot.

4. After a profile similar to that shown in **Fig. 1** is obtained, gate out red and dead cells, and collect 100,000 events. The SP region, of a prevalence of approximately 0.05 %, will appear as indicated in **Fig. 1**. Especially for beginners, it is helpful to confirm this SP by co-staining with antibodies (*see* **Subheading 3.1.3.**) to detect additional surface markers, and/or by treatment with verapamil (*see* **Note 7**) or similar inhibitors to abolish efflux activity, which should result in an absence of the SP.

3.2. Clonal Functional Assay of SP Cells

The following protocols describe single-cell transplantation of BM stem cells isolated through the SP method, as well as assessment of engraftment and reconstitution activity in recipient hosts.

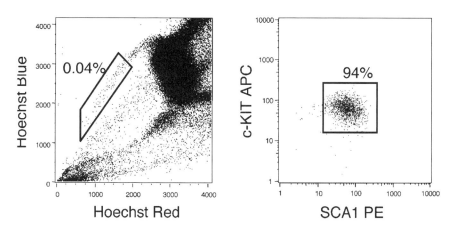

Fig. 1. Murine bone marrow (BM) side population (SP). The left panel shows a typical fluorescence-activated cell-sorting (FACS) profile after a Hoechst 33342 stain as described in this chapter. To visualize the SP population, signals are displayed in a Hoechst Blue versus Hoechst Red dot plot. The SP population (in trapezoid; left panel) comprises between 0.04 and 0.07% of whole BM. The dot plot in the right panel is gated on the SP gate (left), and it demonstrates that most SP cells express high levels of the c-KIT receptor and stem cell antigen-1 (Sca-1).

3.2.1. Single SP Cell Transplantation

1. Using the procedure described above, sort SP cells from a B6-CD45.2 mouse, at one cell per well, into a flat-bottomed 96-well microtiter plate, of which each well contains 100 μl of StemPro medium (*see* **Note 8**).
2. Subject recipient mice (C57Bl/6-CD45.1) to an 11-Gy dose of total body irradiation (*see* **Note 9**).
3. Take recipient mice and the cells to be injected into a biological safety cabinet.
4. Using an insulin syringe, aspirate the entire medium (100 μl) from a single well of the 96-well plate containing each a single SP cell.
5. Wash same well with 100 μl of StemPro-containing 1×10^5 whole BM cells (*carrier* cells) isolated from a C57Bl6 CD45.1 mouse (*see* **Note 10**).
6. Aspirate medium into the syringe containing the single cell.
7. Transfer contents of the syringe (total of 200 μl) into an anesthetized, irradiated mouse through retro-orbital injection.
8. Start antibiotic treatment of transplanted animals.

3.2.2. FACS Analysis of Peripheral Blood

1. Obtain about 70 μl of peripheral blood (PB) from mice to be analyzed; also, bleed B6 CD45.1 and CD45.2 mice for controls (*see* **Note 11**).
2. Add 1 ml of RBC-lysing solution and incubate at room temperature for 5 min.
3. Stop the reaction by diluting the lysis buffer with 10–20 ml of DME+.
4. Spin the cells ($300 \times g$) at 4°C and resuspend the pellet in 100 μl DME+ containing 0.5 μg of anti-CD16/32 antibody. Incubate for 15 min at 4°C.
5. Divide the samples in two tubes, wash with 10 vol of DME+ and spin the samples down at 4°C.
6. Prepare two antibody-staining cocktails, diluting each antibody at 1:100 in DME+. Lymphoid engraftment cocktail: CD45.1, CD45.2, B220, CD3. Myeloid engraftment cocktail: CD45.1, CD45.2, Mac1, Gr1 (*see* **Note 12**).
7. Resuspend pellets in 100 μl of the appropriate antibody cocktail and stain for 15 min at 4°C.
8. Wash with 10–20 vol of DME+ and spin down.
9. Resuspend in 200 μl of DME+ with PI and analyze using flow cytometry.

4. Notes

1. Both the initial and most subsequent reports based on SP methodology have used C57Bl/6 mice, often aged 8–12 weeks. Therefore, it is recommended to use this strain of mice to facilitate comparison of results with published data.
2. Strict adherence to the staining protocol is critical to resolving a useful SP. Special attention should be paid to Hoechst concentration, cell concentration, staining time, and staining temperature. To prohibit further dye efflux, it is also critical that cells be maintained at 4°C after staining.

3. Accuracy of nucleated cell counts is important. Exclusion of enucleate RBCs can be accomplished by an experienced eye or by use of one of several RBC lysis protocols or commercially available reagents.

4. The staining tubes should be both completely submerged in the water bath and shaken several times during the incubation to guard against non-uniform temperature distributions. During the stain, avoid using the water bath for any other cell culture functions to minimize temperature fluctuations; in particular, by no means should tubes of frozen serum or containers of chilled medium be thawed or warmed in this bath.

5. Most murine SP cells will be Sca-1$^+$, c-Kit$^+$, CD45$^+$, and lineage marker$^{-/low}$. To confirm the identities of the SP and MP, it is advisable to stain for at least two markers, one which positively identifies SP cells (e.g., Sca-1 or c-Kit) and another which identifies MP cells (e.g., Gr-1 or B220).

6. We have obtained good results using a MoFlo cell sorter (DakoCytomation, Fort Collins, CO) and LSR FACS analyzers (Becton Dickinson). A very thorough comparison of the SP profiles generated from these machines has been recently reported *(17)*. Note that the analysis of Hoechst fluorescence is performed on a linear scale, and thus, optimal configuration of the flow cytometer is important, including good coefficients of variation (CVs), which can be obtained by relatively high power (50–100 mW) on the UV laser. Nevertheless, methods using less power may suffice and have been described elsewhere *(18)*.

7. Verapamil is used at 50 µM and is included in the cell-dye solution during the duration of the Hoechst-staining incubation.

8. It is important to verify the accuracy of single-cell deposition by the cell sorter used. To do this, we usually visualize under an inverted microscope that only single cells are present in the sorted wells. Sorting of SP cells carrying a fluorescent marker, e.g., GFP, can be performed in order to aid with the visualization. If more than one cell is present in the sorted wells, it might be necessary to recalibrate the FACS device.

9. The dose for irradiation needs to be verified empirically. Currently, we use a split dose of 5.5 Gy each, with 2–4 h in between doses. Mice can also be irradiated the day before transplantation.

10. *Carrier* cells allow for short-term radioprotection of the recipient animals and sustain immune and hematopoietic function, whereas the single donor HSC produces functionally mature progeny. Overall, higher levels of reconstitution derived from the single HSC can be achieved if the *carrier* population is depleted of HSCs. We have achieved this by two different ways: *(1)* transplanting a population of Sca1-depleted cells (usually 3×10^5) cells or *(2)* transplanting a highly purified short-term stem cell population. We have utilized the Lin^{neg}Sca^{pos}c-Kit^{pos}CD34^{pos} population described originally by Nakauchi and colleagues. We usually co-transplant about 500–600 of these cells.

11. We usually begin our PB analyses at 4 weeks after the initial transplant, and do it every month thereafter. To make any conclusions about the single cell–derived

graft being long term, the analysis should extend until at least 6 months post transplant.

12. We utilize the combination of anti-CD45.2 and anti-CD45.1 antibodies in order to obtain a much cleaner and sensitive readout of the usually low contribution by the CD45.2 donor cell. Using this strategy, we have been able to detect multi-lineage contribution of as low as 0.1% of the total PB. Alternatively, if only a three-color FACS analyzer is available, samples can be divided into three and then stained with antibodies against CD45.2, CD45.1, and each of B220, CD3, or a combination of Gr1/Mac1.

References

1. Wolf, N.S., A. Kone, G.V. Priestley, and S.H. Bartelmez. (1993) In vivo and in vitro characterization of long-term repopulating primitive hematopoietic cells isolated by sequential Hoechst 33342-rhodamine 123 FACS selection. *Exp Hematol* **21**, 614–622.
2. Goodell, M.A., K. Brose, G. Paradis, A.S. Conner, and R.C. Mulligan. (1996) Isolation and functional properties of murine hematopoietic stem cells that are replicating in vivo. *J Exp Med* **183**, 1797–1806.
3. Camargo, F.D., S.M. Chambers, E. Drew, K.M. McNagny, and M.A. Goodell. (2006) Hematopoietic stem cells do not engraft with absolute efficiencies. *Blood* **107**, 501–507.
4. Goodell, M.A., M. Rosenzweig, H. Kim, D.F. Marks, M. DeMaria, G. Paradis, S.A. Grupp, C.A. Sieff, R.C. Mulligan, and R.P. Johnson. (1997) Dye efflux studies suggest that hematopoietic stem cells expressing low or undetectable levels of CD34 antigen exist in multiple species. *Nat Med* **3**, 1337–1345.
5. Challen, G.A. and M.H. Little. (2006) A side order of stem cells: the SP phenotype. *Stem Cells* **24**, 3–12.
6. Zhou, S., J.D. Schuetz, K.D. Bunting, A.M. Colapietro, J. Sampath, J.J. Morris, I. Lagutina, G.C. Grosveld, M. Osawa, H. Nakauchi, and B.P. Sorrentino. (2001) The ABC transporter Bcrp1/ABCG2 is expressed in a wide variety of stem cells and is a molecular determinant of the side-population phenotype. *Nat Med* **7**, 1028–1034.
7. Zhou, S., J.J. Morris, Y. Barnes, L. Lan, J.D. Schuetz, and B.P. Sorrentino. (2002) Bcrp1 gene expression is required for normal numbers of side population stem cells in mice, and confers relative protection to mitoxantrone in hematopoietic cells in vivo. *Proc Natl Acad Sci USA* **99**, 12339–12344.
8. Krishnamurthy, P., D.D. Ross, T. Nakanishi, K. Bailey-Dell, S. Zhou, K.E. Mercer, B. Sarkadi, B.P. Sorrentino, and J.D. Schuetz. (2004) The stem cell marker Bcrp/ABCG2 enhances hypoxic cell survival through interactions with heme. *J Biol Chem* **279**, 24218–24225.
9. Bertoncello, I. and B. Williams. (2004) Hematopoietic stem cell characterization by Hoechst 33342 and rhodamine 123 staining. *Methods Mol Biol* **263**, 181–200.

10. Telford, W.G., J. Bradford, W. Godfrey, R.W. Robey, and S.E. Bates. (2007) Side population analysis using a violet-excited cell permeable DNA binding dye. *Stem Cells*, **25**, 1029–1036.

11. Camargo, F.D., R. Green, Y. Capetanaki, K.A. Jackson, and M.A. Goodell. (2003) Single hematopoietic stem cells generate skeletal muscle through myeloid inter-mediates. *Nat Med* **9**, 1520–1527.

12. Uchida, N., B. Dykstra, K.J. Lyons, F.Y. Leung, and C.J. Eaves. (2003) Different in vivo repopulating activities of purified hematopoietic stem cells before and after being stimulated to divide in vitro with the same kinetics. *Exp Hematol* **31**, 1338–1347.

13. Hendrikx, P.J., C.M. Martens, A. Hagenbeek, J.F. Keij, and J.W. Visser. (1996) Homing of fluorescently labeled murine hematopoietic stem cells. *Exp Hematol* **24**, 129–140.

14. van der Loo, J.C. and R.E. Ploemacher. (1995) Marrow- and spleen-seeding efficiencies of all murine hematopoietic stem cell subsets are decreased by prein-cubation with hematopoietic growth factors. *Blood* **85**, 2598–2606.

15. Osawa, M., K. Hanada, H. Hamada, and H. Nakauchi. (1996) Long-term lympho-hematopoietic reconstitution by a single CD34-low/negative hematopoietic stem cell. *Science* **273**, 242–245.

16. Wagers, A.J., R.I. Sherwood, J.L. Christensen, and I.L. Weissman. (2002) Little evidence for developmental plasticity of adult hematopoietic stem cells. *Science* **297**, 2256–2259.

17. Simpson, C., D.J. Pearce, D. Bonnet, and D. Davies. (2006) Out of the blue: a comparison of Hoechst side population (SP) analysis of murine bone marrow using 325, 363 and 407 nm excitation sources. *J Immunol Methods* **310**, 171–181.

18. Cabana, R., E.G. Frolova, V. Kapoor, R.A. Thomas, A. Krishan, and W.G. Telford. (2006) The minimal instrumentation requirements for Hoechst side population analysis: stem cell analysis on low-cost flow cytometry platforms. *Stem Cells* **24**, 2573–2581.

14

Transplantation of Chimeric Fetal Liver to Study Hematopoiesis

Sigrid Eckardt and K. John McLaughlin

Summary

Complementing mutant embryos or embryonic stem cells with normal cells in embryonic chimeras is a valuable tool for investigating phenotypes. Chimera approaches provide a method to examine the phenotype of mutant cells, including hematopoiesis, in mutants with early embryonic lethality. Complementation with normal cells in a chimera can, in most instances, rescue mutant cells to later stages of gestation and beyond, permitting analysis of contribution and function of mutant cells in various organs, both within the chimera, but also by using functional transplantation assays for hematopoietic stem and progenitor cells. This chapter describes principles and methods for the generation of mouse chimeras, for identification and quantitative analysis of cell contribution in chimeras, and for chimeric fetal liver transplantation into adult recipients and analysis of mutant cells in the adult.

Key Words: Developmental chimera; ES cell; fetal liver; fetal liver hematopoietic stem cell.

1. Introduction

Gene targeting and transgenic strategies in the mouse are widely applied tools to determine the relationship between gene products and in vivo function. The derivation and genetic manipulation of embryonic stem (ES) cells and their subsequent introduction into the germline provides the basis for many of these approaches, such as the generation of null mutants. Frequently, however, null mutants exhibit early embryonic lethality or lethality before the stage of interest, thereby preventing analysis of the function of the gene product at later stages

From: *Methods in Molecular Biology, vol. 430: Hematopoietic Stem Cell Protocols*
Edited by: K. D. Bunting © Humana Press, Totowa, NJ

or in certain tissues. If mutants exhibit multiple defects, tissue-specific analysis of defects can also be obscured by cell-extrinsic factors. One way of addressing this problem is the generation of conditional mouse mutants in which embryonic lethality is bypassed by introducing the mutant in a stage or tissue-specific manner [conditional mutant; *(1)*]. For some tissues, including hematopoietic lineages, development from ES cells in vitro has been characterized sufficiently to permit using in vitro differentiation as an alternative approach toward the characterization of mutant ES cell lines *(2)*. The complementation of mutant mouse embryos or ES cells with normal cells in developmental chimeras has also been used as a powerful tool to analyze competence and function of mutant cells in vivo *(3)*. Mouse embryonic chimeras can be made by several approaches, including the combination of cleavage stage embryos, typically at the four- to eight-cell stage, to form aggregation chimeras, or the injection of ES cells into normal host morula- or blastocyst-stage embryos (ES cell injection chimera). In many cases, complementation with normal cells rescues mutant cells to later stages of gestation or to term and postnatally, such that contribution to and function of mutant cells to tissues and organs can be studied in the chimeric organism. One system that is frequently used in studies aimed at the analysis of hematopoiesis is the injection of ES cells into Rag2-deficient blastocysts; $Rag2^{-/-}$ mice do not produce mature B and T lymphocytes, such that lymphocytes in chimeric animals are entirely ES cell derived [Rag complementation; *(4)*]. Examples for the successful application of developmental chimeras to investigate the effect of mutants on hematopoiesis include analyses of the *TEL, Mixed Lineage Leukemia*, and *FOG* genes, which exhibit early embryonic lethality in homozygous null mutants *(5–7)*. For example, analysis of $TEL^{-/-}$ ES cells in chimeras with normal and *Rag2*-deficient cells revealed that *TEL* is not essential for fetal liver hematopoiesis (i.e., $TEL^{-/-}$ ES cells contributed to fetal liver and had normal progenitor activity) but is required for adult bone marrow hematopoiesis, evident from a lack of $TEL^{-/-}$ cells to bone marrow myelopoiesis and erythropoiesis *(5)*.

The ability of fetal liver stem cells to reconstitute hematopoiesis in adults permits analysis of hematopoietic stem cell (HSC) function even in mutants in which chimeras may not survive to term. By transplantation of fetal liver cells from chimeras into lethally irradiated adult mice and subsequent analysis of mutant cell contribution and function in reconstituted animals, the ability of mutant cells to form fetal liver HSC (FL-HSC) can be ascertained and HSC function be studied in respect to self-renewal and multi-lineage differentiation using a variety of approaches (lineage analysis, secondary transplants, number of colony-forming units in bone marrow and spleen; *(8)*].

In this chapter, we describe strategies and methods for the production and analysis of ES cell chimeras. As chimera analysis requires experimental design specifically tailored for the type of embryo/ES cell used, we describe here the experimental design and methods that we applied for the analysis of uniparental ES cell–derived hematopoiesis as an example *(8)*. Mammalian uniparental embryos with genomes of only one parental type, e.g., oocyte-only (partheno-genetic, gynogenetic) or sperm-only (androgenetic) derived, exhibit early embryonic lethality but develop sufficiently to produce ES cells. Uniparental ES cells can contribute to tissues in chimeras, however, with bias in their differentiation into, and exclusion from, certain lineages. Androgenetic cells, even at low levels of contribution, cause severe defects and embryonic and neonatal lethality in chimeras; parthenogenetic cells exhibit proliferation defects. To ascertain the ability of mammalian uniparental cells to undergo hematopoiesis and to form transplantable FL-HSC, we therefore produced uniparental developmental chimeras by injecting uniparental ES cells into normal blastocysts, and transplanted uniparental chimeric fetal liver into lethally irradiated adult mice. As markers to identify contribution of ES cell- and blastocyst-derived cells in chimeras and in recipients of chimeric fetal liver transplants, we combined two markers suitable for chimera analysis: Expression of enhanced green fluorescent protein (eGFP) in ES cells, and presence of distinct isoforms of glucose-phosphate-isomerase 1 (Gpi1) in ES cell-, blastocyst-, and recipient-derived cells. Contribution of uniparental-derived cells to hematopoiesis of recipients was determined by electrophoretic analysis of Gpi1 isoforms. We used standard approaches to analyze differentiation and function of uniparental-derived cells in recipients, including multi-lineage analysis by fluorescence-activated cell sorting (FACS), peripheral hematology, and colony-forming units in bone marrow of uniparental cells. Some recipients developed entirely uniparental cell-derived hematopoiesis, such that bone marrow from these could be used to confirm the presence of uniparental long-term repopulating HSC by transplantation into irradiated secondary recipients. In this chapter, we describe methods for (1) generating fetal stage ES cell chimeras, (2) identifying and quantifying ES cell contribution in fetal chimeras, (3) transplanting fetal liver cells into adult recipients, and (4) identifying and quantifying ES cell contribution in adult recipients. The experimental design, mouse strains, cell identification markers, and methods that were applied to analyze uniparental ES cells are examples; however, these are in many cases applicable to other mutant ES cells.

2. Materials

2.1. Producing Mouse ES Cell Chimeras

2.1.1. ES Cell Maintenance

1. 12-well tissue culture dishes (No. 35-3043; BD-Falcon, San Jose, CA, USA).
2. Feeder cells for ES cell culture [mouse embryonic fibroblasts (MEF) can be made as described *(9,10)* or be purchased commercially. STO fibroblasts are available

from ATCC (CRL-1503). STO cells that have been stably transfected with a neo^r vector and a LIF expression vector (SNL cells) are courtesy of Allan Bradley and Elizabeth Robertson].

3. 500 ml DMEM (Specialty Media/Chemicon EmbryoMax, No. SLM-220-B; without L-glutamine and Na-Pyruvate; with 4500 mg/l glucose, 2250 mg/l Na Bicarb; Specialty Media, Phillipsburg, NJ, USA).

4. 6 ml non-essential amino acids (100×; Gibco, No. 11140-050; Invitrogen, Carlsbad, CA, USA).

5. 6 ml Pen/Strep (100×; Gibco, No. 15140-122).

6. 6 ml L-glutamine (100×; Gibco, No. 25030-081).

7. 0.6 ml β-mercaptoethanol (1000×; Gibco, No. 21958-023).

8. 75 ml fetal bovine serum (Hyclone Defined FBS; No. SH30070.03; Hyclone, Logan, UT, USA).

9. Store at 4°C. After 3 weeks, replenish glutamine and 2-mercaptoethanol from the respective stock solution according to the amount of medium left in bottle. If using mouse primary embryonic fibroblasts (MEF) or STO cells as feeder layers, add LIF (Chemicon Esgro®LIF, No. ESG1106 or 1107)) to the medium (500 U/ml final).

10. GROWTH MEDIUM FOR FEEDER CELLS: Same composition as ES cell growth medium, except: 35 ml FBS per 500 ml (7% FBS); DMEM from Gibco (No. 11965). Do not add Pen/Strep (*see* **Note 1**). Store at 4°C.

11. 1× DULBECCO'S PHOSPHATE-BUFFERED SALINE: Dilute from 10× Stock; Gibco, No. 14200-075; without calcium or magnesium).

12. MITOMYCIN C: Dissolve 2 mg of mitomycin C (MMC) (Sigma, No. M 0503; Sigma, St. Louis, MO, USA) in 5.0 ml of DPBS (40× stock). Store 200 µl aliquots in sterile tubes at –80 °C; add one aliquot to 10 ml medium for treatment (8 µg/ml final).

13. PBS/GELATIN: Add 1 g gelatin (Sigma, No. G 2500) to 1000 ml 1× DPBS in glass bottle, then autoclave. Store at 4°C after opening; can keep for 2 months.

14. TRYPSIN/EDTA: 0.25% Trypsin (Sigma T 4799), 1 mM EDTA, 1× DPBS. For 1000 ml, Trypsin, 2.5 g; EDTA 500 mM Stock, 2 ml; 10× DPBS stock, 100 ml. To 100 ml working solution, add 1 ml of filter-sterilized 5% (w/v) BSA (Sigma, No. A 9647) in water. Store working solution at 4°C for 1–2 weeks, keep stocks at –20 °C.

15. CELL-FREEZING SOLUTIONS:

    a. Solution I: 50% (v/v) FBS in DPBS+ (1×; Gibco, No. 14287-080).
    b. Solution II: mixture of 20 ml DPBS and 5 ml DMSO (Sigma, No. D 2650).
    c. Store at 4°C. Discard after 14 days.

2.1.2. Generation of Chimeras

1. 3.5- and 6-cm tissue culture dishes (Falcon, No. 35-3001, No. 35-3002).
2. Sterile bulb pipettes (3 ml size, dropette brand or similar, bulk, sterilize by irradiation).

3. ES Cell Injection Medium: Iscove's Modified Dulbecco's Medium (IMDM, with L-glutamine and 25 mM Hepes; Gibco, No. 12440-053) supplemented with 5% FBS.

2.2. Identification and Contribution Analysis in Fetal Stage Chimeras

1. Scissors for cutting skin (Fine Science Tools (FST), Foster City, CA, USA, No. 14058-11, or similar).
2. Fine scissors [(FST) No. 14084-08, or similar].
3. 2 pair fine point forceps (No. 5; FST, No. 11254-20, or similar).
4. 6-cm tissue culture dishes (Falcon, No. 35-3002) or sterile Petri dishes.
5. 24- and 48-well culture dishes (Falcon, No. 35-3043; No. 35-3078).
6. PBS supplemented with 3% (w/v) BSA (PBS/BSA; sterile filtered; BSA: Sigma, No. A 9647).
7. Helena Labs Zip-Zone chamber (No. 1283; Helena Labs, Beaumont, TX, 1-800-231-5663, www.helena.com) or normal horizontal electrophoresis tank (*see* **Note 2**).
8. Titan III cellulose acetate plates (Helena, No. 3024).
9. Disposable wicks (Helena, No. 5081, 500 pack).
10. Super Z-12 applicator kit (Helena, No. 4093).
11. Running Buffer: 12 g Tris base and 5.76 g glycine in 1000 ml H_2O. Shake to dissolve, store at room temperature (RT). No need to adjust pH.
12. Stock Solutions for Staining:

	Concentration (in H_2O)	Storage
Staining buffer: 0.5 M Tris–HCl pH 8.0	RT	
D-fructose-6-phosphate (Sigma, No. F 3627; toxic!)	200 mg/ml	–20 °C[a]
beta-NADP (Sigma, No. N 0505)	20 mg/ml	–20 °C[a]
Phenazine methosulfate (Sigma, No. P 9625)	5 mg/ml	–20 °C[a]
Thiazolyl blue tetrazolium bromide (Sigma, No. M 2128)	20 mg/ml	–20 °C[a]
Glucose-6-phosphate dehydrogenase (Sigma No. G 8289)	undiluted	4 °C
Agarose	1.5%	RT

[a] Stock to be stored in aliquots at –20 °C; once thawed, store aliquot at 4 °C and use within 1–2 weeks. Phenazine methosulfate, and thiazolyl blue tetrazolium bromide: Protect from light.

2.3. Fetal Liver Transplantation

1. 70 μM cell strainer (BD, No. 352350; BD Biosciences, San Jose, CA, USA).
2. Mouse restrainer (we use a Broome style mouse restrainer; No. 551-BSRR; Plas Labs, Inc., Lansing, MI; 800-866-7527; www.plas-labs.com).
3. Thirty 1/2-g needles (BD, No. 305106).
4. 1 ml syringes (BD, No. 309602).
5. "Goggles" for identification of eGFP-positive animals (GF-sPectacles, Model gfsP-5, Biological Laboratory Equipment (BLS), Budapest, Hungary; 36 *(1)* 407-2602; bls@t-online.hu; http://www.bls-ltd.com).
6. Hank's balanced salt solution (HBSS, Gibco, No. 14025-076) supplemented with 10% FCS; make fresh.
7. Water bath or warming lamp (40–100 watt bulb).

2.4. Analysis of ES Cell Contribution in Chimeras and Recipients

1. Scissors for tail-clipping (FST, No. 14058-11, or similar).

3. Methods

3.1. Considerations for Experimental Design: Cell Identification and Marker Genes

The identification of ES cell- and blastocyst-derived cells in chimeras and in recipients of chimeric fetal liver transplants requires that these components are genetically distinct while immune compatible with the host. Ideally, all three components should be distinguishable. If this is not possible, the ES cell-derived component must be distinguishable from the normal cells of both blastocyst and recipient origin.

Depending on the overall experimental design including the genetic background of the ES cell lines to be used, the choice of markers can include one or more of the following:

1. Intracellular biochemical markers, such as Gpi1. This marker for identification of chimeric contribution is based on electrophoretic polymorphisms of Gpi1. In mouse laboratory strains, three variants of the enzyme have been identified, encoded by the *Gpi1 a, b,* and *c* alleles. Gpi1 forms homo- and heterodimers, such that cells from a mouse homozygous for the *a* allele will only exhibit the AA homodimer, whereas those from a mouse heterozygous for *a* and *b* alleles will exhibit three variants, the AA and BB homodimers, and the AB heterodimer with intermediate mobility and double density. Mixtures of cells that are homozygous for the *a* and *b* alleles such as those from a chimera will lack the intermediate form. Gpi1 analysis permits quantitative analysis of cell contribution in samples in comparison with standard curves obtained by mixing peripheral blood from mice carrying different *Gpi1* alleles at known ratios. Mouse strains homozygous

for the *b* allele include C57Bl6, C3H, CBA; the *a* allele is present in mice of the 129S1 substrain and many 129 strain-derived ES cell lines (E14, D3, AB1). Mice of the substrain 129SvEv (Taconic) are homozygous for the *c* allele. Advantage: Housekeeping gene, so expressed ubiquitously. Using mice of 129 substrains with different Gpi1 variants permits to detect cellular origin while maintaining immune compatibility. Disadvantage: Cannot be used for FACS analysis; requires disruption of cells.

2. Expression of eGFP [for example, ubiquitously expressed transgene *(11)*]. For analyses involving phenotyping by FACS, eGFP expression can be used, provided that it is expressed in the cells to be analyzed [example: eGFP is expressed in lymphocytes of mice from ref. 11 *(12)*]. This marker also enables immediate detection of chimeras by visualizing eGFP expression in fetuses using "GFP-goggles."

3. The use of eGFP may require complementation with other markers (*see* **ref. *13***). When using hybrid ES cells and recipients, analysis of eGFP expression can be complemented with that of Gpi1 isoforms. Example: ES cells derived from eGFP-transgenic strain *(11)* with C57/Bl6x129S1 background (therefore *Gpi-1 a,b* and eGFP expressing) are injected into C57/Bl6 blastocyst (*Gpi-1 b,b*). Fetal chimeras can be identified using eGFP fluorescence. Contribution of ES cell and blastocyst-derived components to fetal liver and fetus can be quantified by Gpi1 isozyme analysis comparing the ES cell (*a/b*)-derived (AA, AB, BB) isomers versus the BB isoform. Chimeric fetal liver cells are transplanted into irradiated C57/BL6x129 SvEv mice (*Gpi-1 b,c*). ES cell-derived cells in blood and other organs of recipients can be studied using FACS analysis (eGFP expression), and contribution of ES-cell, blastocyst- and host-derived cells quantified by Gpi1 analysis.

4. Extracellular immunological markers present in congenic mouse strains, such as Ly5 (CD45). For example, when using ES cells with a C57BL/6 F1 genotype, the Ly5.2 (*Ly5^b*, Cd45.2, *Ptprc^b* antigen specificity) of the ES cell component can be used to distinguish ES cell-derived cells in the hematopoietic system of recipients by flow cytometry using a CD45.2-specific antibody (BD Pharmingen), provided that both blastocyst and recipient cells have the Ly5.1 allele (*Ly5^a*, *Cd45^a*, Cd45.1, *Ptprc^a* antigen specificity). This allele is available in congenic C57BL/6 mouse models, for example B6.SJL-*Ptprc^a*/BoAiTac (model No. 004007) from Taconic. Advantage: Can be used for FACS analysis. Disadvantages: Only two components can be distinguished. Limited to the hematopoietic system, cannot be used to determine ES cell contribution to fetal chimeras.

3.2. Considerations for Experimental Design: Chimera Approaches

Several approaches are possible for the generation of embryonic chimeras, with the injection of ES cells into blastocysts being the method that is most commonly used. Because this method is typically performed at trans-genic/knockout core facilities, it is likely to be the approach of choice. Because the objective of this chapter is to evaluate the phenotype of ES cell-derived cells in the chimera, it is not always necessary to obtain high contribution or germ line chimeras. Even normal wild-type ES cell lines derived and tested in chimeras at low passage are extremely variable in their capacity form viable term chimeras, an outcome that will not necessarily preclude that studies can be made on earlier developmental stages. Various approaches exist to generate chimeras with different contribution to different tissues *(3)*. Manipulating contribution of cell types can for example be achieved by generating chimeras with cells of different developmental stage (later stage cells in a chimera tend to contribute more to the embryonic lineages) or ploidy (tetraploid cells typically contribute to extraembryonic tissues). These modifications can be effective in directing contribution, but in many cases, chimera contribution is nonetheless a stochastic process. There are several approaches that can be used in a standard knockout blastocyst injection facility to change the contribution of ES cells when injected into diploid blastocysts. Firstly, the number of ES cells injected can increase or reduce contribution. As little as one single ES cell may contribute substantially to the fetus *(14)*. An option to increase the contribution dramatically but often with the consequence of a more variable outcome is to inject 3–4 ES cells subzonally into a morula stage embryo [day 2.5 embryo; at the 8–16 cell stage *(15)*] and transfer them to the recipient the next day as per normal blastocyst injection. An option to decrease the contribution is to co-inject the ES cells of interest with some normal wild-type ES cells to ameliorate the dominance of the ES cells over those of the inner cell mass (ICM) of the blastocyst.

Strain effects have a considerable effect on cell contribution in chimeras where cells are of the same developmental stage. The most commonly used ES cell strains, 129 or C57BL/6x129F1 strains, are typically dominant over other F1 or inbred strains, notably C57, CBAC57, and C3HC57. C57BL/6 ES cells although less dominating are still potent because the ES cell stage tends to dominate over the cells of the ICM.

3.3. Producing Midgestation Chimeras from ES Cells as a Source of Fetal Liver

3.3.1. ES Cell Maintenance

1. Preparation of feeder layers. Grow STO, SNL, or MEF on 15-cm dish to confluency. Aspirate medium and replace with medium containing MMC (1×) and incubate for 2 h at 37°C in CO_2 incubator. Wash plate several times with PBS, trypsinize cells, and determine cell count. Plate 3.5×10^5 STO or SNL or 2×10^5 MEF per each well of a gelatinized 12-well plate (resuspend cells at a concentration of 1×10^7 cells/ml, plate 35 or 20 µl). For gelatin-treatment of

dishes, cover bottom of wells with gelatin/PBS, and incubate at 4°C for 1 h or at RT for 20 min. Aspirate gelatin completely (tilt plates when aspirating) and add feeder medium to well.

2. ES cell maintenance. ES cells are split 1:3 to 1:6 every 2 days, when they reach about 70% confluency. Aspirate medium, rinse with 1× DPBS, rinse briefly with trypsin, then add small amount of trypsin such that the bottom of the well is just covered. Place into incubator for approximately 3–5 min. Cells will lift off and disaggregate. Add ES cell growth medium (for 12 well: 150 µl) and, using Gilson pipette, repeatedly pipet up down to break up clumps and obtain a single cell suspension. Transfer cells onto fresh wells with feeder cells.

3. Freezing and thawing ES cells. *Freezing.* Freeze cells at about 70% confluency. Change medium 2–3 h before freezing. At time to freeze, trypsinize cells as described above, add 2 ml ES cell maintenance medium, and collect by brief centrifugation (400 x g, 3 min) in 15-ml tube. Resuspend in 1/2 of the desired final volume of freeze solution I, then add equal volume of freeze solution II slowly, drop by drop, while mixing carefully by gently flicking tube. Transfer into cryovial and place into –80°C freezer either in styrofoam box or in a cooling device providing a controlled freezing rate (StrataCooler® or similar). Transfer to liquid nitrogen storage after 24 h. Freeze 1 vial with 500 µl volume from 1 well of a 12-well plate, 2 vials from 1 well of a 6-well plate, 4 vials from one 6-cm dish. *Thawing.* Thaw vial quickly in 37°C water bath and transfer contents into 15-ml tube. Slowly add ES cell maintenance medium to about 3–4 ml, collect cells by brief centrifugation, resuspend in ES cell maintenance medium, and distribute onto wells with feeder cells.

3.3.2. Generation of ES Cell Chimeras

1. ES cell preparation for blastocyst injection. For blastocyst injection, cells from one well of a 12-well plate (60–70% confluent) are sufficient. Two days before blastocyst injection, passage or thaw ES cells in various dilutions onto 12-well plate with feeder cells. Distribute one vial of frozen ES cells (frozen as described above: 1 vial from 1 well of a 12-well plate or 2 vials from one well of a 6-well plate) in the following manner onto individual wells of a 12-well tissue culture plate: 1/2 of the thawed cells on one well, 1/4 on the next well, and 1/8 and 1/16 on two other wells. If ES cells are already growing in culture, passage 1:3, 1:4, 1:6 onto wells of a 12-well plate. On the day of blastocyst injection, choose the well in which ES cells are about 70% confluent, with distinct- and medium-sized colonies but no sign of differentiation. Change medium 2–3 h before starting the preparation. For preparation, disaggregate cells by trypsinization (see above: passaging of ES cells) and plate in 4 ml of ES maintenance medium onto a non-gelatinized 6-cm tissue culture dish. Place in CO_2 incubator for 50 min. During this period, feeder cells will attach well, healthy ES cells will slightly attach, and damaged or dead cells will not, thus permitting to separate cells. Remove supernatant gently and wash dish very gently once with 1–2 ml of

injection medium. Using a bulb pipette, use 2–3 ml of injection medium to repeatedly rinse the bottom of the dish to remove lightly attached cells. Avoid touching the surface with the pipette (such that feeder cells are not removed). Rinse for about 1 min, then transfer supernatant with cells into 15-ml tube, spin briefly (400 x g, 3 min) and remove most of the supernatant. Resuspend cells in approximately 10–20 µl medium left in tube. Place 1 µl aliquot onto the inside of the lid of the 6-cm dish used for the preparation or a slide to observe morphology of cells and to ascertain purity of ES cell preparation under a tissue culture microscope.

2. Blastocyst injection and embryo transfer. This part of the methodology requires micromanipulation skills and equipment as well as surgical techniques, and, unless these are already established in the laboratory, will therefore likely be performed in collaboration or at a core facility. The outline of methods above (General Considerations 2: Chimera approaches) provides a guideline for key/pertinent aspects of the technology that need to be discussed with the collaborator or core facility. A detailed description would be beyond the scope of this chapter, requiring a chapter or book on its own. If the technique needs to be established in the laboratory, we refer to several excellent detailed descriptions of the techniques involved *(9,16,17)*.

3.4. Identification and Contribution Analysis in Fetal Stage Chimeras

3.4.1. Dissection of Fetuses and Fetal Liver

Euthanize recipient (13.5–14.5 days post coitum (d.p.c.); i.e., 11–12 days post blastocyst transfer that is equivalent to 2.5 d.p.c.). Open the skin midventrally: Make a small lateral cut at the midline (use skin scissors—not fine scissors for dissection, contact with hair will make these blunt), hold skin above and below incision, and pull skin apart to expose the entire abdomen. The uterus with conceptuses is visible underneath the body wall (*see* **Fig. 1A**). Using small dissection scissors, cut open body wall from the center diagonally toward the rib cage on both sides, whereas holding up the body wall with forceps. Locate the distal end of one uterine horn, cut at oviduct, and holding the tip of the uterine horn with forceps, carefully dissect the uterus by gently tearing the mesometrium with the tips of the scissors (*see* **Fig. 1B**). Cut at cervix and continue dissection for the other horn, transfer uterus to a 6-cm dish with PBS/BSA and, using small scissors, carefully cut open along side on the antimesometrial wall (*see* **Fig. 1C–E**). With forceps, remove each conceptus as a whole, peeling the placenta from the uterine wall and place into PBS/BSA (**Fig. 1F and G**). Rinse and place into sterile PBS/BSA, then open yolk sac to expose fetus (**Fig. 1H and I**). If using eGFP-transgenic ES cells, identify chimeras using eGFP fluorescence (with "GFP-goggles"). If using Gpi1 isozymes for identification of chimeras, take aliquots of each fetal liver

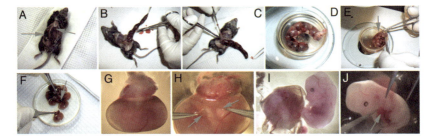

Fig. 1. Dissection of fetal liver from 13.5 days post-conception (d.p.c.) fetus. (A) Pregnant recipient with skin cut open, body wall still intact. Arrows mark the uterus with conceptuses. (B) Dissection of one uterine horn from the mesometrium. The oviduct has been cut. (C) Both horns of the uterus are shown. Embryos were transferred unilaterally such that one horn does not contain any implantations. (D) Uterus in dish before dissection of fetuses. (E) Opening the uterus by cutting the antimesometrial wall. (F) Dissection of conceptus using forceps. (G) Conceptus with intact yolk sac, placenta is oriented toward the top of the image. (H) Opening of the yolk sac using forceps, to expose fetus. (I) Fetus with placenta and yolk sac attached. (J) Fetus with exposed fetal liver (arrows).

for analysis: Under dissection scope, dissect fetal liver from each fetus using sterile forceps (**Fig. 1J**). Dissection is best done in the lid of a 6-cm dish as the lower walls allow better freedom of movement with forceps. With clean forceps, remove small aliquot of fetal liver (2 mm^3) and place into 40 μl of water as a sample for Gpi1 analysis (Eppendorf tube or 48-well plate). Place remaining fetal liver into 1.5-ml Eppendorf tube with 0.5 ml sterile HBSS 10% FCS and keep on ice (to be transplanted). Fetal livers can be kept on ice for several hours while Gpi1 analysis is performed. When using eGFP as a marker, and Gpi1 isoforms are distinct between ES- and blastocyst-derived cells, it is also useful to take a small sample for Gpi1 analysis for quantification of ES cell contribution to the fetal liver. For analysis of ES cell contribution to the fetus as a whole, place entire fetus into 500 μl water (most convenient in a 24-well tissue culture dish).

3.4.2. Gpi1 Isozyme Electrophoresis for Determination of ES Cell Contribution

This method is used to ascertain the contribution of ES cell-derived cells to the fetal liver and is performed on a small aliquot of the fetal liver. Because Gpi1 is a housekeeping gene, this method is more accurate to determine contribution to fetal liver than analysis of eGFP-expressing cells (eGFP transgenes are often not expressed in erythroid cells).

1. Sample preparation. Freeze-thaw fetal liver samples in water three times (accelerate by placing at –80°C, then at 37°C), mash up liver by pipetting after first freeze-thaw cycle. Whole fetuses are processed the same way.

2. Electrophoresis. Soak the Titan III cellulose acetate plate by immersing into a beaker filled with running buffer for about 10 min (*see* **Note 3**). Transfer 5 μl of each fetal liver sample into the wells of the applicator loading block (*see* **Note 4**). As controls, load blood samples (3 μl blood diluted into 30 μl H$_2$O) from mice with known Gpi1 alleles, either pure or mixed at certain ratios as a reference for contribution. Remove the cellulose acetate plate from the beaker and gently remove excess buffer with a paper towel without disrupting the cellulose acetate layer. The plate should be moist but not wet. Place into applicator alignment plate. Load the applicator by pressing down onto the loading block gently three times (**Fig. 2A**). Blot once on a paper towel and reload. Blot onto the cellulose acetate plate and hold very light pressure for 10 s (**Fig. 2B**). Remove plate, invert, and place on moist wicks in electrophoresis chamber (coated side down, reflective plastic side up; **Fig. 2C**). Electrophoresis is performed at 200 V for 1.5 h from anode to cathode.

3. Staining gel. Boil agarose solution and keep in a 55°C water bath. In a 15-ml Falcon tube wrapped in aluminum foil, mix 1 ml staining buffer with 75 μl each of fructose-6-phosphate, beta-nicotinamide adenine dinucleotide phosphate (NADP), phenazine methosulfate (PMS), and Thiazolyl blue tetrazolium bromide (MTT) stock solutions. Collect gel from the tank and place plastic side down on a level surface (can use the lid of 6-cm culture dish). Add about 9–10 ml of agarose solution to stain 1, mix briefly, then add 5 μl of glucose-6-phosphate dehydrogenase and mix thoroughly, then pour evenly over gel (*see* **Note 5**). Place gel in dark (cover with a cardboard box, do not move). Bands begin to appear in a few minutes.

4. Analysis. Take photographs/image while reaction is not saturated such that ratios between isoforms can be quantified in comparison with reference samples with known ratios (standard curves). The Gpi1 B isozyme migrates faster than the A isozyme. Gpi1 forms homo- and heterodimers, such that cells from mice heterozygous for *Gpi-1 a* and *b* alleles contain both AA and BB homodimers as well as heterodimer band of intermediate mobility (**Fig. 2D and E**). Type B mice: C57BL/6, CBA, C3H. Type A mice: BALB/c, 129/Sv.

3.5. Fetal Liver Harvest and Transplant

3.5.1. Recipient Preparation

Treat mice with lethal dose of whole body irradiation (differs depending on mouse strain; 9.5 Gy for C57Bl6x129 F1 hybrid animals) and maintain animals on oral antibiotic (as recommended by respective IACUC guideline,

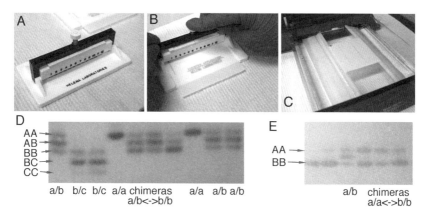

Fig. 2. Gpi1 isozyme electrophoresis. A–C: Setup. (A) Sample loading. (B) Application onto cellulose acetate plate. (C) Electrophoresis setup in a normal horizontal tank. The cellulose acetate plate is placed onto wicks that have been wrapped around pipettes and are joined to wicks that reach into the buffer tanks on either side. (D,E) Gpi1 gel examples. Mono- and heterodimers of the enzyme are indicated on the left (arrows). The genotype of samples is indicated on the bottom. In samples from chimeras between cells with *a/b* and cells with *b/b* alleles, the varying intensity of the BB homodimer is used to determine contribution of the *a/b* and *b/b* components (D). Analysis of chimeras between cells with *a/a* and cells with *b/b* alleles (E) is achieved by comparison of the AA versus the BB dimer (E).

for example, Neomycin 2 mg/ml in drinking water). Transplantation can be performed any time within 24 h after irradiation.

3.5.2. Preparation of Fetal Liver and Transplantation

Dissect fetus and fetal liver as described above; place fetal liver into Eppendorf tube with 0.5 ml sterile HBSS 10% FCS on ice. In tissue culture hood, place 70 µM cell strainer into 50-ml tube and moisten with HBSS/FCS. Place fetal liver on cell strainer. Remove plunger of a 1-ml syringe and use the broad flat end of the plunger to gently dissociate liver on strainer surface, while rinsing with cold HBSS/FCS. Let cells sit on ice for 2 min, then transfer suspended cells into fresh tube, leaving clumps that have settled to the bottom of the dish behind. Determine volume of cell suspension while transferring and remove aliquot for cell count. Collect cells by brief centrifugation and resuspend at concentration needed for injection, typically $0.5–1 \times 10^7$ cells/ml. Place mouse into restrainer, locate lateral tail vein, and inject 0.2 ml (i.e., 1 ×

10^6 cells when at 0.5×10^7 cells/ml) of the fetal liver cell suspension using a 33 g 1/2 needle (*see* **Note 6**).

3.6. Contribution Analysis in Transplants and Recipients

3.6.1. Gpi1 Analysis (Tail Blood)

Overall contribution of donor- (ES- and blastocyst-) and recipient-derived cells to the peripheral blood can be determined by Gpi1 isoform electrophoresis, provided that each component has distinct *Gpi1* alleles. Example: ES cells of C57/Bl6x129S1 origin (*Gpi-1 a/b*), host blastocyst C57/Bl6 (*Gpi-1 b/b*), recipient: C57BL/6x129 SvEv (*Gpi1 b/c*). For each blood sample, dilute 1–3 μl of peripheral blood 1:10 into water. Per electrophoresis, only 5 μl of sample is required; using a 3-μl sample that is diluted into 30 μl permits repeated runs, for example several months side-by-side. Samples can be stored in a –20°C freezer for >1 year. Blood can either be obtained from the saphenous vein, or from the tail tip, by cutting 1–1.5 mm off the very tip with sterile scissors, applying gentle pressure to the tail such that one droplet of blood is produced at the tip, collect blood with a Gilson pipette set to 3 μl, transfer directly into a 1.5-ml Eppendorf tube with 30 μl water and mix. Load and run samples as described above.

To analyze ES-cell-derived contribution in nucleated blood cells, lyse red blood cells of a 20-μl blood sample by osmotic lysis, collect cells by brief centrifugation, wash once with PBS/2% FCS, and add 15 μl of water to the sample. Load 5 μl on a gel and analyze as described (*see* **Note 7**).

4. Notes

1. No need to use EmbryoMax media, high glucose DMEM of any source is fine. We do not add antibiotics to feeder medium to ensure proper tissue culture handling—any infection becomes readily visible before feeders are used for ES cell culture.

2. When using a standard electrophoresis tank, two plastic or Pasteur pipettes on either side can serve as a holder for the plate. Wrap one moistened wick around each pipette and connect to a second wick on each side that is half submerged into buffer. Place plate on top of pipettes and ensure good contact, for example by placing a glass plate on top (*see* **Fig. 1C**). If analyzing Gpi1C, cool chamber with ice; Gpi1 C is unstable.

3. Mark glossy side of plate, top right side with a black marker pen (such that you can see where you are starting from on the other side of gel). For soaking, lower plate into buffer with forceps. Be careful and lower at a steady pace to avoid the forming of air bubbles between the holding film and the cellulose acetate coating.

4. It is important not to overload the cellulose acetate plate with sample. For fetal liver, diluting a 2-mm^3 piece into 40 µl of water works well; for blood, 3 µl of blood are diluted into 30 µl water. Guidelines for the dilution of other tissues can be found here *(18)*. Typically, a 1:10 dilution is a good place to start.

5. Keep solutions and stain mix in dark as much as possible, work quickly when preparing staining mix, best at dim light to avoid background staining. Do not move plate once stain is poured on until the agarose has hardened. An alternative staining method not involving agarose is described in *(9)*.

6. The vein is more difficult to see in strains with pigmented tails, and as this technique requires a lot of practice, it is advisable to first practice injection of sterile saline into non-irradiated mice with non-pigmented tails. Location and injection of the vein are easier if the vein is dilated, either by immersing the tail in warm water for 5–10 s or by warming the mouse 5–10 min in the cage with a warming lamp with a 40- to 100-watt bulb. Place the mouse into restrainer device with the tail first, use the movable white ring to restrain mouse by applying gentle pressure while ensuring that the nose is placed into the center of the ring such that the mouse can breathe. Locate the lateral tail vein—the veins can be seen on either side of the tail. Inject at a location at about one-third from the tail tip, holding the tail steady with one hand and supporting the injecting hand on the holding hand. When the needle is placed correctly, the vessel is visually flushed when the liver cell suspension (or sterile saline for practice) is administered. There should be no resistance to the fluid when injected; inject slowly. The formation of a bleb at the site indicates that the needle is not placed in the vein. Remove the needle and try a second site on the same vessel in a more proximal location on the tail. Needles differ in their quality and sharpness and also become blunt upon contact with skin/hair, so it is best to change the needle after a failed attempt. When finished, hold slight pressure on the site of injection to stop break bleeding.

7. Gpi1 isozyme analysis can be performed subsequent to FACS analysis, for example, after lineage phenotyping. Depending on the initial number of cells that were stained, it may be required to pool several FACS samples from the same recipient. Collect cells by centrifugation and dilute pellet approximately 1:10 in water; perform electrophoresis as described.

References

1. Feil R. (2007) Conditional somatic mutagenesis in the mouse using site-specific recombinases. *Handb Exp Pharmacol* **178**, 3–28.
2. Matsumoto N., Kubo A., Liu H., Akita K., Laub F., Ramirez F., Keller G., Friedman S. L. (2006) Developmental regulation of yolk sac hematopoiesis by Kruppel-like factor 6. *Blood* **107**, 1357–1365.

3. Tam P. P., Rossant J. (2003) Mouse embryonic chimeras: tools for studying mammalian development. *Development* **130**, 6155–6163.

4. Chen J., Lansford R., Stewart V., Young F., Alt F. W. (1993) RAG-2-deficient blastocyst complementation: an assay of gene function in lymphocyte development. *Proc Natl Acad Sci USA* **90**, 4528–4532.

5. Wang L. C., Swat W., Fujiwara Y., Davidson L., Visvader J., Kuo F., Alt F. W., Gilliland D. G., Golub T. R., Orkin S. H. (1998) The TEL/ETV6 gene is required specifically for hematopoiesis in the bone marrow. *Genes Dev* **12**, 2392–2402.

6. Ernst P., Fisher J. K., Avery W., Wade S., Foy D., Korsmeyer S. J. (2004) Definitive hematopoiesis requires the mixed-lineage leukemia gene. *Dev Cell* **6**, 437–443.

7. Tsang A. P., Fujiwara Y., Hom D. B., Orkin S. H. (1998) Failure of megakaryopoiesis and arrested erythropoiesis in mice lacking the GATA-1 transcriptional cofactor FOG. *Genes Dev* **12**, 1176–1188.

8. Eckardt S., Leu N. A., Bradley H. L., Kato H., Bunting K. D., Mclaughlin K. J. (2007) Hematopoietic reconstitution with androgenetic and gynogenetic stem cells. *Genes Dev* **21**, 409–419.

9. Nagy A., Gertsenstein M., Vintersten K., Behringer R. (2003) *Manipulating the Mouse Embryo*, 3rd ed. Cold Spring Harbor Laboratory Press, Cold Spring Harbor, NY.

10. Abbondanzo S. J., Gadi I., Stewart C. L. (1993) Derivation of embryonic stem cell lines, in *Methods Enzymol 225: Guide to Techniques in Mouse Development* (Wassarman, P.M, DePamphilis, M.L., eds.), Academic Press, San Diego, CA, pp. 803–823.

11. Okabe M., Ikawa M., Kominami K., Nakanishi T., Nishimune Y. (1997) 'Green mice' as a source of ubiquitous green cells. *FEBS Lett* **407**, 313–319.

12. Kawakami N., Sakane N., Nishizawa F., Iwao M., Fukada S. I., Tsujikawa K., Kohama Y., Ikawa M., Okabe M., Yamamoto H. (1999) Green fluorescent protein-transgenic mice: immune functions and their application to studies of lymphocyte development. *Immunol Lett* **70**, 165–171.

13. Spangrude G. J., Cho S., Guedelhoefer O., Vanwoerkom R. C., Fleming W. H. (2006) Mouse models of hematopoietic engraftment: limitations of transgenic green fluorescent protein strains and a high-performance liquid chromatography approach to analysis of erythroid chimerism. *Stem Cells* **24**, 2045–2051.

14. Wang Z., Jaenisch R. (2004) At most three ES cells contribute to the somatic lineages of chimeric mice and of mice produced by ES-tetraploid complementation. *Dev Biol* **275**, 192–201.

15. Yagi T., Tokunaga T., Furuta Y., Nada S., Yoshida M., Tsukada T., Saga Y., Takeda N., Ikawa Y., Aizawa S. (1993) A novel ES cell line, TT2, with high germline-differentiating potency. *Anal Biochem* **214**, 70–76.

16. Stewart C. L. (1993) Production of chimeras between embryonic stem cells and embryos, in *Methods Enzymol 225: Guide to Techniques in Mouse Development* (Wassarman, P.M, DePamphilis, M.L., eds.), Academic Press, San Diego, CA, pp. 823–855.

17. Mann J. R. (1993) Surgical Techniques in Production of Transgenic Mice, in *Methods Enzymol 225*: *Guide to Techniques in Mouse Development* (Wassarman, P.M, DePamphilis, M.L., eds.), Academic Press, San Diego, CA, pp. 782–793.

18. Nagy A., Rossant J. (1993) Production of completely ES cell derived fetuses, in *Gene Targeting*, 1 ed. IRL (Joyner A.L., ed.), Oxford, pp. 147–178.

15

Immunodeficient Mouse Models to Study Human Stem Cell-Mediated Tissue Repair

Ping Zhou, Sarah Hohm, Ben Capoccia, Louisa Wirthlin, David Hess, Dan Link, and Jan Nolta

Summary

Hematopoietic stem cell transplantation has traditionally been used to reconstitute blood cell lineages that had formed abnormally because of genetic mutations, or that had been eradicated to treat a disease such as leukemia. However, in recent years, much attention has been paid to the new concept of "stem cell plasticity," and the hope that stem cells could be used to repair damaged tissues generated immense excitement. The field is now in a more realistic and critical period of intense investigation and the concept of cell fusion to explain some of the observed effects has been shown after specific types of damage in liver and muscle, both organs that contain a high number of multinucleate cells. The field is still an extremely exciting one, and many questions remain to be answered before stem cell therapy for tissue repair can be used effectively in the clinic. Immune deficient mouse models of tissue damage provide a system in which human stem cell migration to sites of damage and subsequent contribution to repair can be carefully evaluated. This chapter gives detailed instructions for methods to study human stem cell contribution to damaged liver and to promote repair of damaged vasculature in immune deficient mouse models.

Key Words: Human stem cells; immune deficient mice; tissue repair; revascularization; liver.

1. Background
1.1. Stem Cells and Tissue Repair

Hematopoietic stem cells from adult sources such as bone marrow (BM), mobilized peripheral blood (MPB), adipose tissue, and umbilical cord blood

From: *Methods in Molecular Biology, vol. 430: Hematopoietic Stem Cell Protocols*
Edited by: K. D. Bunting © Humana Press, Totowa, NJ

(UCB) have been shown to promote tissue repair. Different populations of stem cells have been described to contribute to the regeneration of muscle, liver, and heart, among other tissues, although the mechanisms by which they accomplish this are still not well understood. Stem cells are known, however, to secrete various cytokines and growth factors that have both paracrine and autocrine activities. One currently accepted theory of tissue repair and regeneration by adult stem cells is that the mechanism of action is based on the innate functions of the stem cells. It appears that the injected stem cells may home to the injured area and release trophic factors that hasten endogenous repair. These secreted bioactive factors suppress the local immune system, inhibit fibrosis (scar formation) and apoptosis, enhance angiogenesis, and stimulate recruitment, retention, mitosis, and differentiation of tissue-residing stem cells. These effects, which are referred to as *trophic effects*, are distinct from the direct differentiation of stem cells into repair tissue, and have been shown to be significant in cardiac repair *(1,2)*.

Although it is becoming apparent that stem cells can hasten repair of endogenous tissues, they are usually not found in the tissue in large quantities after the repair. In a randomized, large animal model study using cynomolgus monkeys, acute myocardial infarction was generated by ligating the left anterior descending artery, after which autologous CD34$^+$ cells were transplanted to the peri-ischemic zone *(3)*. The group receiving cells demonstrated improved regional blood flow and cardiac function, as compared with the saline-treated group 2 weeks after transplantation. However, very few donor marker-positive cells were found incorporated into the vascular structure, or in the repaired tissue. These results suggest that the cardiac improvement was not the result of generation of transplanted cell-derived endothelial cells or cardiomyocytes, and furthermore the possibility that angiogenic cytokines secreted from transplanted cells potentiate angiogenic activity of endogenous cells *(3)*. In summary, whereas the detailed mechanisms have yet to be established, current data indicate that marrow-derived stem cells do not become a part of the damaged heart, but persist temporarily in the injured tissue to exert significant trophic effects on cardiac repair.

After working with the cardiac infarction system using human stem cell transplantation into immune deficient mice for several years, our group also did not find significant contribution of the adult stem cells to the cardiac tissue at the site of injury 2 or more weeks after injection. Instead, they appeared to promote more rapid angiogenesis to the damaged ischemic tissue, which could be a very worthwhile repair. However, to examine the finer details of this phenomenon, we find the hind limb ischemia model to be far more reproducible, and to have higher throughput; thus, it is discussed in detail below, with protocols.

1.2. Stem Cell Selection

Methods to identify the most primitive hematopoietic stem cells are constantly sought. The CD34 protein is frequently used as a marker for positive selection of human hematopoietic stem and progenitor cells. Goodell *et al.* characterized the Hoechst dye-excluding side population (SP) cells in mouse that lacked CD34 expression and had reconstitution capacity *(4,5)*. Bhatia et al. *(6)* demonstrated a low level of engraftment activity in human CD34⁻cells. We showed that highly purified human CD34⁺ cells generated CD34⁻ cells after 1 year in vivo, which retained the capacity to regenerate CD34⁺ cells upon secondary transplantation into immune deficient mice, demonstrating that expression of CD34 is reversible on human multilineage engrafting stem/progenitor cells *(7,8)*. Phenotype was found to vary with activation state *(7,9–11)*. This calls for different purification strategies, based on conserved stem cell function as well as phenotype. The use of metabolic markers such as rhodamine and Hoechst 33342 dye efflux separates cells based on high expression of membrane pumps encoded by the multiple drug resistance (MDR) genes; a characteristic believed to be associated with very immature stem/progenitor cells *(4,5,12,13)*. Another very interesting metabolic marker indicative of early, immature cells is the enzyme aldehyde dehydrogenase (ALDH), shown by our group and others to correlate with high stem cell activity in vivo *(12,14,15)*.

We characterized a population of primitive cells isolated from UCB by lineage depletion (Lin⁻) followed by selection of cells with high ALDH activity (ALDH^hi^Lin⁻) *(15)*. Conventionally, Lin⁻ cells are sorted based on CD34 expression to purify hematopoietic NOD/SCID-repopulating cells. Rather than isolating cells based on a hematopoietic restricted marker, we have purified based on an enzymatic function known to confer stem cell resistance to cytotoxic drugs such as cyclophosphamide *(12,16–18)*. The advantage of this strategy is the ability to obtain a purified stem cell population from UCB that is not restricted to the hematopoietic lineage.

ALDH^hi^Lin⁻ cells co-expressed the primitive hematopoietic stem cell (HSC) markers CD34 and CD133 at high levels. Nearly all ALDH^hi^Lin⁻ cells expressed CD31. Human hematopoietic progenitor function was enriched in the ALDH^hi^Lin⁻ population whereas ALDH^lo^Lin⁻ cells had little in vitro colony-forming ability. ALDH^hi^Lin⁻ cells were characterized for hematopoietic repopulation in vivo using immune deficient, sublethally irradiated mice *(15)*. Human cells were detected in the BM, spleen, or peripheral blood in NOD/SCID β2M null mice transplanted with as few as 5×10^2 ALDH^hi^Lin⁻ cells, whereas 10^7 ALDH^lo^Lin⁻ cells demonstrated no repopulation. Equivalent doses of ALDH^hi^Lin⁻ demonstrated 10-fold less repopulating capacity in the NOD/SCID model, as compared with the more permissive NOD/SCID β2M null mouse *(15)*, indicating that the NOD/SCID β2M null mouse is a superior model to study the engraftment of ALDH-expressing cells. These data indicate that human UCB ALDH^{hi}Lin$^-$ cells are highly enriched for clonogenic progenitors and primitive hematopoietic repopulating cells. They are currently being assessed in the tissue repair models discussed in this chapter. The cells appear to have a unique and robust capacity for distribution throughout multiple tissues in the mice, so we recommend them for tissue repair studies.

1.3. Immune Deficient Mouse Strains for Transplantation of Human Stem Cells to Assess Tissue Repair

NOD/SCID β2M null mice exhibit enhanced immunodeficiency compared with the NOD/SCID strain because of reduced NK cell function. Consequently, these mice are very permissive for xenograft acceptance and provide the highest and most reproducible engraftment with human cells to date *(19)*. Both committed progenitors and primitive cells engraft NOD/SCID β2M null mice *(20,21)*. We recommend them for tissue repair studies, because of their more robust engraftment capacity, as compared with the standard NOD/SCID strain *(19)*.

NOD/SCID/MPSVII mice provide a murine xenotransplantation model to visualize transplanted cells without the use of cell-surface markers or FISH. Mucopolysaccharidosis Type VII (MPSVII) is a lysosomal storage disease caused by a deficiency in B-glucuronidase (GUSB) activity. The MPSVII mutation was backcrossed onto the NOD/SCID strain, and human CD34$^+$ cells engraft in the resultant NOD/SCID/MPSVII strain to levels equivalent to the NOD/SCID parent *(22)*. Tissue slides prepared from this strain allow rapid visualization of human cells that carry normal levels of GUSB, against the background mouse tissues that are null for the enzyme (*see* **Fig. 1**). Following the enzymatic reaction, slides can be counterstained with antibodies to a tissue-specific protein marker. The enzymatic stain is quite specific, and although the released enzyme can be taken up by neighboring cells, it is in levels too low to be detectable by the histochemical analysis *(22)*. Thus, the individual transplanted human cells stand out vividly against the background, GUSB null murine tissues. Most notably, human cells expressing GUSB can be detected without reliance on the expression of cell-surface markers. The use of the GUSB enzyme staining is proving to be more fast, efficient, and definitive than FISH, and makes assessment of stem cell homing and repair in tissues much more rapid than had previous techniques *(22,23)*.

2. Materials
1. Ficoll (Amersham Biosciences, Piscataway, NJ, USA)
2. Negative selection human progenitor enrichment cocktail (StemCell Technologies, Vancouver, BC, Canada)

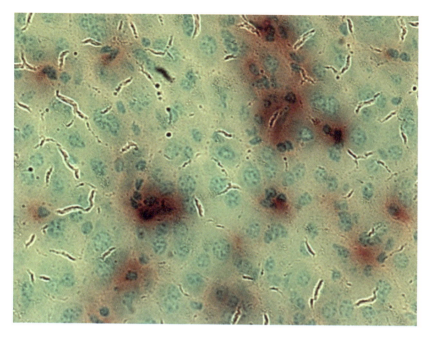

Fig. 1. Human cells in NOD/SCID/MPSVII mouse liver, detected by enzymatic reaction for B-glucuronidase.

3. Anti-human CD41 tetramer antibody (StemCell Technologies)
4. Magnetic colloid (StemCell Technologies)
5. Negative selection column and pump tubing (StemCell Technologies)
6. PBS (HyClone)
7. ALDEFLUOR assay kit (StemCell Technologies)
8. 10× human red blood cell (HRBC) lysis buffer: 400 ml dI H_2O, 40.1 g ammonium chloride (NH_4Cl), 4.2 g sodium bicarbonate ($NaHCO_3$), 10 ml 0.5 M ethylenediaminetetraacetic acid (EDTA), Scale to 500 ml and keep at 4°C, Dilute to 1× with H_2O before using
9. 50-ml tubes (Corning, Corning, NY, USA)
10. 10- and 35-ml pipets (Falcon, Franklin Lakes, NJ, USA)
11. CellTrics 50 μM filter (Partec, Münster, Germany)
12. Shaker
13. Centrifuge (Beckman, Fullerton, CA, USA)
14. Peristatic pump (Rainin, Oakland, CA, USA)
15. StemSep (StemCell Technologies)
16. CCl_4 (Sigma, St. Louis, MO, USA)
17. Corn oil (Sigma)

18. 1/2-CC Syringes with 29-G needles (Terumo, Tokyo, Japan)
19. Tailveiner (Braintree Scientific, Braintree, MA, USA)
20. 6-0 18" Taper C-1 Silk Black Braided 3/8 13 mm by Ethicon
21. Nexaband liquid topical tissue adhesive list 5295 by Abbott Laboratories, Abbott Park, IL, USA
22. Ethanol—70%
23. Eye ointment (Neosporin)
24. Povidone iodine scrub solution Antiseptic germicide—7.5% topical solution by Triadine
25. Ketamine (100 mg/ml)
26. Xylazine (100 mg/ml)
27. Sterile H_2O
28. Electric shaver
29. Cotton Swabs
30. Kim-wipes
31. Needle
32. Tape
33. Dissection microscope
34. Scissors
35. Forceps

3. Methods

3.1. Isolation of Lineage-Depleted ALDH-Expressing Cells from Human Umbilical Cord Blood

1. Add 15 ml Ficoll to a 50-ml tube. Prepare four tubes for each cord. Spray cord blood bags (*see* **Note 1**) with 70% ethanol and water solution (*see* **Note 2**). Cut open a bag at the top.
2. Bring volume to 150 ml with PBS in the bag. Load 30 ml cord blood slowly on the top of Ficoll in the 50-ml tube using a 35-ml pipet at a 45° angle against the wall of the tube (4 tubes for 1 bag). Do not mix the layers.
3. Centrifuge at 320 × *g* with brake off for 30 min (*see* **Note 3**).
4. The mononuclear cells (MNC) are at the middle of the tube, in a layer termed the "buffy coat," which will appear opaque. Remove and discard serum form the top of the tube, then Transfer MNC layer to a new 50-ml tube with a 10-ml pipet. Combine cells from two tubes and bring volume up to 50 ml with PBS (*see* **Note 4**).
5. Centrifuge at 320 × *g* for 5 min with brake (*see* **Note 3**).
6. Discard supernatant (*see* **Note 5**). Resuspend cell pellet in 5 ml PBS and combine cells from the same cord. Bring volume to 50 ml with 1× HRBC buffer and incubate for 7 min to lyse red blood cells.
7. Centrifuge at 320 × *g* for 5 min.

8. Discard supernatant (*see* **Note 5**). Resuspend cell pellet in appropriate PBS and combine cells from all cords (*see* **Note 6**). Count the number of cells and adjust cell density to 1×10^8/ml with cold PBS.

9. Add 100 µl of negative selection human progenitor enrichment cocktail and 10 µl of anti-human CD41 tetramer antibody for every 1×10^8 cells.

10. Incubate at 4°C for 30 min on a shaker with slow shaking.

11. Add 60 µl of magnetic colloid per milliliter cell suspension to the tube and incubate at 4°C for 30 min on a shaker with slow shaking (*see* **Note 7**).

12. Set up negative selection column on StemSep and connect it to a peristaltic pump with pump tubes. Separate non-binding cells according to manufacturer's specifications.

13. Centrifuge the collected cell suspension at $320 \times g$ for 5 min.

14. Discard supernatant. These cells are Lin⁻ cells and can be used for transplantation.

15. For ALDH fluorescence staining, resuspend cells from all eight cords in 1 ml of ALDEFLUOR Assay Buffer (Aldagen, Durham, NC, USA). Add 7 µl of ALDEFLUOR staining reagent. Incubate at 37°C in an incubator for 30 min.

16. Add 4 ml of PBS. Centrifuge $320 \times g$ for 5 min.

17. Discard supernatant. Resuspend cells in 1 ml ALDEFLUOR assay buffer.

18. Filter through CellTrics 50 µM filter to a fluorescence-activated cell sorting (FACS) tube. Put the tube on ice. Add 500 µl PBS to collection tubes (*see* **Note 8**).

19. Sort cells by using FACS, using the collection gates shown in **Fig. 2**.

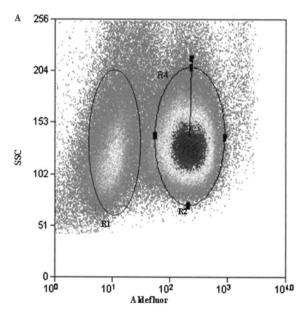

Fig. 2. Aldehyde dehydrogenase (ALDH) sort gates, to define the ALDH lo (R1) and ALDH hi (R2) populations.

20. Centrifuge sorted cells at 320 × *g* for 5 min. Remove fluid, leaving one drop, and tap tube with finger to dislodge pellet. This reduces clumping.
21. Proceed to the following sections on liver damage or hind limb ischemia for details on cell resuspension for injection into immune deficient mice.

3.2. Transplantation of Human Stem Cells to Examine Mechanisms of Liver Repair Following Damage by CCl4

1. To induce liver damage, dilute an appropriate amount of CCl_4 in corn oil and inject 100 µl to a mouse intraperitoneally 1–2 days before transplantation. For guidelines on the amount of CCl_4 to inject per mouse, please see **ref. 24**. The amount can vary by strain. As a general guideline, in our studies, NOD/SCID mice (Jackson labs, Bar Harbor, ME, USA) are treated with a single intraperitoneal dose of 0.4 ml/kg CCl_4, and this dose is associated with a 50% mortality by 2 weeks. The NOD/SCID/MPSVII strain requires one-tenth that amount to achieve the same degree of liver damage and mortality. As with radiation, the amount of CCl_4 to be used will need to be titered for each strain and colony.
2. Count cells and adjust cell density with PBS according to the number of cells needed for transplantation. Cell number can be up to 10 million. The liquid volume injected through mouse tail vein is between 50 to 500 µl.
3. Pull a mouse into a restraint device by dragging the mouse tail through. Inject appropriate volume of cell suspension into tail vein using a syringe with 29-G needle.
4. Cohorts: Each study of liver damage and recovery should contain the experimental group, plus control mice transplanted with BM cells *without* prior CCl_4-mediated liver injury or other conditioning. As another important control group, a cohort of mice should receive CCl_4 but no BM cells. This control group is crucial not only to study the level of recovery without the marrow infusions but to provide injured tissue that lacks donor cells, for histological analyses. It has become apparent in the field of plasticity that antibodies and probes will bind non-specifically to injured tissues, and stringent measures must be taken to rule out this degree of background staining, which could be misinterpreted as positive signals.
5. At the end of the designated experimental period, mice will be euthanized, and tissues prepared and assessed as described (*24*).

3.3. Hind limb Ischemia Model to Assess Ability of Stem Cells to Contribute to Revascularization

These experiments will characterize the population of human cells with the best angiogenic potential, in a hind limb ischemic xenograft model after femoral artery ligation.

3.3.1. Preparation before Mouse Surgery

1. Place a heating pad beneath the surgical drapes to maintain euthermia in the mice.
2. Anesthetize mice with ketamine/xylazine (80/15 mg per kg) IP.

    a) 1-ml ketamine (100 mg/ml)
    b) 0.5 ml xylazine (100 mg/ml)
    c) 4.6 ml sterile H_2O

    Before continuing, use the footpad test to ensure mouse is fully anesthetized.

3. Using a sterilized electric shaver, shave the anterior portion of both limbs up to the stomach, from the hip to the ankle, and prep the area with betadine scrub, then rinse.
4. To prevent the eyes from drying out, apply eye ointment (Vetromycin).
5. Using a cotton swab, apply povidone iodine to entire right limb, up to the stomach.
6. Position mouse with limbs extended underneath dissection microscope. Tape feet onto a rotational surface.

3.3.2. Surgery

1. Directly above the knee, make a small, lateral incision using scissors. The incision should be no larger than 1–2 mm.
2. Using this notch, make straight incision through skin to midline, over the location where the femoral artery is located. A longitudinal incision will be made, extending inferiorly from the inguinal ligament to a point just proximal to the patella.
3. Avoiding blood vessels, remove fatty tissue covering the artery and four main veins. Use forceps and scissors, but be careful not to cut veins entrapped within fatty tissue.
4. With the blunt end of the needle, carefully pull the nerve attached to the femoral artery off to the left side. This prevents the nerve from being sutured in with the vein.
5. There are four main veins leading to the artery that are used for placement of the sutures (6-0 18" Taper C-1 Silk Black Braided 3/8 13 mm by Ethicon, Cornelia, GA, USA). Placing a suture between each of these four veins and the main artery will close off any incoming blood supply. The left femoral artery as well as all the main branches will be dissected free and ligated with 4-O silk, preventing blood flow distal to the bifurcation of the internal iliac artery. The superior and inferior sutures will be tied directly onto the artery in front of the protruding vein. The remaining sutures will be tied onto the veins themselves. After each suture, wipe thread with ethanol to prevent infection.
6. Excise the artery between the four sutures.
7. The femoral artery will be completely excised from its proximal origin as a branch of the external iliac to the point distally where it bifurcates into the saphenous and popliteal arteries. Excision of the femoral artery is required to prevent rapid revascularization secondary to collateral vessels bridging the ligature.
8. Glue the incision (Nexaband by Abbott Laboratories).
9. Mice should be observed continuously until they recover fully from the anesthesia. In addition, mice should be examined on a daily basis for signs of wound infection or distress.

It is expected that the majority of mice will recover near-normal blood flow to the ischemic limb within 3 weeks. Approximately 10% of mice will develop toe necrosis during the first post-operative week but then heal rapidly within the following weeks without specific treatment. If toe necrosis is observed, the mice will be examined daily for evidence of extension of the necrosis to the foot and, if present, the mouse euthanized. It also is possible that mice may develop an infection at the surgical site. Careful aseptic technique should reduce this possibility. However, if wound infection is observed, the mice must be euthanized.

3.4. Stem Cell Injection

3.4.1. Flowchart: Hind limb Ischemia Model

1. Perform femoral artery ligation in 15 NOD/SCID/B2M or NOD/SCID/MPSVII mice to induce right hind limb ischemia. Laser Doppler Imaging (LDPI) will quantitate the reduction in hind limb blood flow immediately following surgery. Recovery from ischemia will be quantified by the ratio of blood flow in the ischemic versus the contralateral (unligated) limb.
2. Twenty-four hours post-ligation, tail vein inject mice with PBS ($n=5$), 10^5 purified ALDH hi lin$^-$ cells ($n=5$), or 10^5 cells from other purified HSC or Mesenchymal Stem Cells (MSC) populations ($n=5$). Mice will be transplanted without prior irradiation.
3. Revascularization is assessed by LDPI, a non-invasive method to assess blood flow. LDPI is performed with an LDR-IRS laser Doppler imager system (Moor Instruments, Inc., Glasgow, UK). Mice are anesthetized and placed prone on a black felt surface. LDPI, performed as recommended by the manufacturer of the device, takes approximately 5 min. Mice should be observed continuously until they recover fully from the anesthesia and should then be returned to a clean cage. LDPI should be performed twice weekly to quantitate recovery of blood flow to ischemic limb over 4 weeks in transplanted mice versus mock-injected controls.
4. At 4 weeks, vascularization of ischemic versus contralateral muscle can be quantitated by capillary density using H+E with CD31 immunostaining (capillary density/mm^2).
5. Human cell recruitment to ischemic muscle can be quantitated on slides by immunohistochemical analysis or by FACS analysis following collagenase-free mechanical digestion and flow cytometry. The GUSB enzyme reaction for NOD/SCID/MPSVII mice will be used, in conjunction with HLA A,B,C (human nucleated cells), CD45 (human hematopoietic), CD14 (human monocytes) and PECAM-1 (human endothelial/macrophage), and VE-cadherin or CD144 (mature endothelial) expression. The frequency of human cells in ischemic muscle can be compared with human reconstitution in BM and spleen.
6. Human cells in recovering muscle can also be visualized by immunohistochemistry of frozen muscle tissue for CD45, CD31, CD144, and human von Willebrand factor (vWF, mature endothelial cells).

4. Notes

1. This protocol can also be used to isolate human stem/progenitor cells from BM or MPB samples. The volumes should be scaled up appropriately, based on cell counts, as recommended by the Aldefluor manufacturer, Aldagen.
2. We commonly use a 70% ethanol and H_2O solution to sterilize blood bags and tools. It is important to dry off the bag or tool with a clean Kimwipe™ and then let the remainder of the alcohol evaporate before contact with cells, however, because ethanol can have negative effects on cell survival.
3. Centrifuge brakes will disrupt the Ficoll layers, dispersing the buffy coat that contains the white blood cells. For optimal results, let the spin come to a natural stop, without braking, during the Ficoll spin. The brake can be on, to save time, for all subsequent centrifugations.
4. To ensure that each cord blood sample is not contaminated with bacteria or yeast, which can happen during normal delivery, plate 10 μl from each Ficolled sample into a 60-mm dish in any cell culture medium and place in the incubator while you start manipulations. By the time you are ready to sort or to transplant mice, outgrowth of contaminants will already be visible. Do not transplant contaminated samples, because the immune deficient mice will have no way to eradicate the pathogen.
5. Do not leave the cells pelleted for long after centrifuging, but remove liquid and tap the bottom of the tube with your finger. This reduces cell clumping and loss because of extruded DNA.
6. We have not observed adverse events or cell loss, from mixing cord samples derived from different donors before selection of progenitors. Because samples are maintained at 4° after this step, until sorting, the opportunities for immune reactions form one donor to the next are minimized. The lineage depletion step will eliminate cells of the immune system, and the ALDH-sorting procedure further depletes them. Pooling several cord blood samples reduces donor-to-donor variability.
7. This rotation step should optimally be done in a cold room or a deli fridge with electrical outlet, to keep cells cold. If this is not available, keep the tube on wet ice and tap with finger, rotating tube, each 2–3 min during the incubation step.
8. If the cells will not be taken for further processing immediately, add 2% fetal bovine serum to the collection tubes. This will increase viability.

References

1. Caplan, A. I. & Dennis, J. E. (2006) Mesenchymal stem cells as trophic mediators. *J Cell Biochem* **98**, 1076–1084.
2. Schatteman, G. C. (2004) Non-classical mechanisms of heart repair. *Mol Cell Biochem* **264**, 103–117.
3. Yoshioka, T., Ageyama, N., Shibata, H., Yasu, T., Misawa, Y., Takeuchi, K., Matsui, K., Yamamoto, K., Terao, K., Shimada, K., Ikeda, U., Ozawa, K. & Hanazono, Y. (2005) Repair of infarcted myocardium mediated by transplanted

bone marrow-derived CD34+ stem cells in a nonhuman primate model. *Stem Cells* **23**, 355–364.

4. Goodell, M. A., Brose, K., Paradis, G., Conner, A. S. & Mulligan, R. C. (1996) Isolation and functional properties of murine hematopoietic stem cells that are replicating in vivo. *J Exp Med* **183**, 1797–1806.

5. Goodell, M. A., Rosenzweig, M., Kim, H., Marks, D. F., DeMaria, M., Paradis, G., Grupp, S. A., Sieff, C. A., Mulligan, R. C. & Johnson, R. P. (1997) Dye efflux studies suggest that hematopoietic stem cells expressing low or undetectable levels of CD34 antigen exist in multiple species. *Nat Med* **3**, 1337–1345.

6. Bhatia, M., Bonnet, D., Murdoch, B., Gan, O. I. & Dick, J. E. (1998) A newly discovered class of human hematopoietic cells with SCID- repopulating activity [see comments]. *Nat Med* **4**, 1038–1045.

7. Dao, M. A., Arevalo, J. & Nolta, J. A. (2003) Reversibility of CD34 expression on human hematopoietic stem cells that retain the capacity for secondary reconstitution. *Blood* **101**, 112–118.

8. Dao, M. A. & Nolta, J. A. (2000) CD34: to select or not to select? That is the question. *Leukemia* **14**, 773–776.

9. Hess, D. A., Karanu, F. N., Levac, K., Gallacher, L. & Bhatia, M. (2003) Coculture and transplant of purified CD34(+)Lin(-) and CD34(-)Lin(-) cells reveals functional interaction between repopulating hematopoietic stem cells. *Leukemia* **17**, 1613–1625.

10. Sato, T., Laver, J. H. & Ogawa, M. (1999) Reversible expression of CD34 by murine hematopoietic stem cells. *Blood* **94**, 2548–2554.

11. Zanjani, E. D., Almeida-Porada, G., Livingston, A. G., Porada, C. D. & Ogawa, M. (1999) Engraftment and multilineage expression of human bone marrow CD34-cells in vivo. *Ann N Y Acad Sci* **872**, 220–231; discussion 231–232.

12. Storms, R. W., Goodell, M. A., Fisher, A., Mulligan, R. C. & Smith, C. (2000) Hoechst dye efflux reveals a novel CD7(+)CD34(-) lymphoid progenitor in human umbilical cord blood. *Blood* **96**, 2125–2133.

13. Cai, J., Weiss, M. L. & Rao, M. S. (2004) In search of "stemness". *Exp Hematol* **32**, 585–598.

14. Fallon, P., Gentry, T., Balber, A. E., Boulware, D., Janssen, W. E., Smilee, R., Storms, R. W. & Smith, C. (2003) Mobilized peripheral blood SSCloALDHbr cells have the phenotypic and functional properties of primitive haematopoietic cells and their number correlates with engraftment following autologous transplantation. *Br J Haematol* **122**, 99–108.

15. Hess, D. A., Meyerrose, T. E., Wirthlin, L., Craft, T. P., Herrbrich, P. E., Creer, M. H. & Nolta, J. A. (2004) Functional characterization of highly purified human hematopoietic repopulating cells isolated according to aldehyde dehydrogenase activity. *Blood* **104**, 1648–1655.

16. Takebe, N., Zhao, S. C., Adhikari, D., Mineishi, S., Sadelain, M., Hilton, J., Colvin, M., Banerjee, D. & Bertino, J. R. (2001) Generation of dual resistance to 4-hydroperoxycyclophosphamide and methotrexate by retroviral transfer of the human aldehyde dehydrogenase class 1 gene and a mutated dihydrofolate reductase gene. *Mol Ther* **3**, 88–96.

17. Jones, R. J., Barber, J. P., Vala, M. S., Collector, M. I., Kaufmann, S. H., Ludeman, S. M., Colvin, O. M. & Hilton, J. (1995) Assessment of aldehyde dehydrogenase in viable cells. *Blood* **85**, 2742–2746.

18. Jones, R. J., Collector, M. I., Barber, J. P., Vala, M. S., Fackler, M. J., May, W. S., Griffin, C. A., Hawkins, A. L., Zehnbauer, B. A., Hilton, J., Colvin, O. M. & Sharkis, S. J. (1996) Characterization of mouse lymphohematopoietic stem cells lacking spleen colony-forming activity. *Blood* **88**, 487–491.

19. Meyerrose, T. E., Herrbrich, P., Hess, D. A. & Nolta, J. A. (2003) Immune-deficient mouse models for analysis of human stem cells. *Biotechniques* **35**, 1262–1272.

20. Glimm, H., Eisterer, W., Lee, K., Cashman, J., Holyoake, T. L., Nicolini, F., Shultz, L. D., von Kalle, C. & Eaves, C. J. (2001) Previously undetected human hematopoietic cell populations with short-term repopulating activity selectively engraft NOD/SCID-beta2 microglobulin-null mice. *J Clin Invest* **107**, 199–206.

21. Christianson, S. W., Greiner, D. L., Hesselton, R. A., Leif, J. H., Wagar, E. J., Schweitzer, I. B., Rajan, T. V., Gott, B., Roopenian, D. C. & Shultz, L. D. (1997) Enhanced human CD4+ T cell engraftment in beta2-microglobulin-deficient NOD-scid mice. *J Immunol* **158**, 3578–3586.

22. Hofling, A. A., Vogler, C., Creer, M. H. & Sands, M. S. (2003) Engraftment of human CD34+ cells leads to widespread distribution of donor-derived cells and correction of tissue pathology in a novel murine xenotransplantation model of lysosomal storage disease. *Blood* **101**, 2054–2063.

23. Meyerrose, T. E., De Ugarte, D. A., Hofling, A. A., Herrbrich, P. E., Cordonnier, T. D., Shultz, L. D., Eagon, J. C., Wirthlin, L., Sands, M. S., Hedrick, M. A. & Nolta, J. A. (2006) In vivo distribution of human adipose-derived MSC. *Stem Cells* **25**, 220–227.

24. Wang, X., Ge, S., McNamara, G., Hao, Q. L., Crooks, G. M. & Nolta, J. A. (2003) Albumin expressing hepatocyte-like cells develop in the livers of immune-deficient mice transmitted with highly purified human hematopoietic stem cells. *Blood*, **101**, 4201–4208.

V

GENETIC MODIFICATION OF HSCs AND IMAGING ENGRAFTMENT

16

Retroviral Transduction of Murine Hematopoietic Stem Cells

Peter Haviernik, Yi Zhang, and Kevin D. Bunting

Summary

Hematopoietic stem cells (HSC) are inherently rare cell types that cannot be obtained in sufficient amounts for classical biochemical characterization. To facilitate functional studies of murine HSC and hematopoietic development, the technique of retroviral-mediated gene transfer provides a useful tool. The generation of high titer retroviral vectors permits transduction of stem cells with a variety of genes and leads to long-term marking in the blood of recipient mice. Optimized promoter/enhancers facilitate high-level transgene expression in mice transplanted with transduced bone marrow (BM) cells. The co-expression of reporter genes along with a gene of interest greatly facilitates tracking donor engraftment of transduced hematopoietic progeny following stem cell transplantation. This methodology can be used to reconstitute defective function in a mutant background or to study protein function during hematopoiesis by overexpression. Despite limitations such as integration site variegation and copy number-dependent effects, this approach is rapid and efficient compared with transgenic mouse technology. In this chapter, we review this broadly applicable technique for achieving high-level murine BM stem cell transduction. We also describe methods for transplantation and subsequent analysis of transplanted mice as a bona fide assay for the stem cell transduction efficiency.

Key Words: Retroviral vector; MSCV vector; producers; GFP marking; HSC transduction; gene transfer; engraftment; flow cytometry.

1. Introduction

The murine hematopoietic system is one of the best-studied mammalian models with very well-described development of all lineages. Hematopoietic stem cell (HSC) transplant into lethally irradiated hosts replenishes the entire blood system. In this regard, manipulation of HSC residing in the bone marrow

From: *Methods in Molecular Biology, vol. 430: Hematopoietic Stem Cell Protocols*
Edited by: K. D. Bunting © Humana Press, Totowa, NJ

(BM) can be used to genetically modify the circulating cells of various lineages in the peripheral blood. Nevertheless, it is a very complex system comprising many regulatory genes acting at various levels of differentiation. Gene transfer techniques are great tools to study the function of a gene. The gene in question is usually delivered into target cells under control of a retroviral (unregulated) promoter allowing its overexpression. This may cause dysregulation of the natural balance resulting in partial or complete impairment of hematopoiesis. If a specifically mutated gene is delivered into target cells, its characterization of function can be informative. It is also possible to use these techniques in sequential analysis of signaling cascades. For example, if mutation blocks a signaling cascade and the delivery of a gene allows the signal to be restored, it complements the defect and may read out functionally.

Among the vehicles that have emerged for gene transfer, retroviruses (RV) became the most widely utilized despite their limitations and inherent risk *(1)*. They are naturally very efficient in transferring their own genetic material into host cells, and this has made them the most popular candidates. The majority of retroviral vectors are based on Moloney murine leukemia virus (Mo-MLV). Deciphering the RV genome as well as its life cycle allowed intervention and directed manipulation resulting in development of several generations of recombinant RV vectors. The RV life cycle consists of several steps, and it is quite complex. It includes binding of the virion to the receptor on the surface of the target cell, penetration inside the cell, and reverse transcription (RT) of the viral genome resulting in proviral dsDNA flanked with the long-terminal repeat (LTR) at both ends. This DNA is then translocated into the cell nucleus and integrated into chromosomal DNA where it becomes a source of new progeny virions through transcription, protein synthesis and processing, and assembly of new particles that bud off the cell surface and infect the other cells. To direct this process, the viral genome encodes just three genes – gag, pol, env and a few regulatory sequences. Only the first part of the cycle until the proviral DNA is integrated into the host's genome is essential for efficient delivery of a transgene into a target cell. This allows replacement of the part of the viral genome coding for these genes by a transgene-coding segment.

The authentic vehicle is a viral particle containing the viral genomic ssRNA comprising a gene of interest to be transduced into the host cell. The virion carries all tools necessary for its efficient incorporation into the cell genome including a primer tRNA for initiating RT and enzymes for catalyzing RT and subsequent integration. Once integrated into target cell chromosomes, the proviral DNA serves as a template for transgene expression allowing complementation of the defective natural allele of the gene or overexpression of a gene in a wild-type background. Unlike the natural gene structure, the transgene is composed of its cDNA under the control of viral or some other foreign promoter, which prevents natural regulation of transgene expression.

The source of transgenic viral particles is the producer cell line. It constitutively produces viral particles because it has the recombinant retroviral vector comprising the packaging, RT, and integration signals and the transgene integrated. In addition, producers contain other viral genes inserted at other parts of their genome, and their expression provides all components necessary for assembly of the functional infective virion *(2–4)*. A gene transfer protocol itself employs transduction through either co-culture or viral supernatant infection. The former utilizes culturing both producer and target cells in the same dish allowing simultaneous production of virus and infection. On the other hand, after completion of transduction, the target cells need to be separated from producers. This harvesting can cause a lower yield of transduced cells. The latter method utilizes a cell-free viral supernatant that eliminates the need for cell separation. However, there is a delay between viral production and infection that could cause diminished viability of virions resulting in a drop in viral titer and lower transduction efficiency. This problem may be overcome by concentrating the viral particles to increase viral titer *(5)*.

Although an RV vector comprises minimal sequences required for efficient packaging, this approach still brings certain restrictions to design of an effective RV vector. The size of the RV vector is limited by the packaging capacity of the virion. In addition, there are some safety issues regarding generation of replication competent retroviruses (RCR) as well as potential adverse effects of integrated proviral DNA causing dysregulated expression of neighboring genes and/or insertional mutagenesis *(6)*, although there is an effort to construct non-integrative vectors *(7)*.

Because efficiency of gene transfer varies, it is suitable to have a means to assess transduction efficiency in target cells. Even though RV particles are highly efficient vehicles for gene transfer, not all target cells become infected, resulting in mixed population of transduced and untransduced cells. A simple solution is to tag vector with a marker, which allows tracking the transduced cells easily based on a phenotypic trait (*see* **Table 1**). The concept of a the tracer gene being co-expressed along with a gene in question from a single vector was originally designed using two independent expression units (two different promoters). Because of independent expression or promoter interference, however, there was a little coordination in their expression. Promoter/enhancer regulation might be very different, and so, the expression level of tracer did not reflect that of the investigated gene. Bringing both genes under control of a single promoter helped to bypass this problem. The same level of transcription was achieved by a single mRNA encoding both genes transcribed from the same promoter. Protein synthesis is split between

Table 1
Selectable Marker Genes Useful for Retroviral Gene Transfer Studies

Gene	Detection/Selection*	References
Green fluorescent protein (GFP) (and other fluorescent markers)	Fluorescence	*(14–16)*
Neo (Hygro, Puro)	G418 (hygromycin B, puromycin)	*(17–20)*
Multidrug resistance 1 (MDR1)	Anthracyclines	*(21,22)*
Methylguanine DNA methyltransferase (MGMT)	BCNU, TMZ	*(23,24)*
Dihydrofolate reductase (DHFR^{L22Y})	MTX, TMTX	*(25–27)*
Glutathione S-transferase (GST)	Cyclophosphamide	*(28)*
Cytidine deaminase (CDD)	Ara-C	*(29–31)*
Aldehyde dehydrogenase (ALDH)	Cyclophosphamide, mafosfamide, ifosfamide	*(32,33)*

cap-driven translation of the first gene and Internal ribosomal entry sequence (IRES)-driven translation from the second cistron. Because there is a difference in efficiency of translation initiation between cap and IRES, the tracking gene is usually placed under IRES translation. Modification of IRES structure to accommodate some cloning features caused variable relatively inefficient IRES-mediated translation. Respecting the original EMCV IRES structure maintained a high level of the second cistron expression *(8)*. As an alternative to the rather long IRES sequence, a short FMDV-2A sequence allows the production of two separate polypeptides during co-translational processing *(9)*. Another approach was construction of a synthetic bidirectional promoter, which would mediate coordinate transcription of two genes in opposite directions *(10)*.

Finally, selectable markers eventually provide a means for enrichment of transduced cells from untransduced in addition to providing protection from chemotherapy-induced myelosuppression *(11–13)*. On the other hand, the inclusion of such a tag into RV vectors diminishes the actual capacity of the vector for accommodation of the transgene.

2. Materials

2.1. Media or Buffers

1. Producer cell medium: Dulbecco's modified eagle's medium (DMEM), 10% calf serum (CS, HyClone, Logan, Utah, USA), 1% penicillin/streptomycin/ amphotericin B (Gibco-BRL, Carlsbad, California, USA).
2. Bone marrow cell (BMC)-flushing buffer: PBS, 2% FBS.

3. BMC culture medium: Iscove's modified Dulbecco's medium (IMDM), 15% FBS (HyClone), 2% penicillin/streptomycin/amphotericin B supplemented with 50 ng/ml recombinant murine stem cell factor (rmSCF), 20 ng/ml recombinant murine interleukin-3 (rmIL-3), 50 ng/ml recombinant human interleukin-6 (rhIL-6) (R&D System)

4. Methylcellulose for colony-forming unit-cell (CFU-C) assay: MethoCult M3334 (StemCell Technologies, No. 3334) supplemented with 50 ng/ml rmSCF, 20 ng/ml rmIL-3, 50 ng/ml rhIL-6 (R&D System)

5. Polybrene (Sigma, St. Louis, MO, USA): stock solution of 6 mg/ml, working concentration of 6 μg/ml.

2.2. Flow Cytometry Analysis

1. Red blood cell (RBC)-lysing solution: 0.155 M NH_4Cl, 10 mM $KHCO_3$, 0.1 mM EDTA

2. Blocking buffer: 5% normal mouse serum (NMS) in PBS.

3. Washing buffer: PBS supplemented with 2% heat-inactivated FBS.

4. Staining buffer: PBS, 2% FBS

5. Monoclonal antibodies (BD-Pharmingen, San Jose, CA, USA): Lineage antibodies include Gr-1~PE (No. 553128), B220~PE (No. 553089), Ter119~PE (No. 553673), and CD4~PE (No. 553652) antibodies, CD45.2~bio (No. 553771) antibody; Streptavidin~AF647 (No. S32357, Invitrogen–Molecular Probes, Carlsbad, CA, USA).

2.3. Mice

1. C57BL/6J (No. 000664): purchased from the Jackson Laboratory (JAX, Bar Harbor, ME, USA)

2. Boy J (B6.SJL-Ptprc[a]Pep3[b]/BoyJ, No. 002014): purchased from JAX

3. Acidified water: 5 mM HCl in drinking water.

3. Methods

The initial transduction procedure takes a little over a week. Additional analyses of transplanted mice are required to assess transduction into long-term repopulating HSC. This approach is summarized as a time-line in **Fig. 1**.

3.1. 5-FU Treatment of Donor Mice

To induce the quiescent HSC into proliferation, donor mice (C57BL/6J) are treated with 5-FU at 150 mg/kg. 5-FU is administered as i.p. injection of 10 mg/ml solution (*see* **Note 1**). After 3 days, the mice are euthanized by cervical dislocation under anesthesia (*see* **Note 2**).

D(-7)	Friday	5-FU treatment of donor mice
D(-4)	Monday	BMC harvesting, setting up pre-culture
D(-3)	Tuesday	Irradiation and plating of producer cells
D(-2)	Wednesday	Harvesting of pre-stimulated BMC, setting up transduction
		(co-culture)
D0	Friday	Harvesting transduced BMC
		BMT (injecting transduced BMC into irradiated recipient mice)
		plating transduced BMC on methylcellulose media for a CFU-C assay
		culturing transduced BMC for a FCM analysis
D(+3)	Monday	FCM analysis of transduced BMC
D(+7)	Friday	scoring colonies for CFU-C assay
W(+8-10)		First FCM analysis of PBL in recipient mice (donor engraftment/
		GFP marking)
M(+3-4)		Second FCM analysis of PBL in recipient mice (multilineage)
M(+4-6)		Secondary BMT

D = day
W = week
M = month

Fig. 1. Experimental time line for retroviral-mediated gene transfer into hemato-poietic stem cells, transplantation, and analysis. The time line provides a day-by-day schedule from preparation of the donor mice with 5-FU, bone marrow harvesting, and retroviral transduction. Following transplant, analyses are done on weekly or monthly schedules beginning 8 weeks post-transplant when the hematopoietic system has fully recovered from the myelosuppression. The busiest days in the schedule can be day-4 for the bone marrow harvest and day 0 for the intravenous injections into irradiated recipients.

3.2. BMC Harvesting and Pre-Culture

1. Kill the mouse and then wet the pelt thoroughly with 70% ethanol.
2. Strip skin from hind limbs, remove excess tissue; using sterile sharp scissors, cut off the legs at the hip joints, separate femurs and tibias at the knee joints.
3. Flush the marrow using 21-G (femurs) and 25-G (tibias) needles attached to a syringe. Collect the BM cells in PBS containing 2% FBS.

4. Make a single cell suspension using 21-G needle and syringe and spin them down for 5 min at $1,000 \times g$.
5. Resuspend the cell pellet in RBC-lysing buffer (1–2 ml/mouse) and incubate on ice for 10 min. Spin them down for 5 min at $1,000 \times g$.
6. Resuspend the cells in PBS/2% FBS (1 ml/mouse) and count the cell number using hemocytometer after dilution in 2% acetic acid (HOAc) (1:1) and Trypan blue (1:1) and calculate the cell yield (*see* **Note 3**).
7. Spin the cells down for 5 min at $1,000 \times g$ and then resuspend them in BMC culture medium at 2×10^6 cells/ml.
8. Plate BMC in 15-cm suspension dishes and pre-stimulate them for 2 days (*see* **Note 4**).

3.3. Producer Cell Line Preparation

Because of specific focus of this chapter, generation of retroviral producer cell lines is not discussed here.

1. A day before co-culture, harvest the producer cells and count them.
2. Irradiate (15 Gy, γ-rays ^{137}Cs) the required number of cells; spin them down for 5 min at $1,000 \times g$.
3. Resuspend irradiated producer cells in producer medium at 8×10^5 cells/ml and plate them on gelatin-coated (0.1% gelatin in PBS) 10-cm tissue culture dishes (10 ml/dish) (*see* **Notes 5** and **6**).

3.4. Retroviral Transduction using Co-Culture Method

1. Harvest pre-stimulated BMC, count them using a hemocytometer as before, and calculate the cell yield.
2. Spin the cells down for 5 min at $1,000 \times g$. Resuspend the cell in BMC culture medium at a density of 1×10^6 cells/ml.
3. Spare 1-ml cells (non-transfected control-mock); seed them into a 24-well plate. These will be used later to assess transduction efficiency of the total cell population.
4. Add polybrene to the rest of cells at 6 μg/ml and split them according to the number of different transduction groups.
5. Remove the medium from irradiated producer cells and plate BMC over them (10 ml/dish). Continue co-culture for 2 days (*see* **Note 7**).

3.5. BMC Harvest, Recipient Preparation, and Transplantation

1. Irradiate recipient mice (11 Gy; γ-rays ^{137}Cs) (*see* **Note 8**).
2. Harvest transduced BM cells: carefully wash them out of producers, rinse them with 5 ml PBS/2% FBS per five plates.
3. Spin the cells down for 5 min at $1,000 \times g$. Resuspend them in 10 ml PBS/2% FBS, count them (use 2% HOAc).

4. Spare 1×10^5 cells for transduction efficiency and resuspend them in 1 ml of BMC culture medium.
5. Spin the rest of the cells down for 5 min at $1000 \times g$, resuspend them in PBS/2% FBS.
6. Adjust the cell density to $4–5 \times 10^6$ BM cells/ml. Add heparin to 0.1 mg/ml (*see* **Note 9**).
7. Filter the cells through a 40-μm mesh to remove aggregates of producer cells.
8. Inject (i.v., lateral tail vein) 500 μl ($\sim 2 \times 10^6$ cells) into irradiated recipient mice (*see* **Note 10**).

The first 3 weeks following transplantation, mice are put on acidified water (*see* **Note 11**).

3.6. Transduction Efficiency Evaluation by Flow Cytometry and Colony-Forming Unit in Culture Assay

The design of the RV vector determines the tracking option. If a fluorescence protein gene is included, then transduced cells and their progeny can be tracked based on its fluorescence.

1. Flow cytometry analysis.
   Take 1×10^5 transduced BMC, perform flow cytometry analysis for GFP (or other fluorescent marker) expression. Whereas this analysis does not tell much about transduction of HSC, it is generally informative regarding the overall transduction efficiency of total BMC.

2. CFU-C assay.
   a. Seed 1×10^4 transduced BMC in 3 ml methylcellulose medium containing Epo (3 U/ml), supplemented with SCF (50 ng/ml), IL-3 (20 ng/ml), and IL-6 (50 ng/ml) for assay of the myeloid colonies.
   b. After 7 days culture, score the total number of colonies as well as the number of green ones under a fluorescent microscope using a blue filter.
   c. Calculate the frequency of total myeloid progenitor and the transduction efficiency of progenitors. However, it is important to remember that this does not reflect transduction efficiency of HSC but rather of committed myeloid CFU-C.

3.7. Assessment of Donor BMC Engraftment and Marking

In order to contribute to host hematopoiesis, transplanted BMC have to engraft in the BM. The criteria for detecting donor HSC are long-term multilineage contribution and sustained reconstitution after serial transplantation.

3.7.1. GFP Marking

To evaluate GFP marking in recipient mice, flow cytometry analysis of GFP is performed in total PBL in circulating white blood cells (WBC) after 8, 12, and 16 weeks transplantation.

1. Bleed the transplanted mice from retro orbital sinus vein under isoflurane anesthesia through a heparinized micro-capillary tube (~200 μl) (*see* **Notes 12** and **13**).
2. Add 1 ml of hypotonic RBC-lysing solution and incubate at room temperature for 10 min to lyse RBC; spin them down for 2 min at 2000 × *g*.
3. Resuspend the cell pellet in 1 ml of RBC-lysing solution and spin them down for 2 min at 2000 × g.
4. Wash the cells with 1 ml of PBS; spin them down for 2 min at 2,000 × *g*.
5. Resuspend the cell pellet in 100 μl of blocking buffer to block unspecific binding sites and incubate for 15 min on ice.
6. Add 1 μl of CD45.2~PE antibody and incubate on ice for 15 min.
7. Wash the cells with 1 ml of PBS/2% FBS; spin them down for 2 min at 2,000 × *g*.
8. Resuspend the cell pellet in 500 μl of PBS/2% FBS for a direct flow cytometry measurement (FCM) analysis of GFP expression in donor mice BMC (*see* **Note 14**).

3.7.2. Lineage-Specific Antibody Staining

To test for long-term multilineage hematopoietic reconstitution, the peripheral blood is collected from recipient mice at least 4 months after transplantation.

1. Bleed the mice, lyse RBC, and block unspecific binding as above.
2. After blocking, split WBC suspension into four aliquots; spin them down for 2 min at 2000 × *g*.
3. Resuspend the cell pellet of each aliquot in 100 μl of one of four lineage-specific antibody cocktails: CD45.2~bio + Gr-1~PE, or CD45.2~bio + B220~PE, or CD45.2~bio + Ter119~PE, or CD45.2~bio + CD4~PE, and incubate them on ice for 15 min.
4. Wash out unbound antibodies with 1 ml PBS/2% FBS; spin the cell down for 2 min at 2,000 × *g*.
5. Resuspend the cell pellet in 100 μl of PBS/2% FBS containing streptavidin~AF647 and incubate on ice for 15 min.
6. Wash the cell with 1 ml PBS/2% FBS again, resuspend the cell in 500 μl PBS/2% FBS for flow cytometry analysis.

3.8. The Secondary BMT

Because HSC are rare in normal BMC and ex vivo culture can alter their functional hematopoietic repopulation capacity, the transplanted cell dose can be depleted of HSC by the procedure. Serial transplantation can be done to define the self-renewal capacity of the donor graft and to validate gene transfer into HSC. BMC harvested from primary recipients are typically transplanted into five secondary recipients (lethally irradiated, as above), and after full engraftment, long-term contribution of transduced donor cells indicates successful HSC level transduction. All the test procedures are the same as for the primary BMT.

4. Notes

1. To facilitate injection of a large number of mice, we replaced i.v. injection of 5-FU with an i.p. injection. This was found to give similar myelosuppression and BM cell yield as i.v. and speeds up the procedure. Special attention should be made to avoid hitting an internal organ or the bladder. Direct excretion of 5-FU injected into the bladder can result in sub-optimal HSC enrichment. 5-FU is a biohazard and care must be taken to discard bedding materials from the mouse cages into a biohazard bag for incineration.
2. The use of isoflurane is the preferred method of anesthesia because of simple administration (breathing vapors) without any noticeable adverse effects on normal function. Hypoxia caused by using CO_2 anesthesia might affect this function.
3. The use of HOAc causes additional lysis of RBC and provides more accurate WBC count.
4. Although various and more complex combinations of cytokines could be used, this minimal cocktail seems to be sufficient to stimulate and support HSC in short-term culture while still preserving engraftment ability.
5. Different amounts of irradiated producers were seeded and tested for transduction efficiency. 8×10^6 producers on 10-cm dishes gave a transduction efficiency that was almost double that of 4×10^6 producers. We tested even higher amounts of irradiated producers (1×10^7), but these grew too dense causing the producer cells to detach.

	% of GFP$^+$ BMC transduced with vector	
Number of producers	MSCV-irGFPx	MSCV-TIMP1-irGFP
4×10^6	26.1	27.6
8×10^6	44.9	51.8

6. The irradiated producers tend to detach (especially at higher densities). Coating the plates with gelatin facilitates firmer attachment. This is very helpful during harvesting BMC off the producers after transduction and helps to get a more complete harvest of BM cells.
7. The culture of cells should be monitored. If the media gets depleted (indicated by yellow color), the cells need to be re-fed with 3 ml of fresh media (including cytokines and polybrene).
8. Heparin prevents aggregation of cells. This improves mouse survival for higher cell doses, especially if some producer cells have dislodged from the plates during BMC harvest after co-culture.

9. We usually irradiate recipient mice between 15 min and 2 h before transplant, but the mice can be irradiated up to 24 h in advance. The irradiation rate can also be varied. Our experience with delivery of 11 Gy has been the same with as few as 4 min and as long as 13 min, depending on the source intensity. If a higher source dose rate is being used, the total dose can be split into two halves and the mice can be irradiated in two cycles approximately 4 h apart. This is believed to lessen some of the non-hematopoietic toxicities associated with radiation sickness in mice.

10. If the lateral veins are obscured, they can be dilated by heating mice with a heat lamp or rubbing the tail with wintergreen oil.

11. The irradiated mice have their immune defense weakened, and even their own gut micro flora can overgrow leading to sepsis. Special drinking water helps to maintain balance and promote recovery during the nadir that occurs 2–3 weeks after irradiation. Some immune deficient mouse strains require special additives (antibiotics, e.g., 1% Baytril in drinking water).

12. Some other anticoagulants may be used to prevent blood clotting.

13. A part of the blood may be spared for basic hematology and/or blood smears; the remaining blood may be used for collecting plasma after centrifugation for 10 min at 2000 × g.

14. We usually use 5% NMS in PBS for blocking some unspecific binding sites on the cell surface. If this blocking is not sufficient, the purified CD16/CD32 (FcγIII/II R, clone 2.4G2) mAb can be used.

References

1. Nienhuis A. W., Walsh C. E., and Liu J. Viruses as therapeutic gene transfer vectors. In: Young N. S., ed. (1993) *Viruses and Bone Marrow: Basic Research and Clinical Practice*. New York: Dekker, 353–414.
2. Markowitz, D., Goff, S., and Bank, A. (1988) Construction and use of a safe and efficient amphotropic packaging cell line. *Virology* **167,** 400–406.
3. Markowitz, D. G., Goff, S. P., and Bank, A. (1988) Safe and efficient ecotropic and amphotropic packaging lines for use in gene transfer experiments. *Trans. Assoc. Am. Physicians* **101,** 212–218.
4. Miller, A. D. (1990) Retrovirus packaging cells. *Hum. Gene Ther.* **1,** 5–14.
5. Kohno, T., Mohan, S., Goto, T., Morita, C., Nakano, T., Hong, W., Sangco, J. C., Morimatsu, S., and Sano, K. (2002) A new improved method for the concentration of HIV-1 infective particles. *J. Virol. Methods* **106,** 167–173.
6. Baum, C., Dullmann, J., Li, Z., Fehse, B., Meyer, J., Williams, D. A., and von Kalle, C. (2003) Side effects of retroviral gene transfer into hematopoietic stem cells. *Blood* **101,** 2099–2114.
7. Philippe, S., Sarkis, C., Barkats, M., Mammeri, H., Ladroue, C., Petit, C., Mallet, J., and Serguera, C. (2006) Lentiviral vectors with a defective integrase allow efficient and sustained transgene expression in vitro and in vivo. *Proc. Natl. Acad. Sci. U.S.A.* **103,** 17684–17689.

8. Martin, P., Albagli, O., Poggi, M. C., Boulukos, K. E., and Pognonec, P. (2006) Development of a new bicistronic retroviral vector with strong IRES activity. *BMC. Biotechnol.* **6**, 4.
9. Klump, H., Schiedlmeier, B., Vogt, B., Ryan, M., Ostertag, W., and Baum, C. (2001) Retroviral vector-mediated expression of HoxB4 in hematopoietic cells using a novel coexpression strategy. *Gene Ther.* **8**, 811–817.
10. Amendola, M., Venneri, M. A., Biffi, A., Vigna, E., and Naldini, L. (2005) Coordinate dual-gene transgenesis by lentiviral vectors carrying synthetic bidirectional promoters. *Nat. Biotechnol.* **23**, 108–116.
11. Zaboikin, M., Srinivasakumar, N., and Schuening, F. (2006) Gene therapy with drug resistance genes. *Cancer Gene Ther.* **13**, 335–345.
12. Milsom, M. D., Fairbairn, L. J. (2004) Protection and selection for gene therapy in the hematopoietic system. *J. Gene Med.* **6**, 133–146.
13. Flasshove, M., Moritz, T., Bardenheuer, W., and Seeber, S. (2003) Hematoprotection by transfer of drug-resistance genes. *Acta Haematol.* **110**, 93–106.
14. Persons, D. A., Allay, J. A., Allay, E. R., Smeyne, R. J., Ashmun, R. A., Sorrentino, B. P., and Nienhuis, A. W. (1997) Retroviral-mediated transfer of the green fluorescent protein gene into murine hematopoietic cells facilitates scoring and selection of transduced progenitors in vitro and identification of genetically modified cells in vivo. *Blood* **90**, 1777–1786.
15. Hawley, T. S., Telford, W. G., and Hawley, R. G. (2001) "Rainbow" reporters for multispectral marking and lineage analysis of hematopoietic stem cells. *Stem Cells* **19**, 118–124.
16. Persons, D. A., Allay, J. A., Riberdy, J. M., Wersto, R. P., Donahue, R. E., Sorrentino, B. P., and Nienhuis, A. W. (1998) Use of the green fluorescent protein as a marker to identify and track genetically modified hematopoietic cells. *Nat. Med.* **4**, 1201–1205.
17. Dick, J. E., Magli, M. C., Huszar, D., Phillips, R. A., and Bernstein, A. (1985) Introduction of a selectable gene into primitive stem cells capable of long-term reconstitution of the hemopoietic system of W/Wv mice. *Cell* **42**, 71–79.
18. Szilvassy, S. J., Fraser, C. C., Eaves, C. J., Lansdorp, P. M., Eaves, A. C., and Humphries, R. K. (1989) Retrovirus-mediated gene transfer to purified hemopoietic stem cells with long-term lympho-myelopoietic repopulating ability. *Proc. Natl. Acad. Sci. U.S.A.* **86**, 8798–8802.
19. Fraser, C. C., Szilvassy, S. J., Eaves, C. J., and Humphries, R. K. (1992) Proliferation of totipotent hematopoietic stem cells in vitro with retention of long-term competitive in vivo reconstituting ability. *Proc. Natl. Acad. Sci. U.S.A.* **89**, 1968–1972.
20. Hawley, R. G., Lieu, F. H., Fong, A. Z., and Hawley, T. S. (1994) Versatile retroviral vectors for potential use in gene therapy. *Gene Ther.* **1**, 136–138.
21. Bunting, K. D., Galipeau, J., Topham, D., Benaim, E., and Sorrentino, B. P. (1998) Transduction of murine bone marrow cells with an MDR1 vector enables ex vivo stem cell expansion, but these expanded grafts cause a myeloproliferative syndrome in transplanted mice. *Blood* **92**, 2269–2279.

22. Sorrentino, B. P., McDonagh, K. T., Woods, D., and Orlic, D. (1995) Expression of retroviral vectors containing the human multidrug resistance 1 cDNA in hematopoietic cells of transplanted mice. *Blood* **86,** 491–501.

23. Allay, J. A., Dumenco, L. L., Koc, O. N., Liu, L., and Gerson, S. L. (1995) Retroviral transduction and expression of the human alkyltransferase cDNA provides nitrosource resistance to hematopoietic cells. *Blood* **85,** 3342–3351.

24. Allay, J. A., Davis, B. M., and Gerson, S. L. (1997) Human alkyltransferase-transduced murine myeloid progenitors are enriched in vivo by BCNU treatment of transplanted mice. *Exp. Hematol.* **25,** 1069–1076.

25. Corey, C. A., DeSilva, A. D., Holland, C. A., and Williams, D. A. (1990) Serial transplantation of methotrexate-resistant bone marrow: protection of murine recipients from drug toxicity by progeny of transduced stem cells. *Blood* **75,** 337–343.

26. Zhao, S. C., Li, M. X., Banerjee, D., Schweitzer, B. I., Mineishi, S., Gilboa, E., and Bertino, J. R. (1994) Long-term protection of recipient mice from lethal doses of methotrexate by marrow infected with a double-copy vector retrovirus containing a mutant dihydrofolate reductase. *Cancer Gene Ther.* **1,** 27–33.

27. Allay, J. A., Galipeau, J., Blakley, R. L., and Sorrentino, B. P. (1998) Retroviral vectors containing a variant dihydrofolate reductase gene for drug protection and in vivo selection of hematopoietic cells. *Stem Cells* **16 Suppl 1,** 223–233.

28. Matsunaga, T., Sakamaki, S., Kuga, T., Kuroda, H., Kusakabe, T., Akiyama, T., Konuma, Y., Hirayama, Y., Kobune, M., Kato, J., Sasaki, K., Kogawa, K., Koyama, R., and Niitsu, Y. (2000) GST-pi gene-transduced hematopoietic progenitor cell transplantation overcomes the bone marrow toxicity of cyclophosphamide in mice. *Hum. Gene Ther.* **11,** 1671–1681.

29. Momparler, R. L., Eliopoulos, N., Bovenzi, V., Letourneau, S., Greenbaum, M., and Cournoyer, D. (1996) Resistance to cytosine arabinoside by retrovirally mediated gene transfer of human cytidine deaminase into murine fibroblast and hematopoietic cells. *Cancer Gene Ther.* **3,** 331–338.

30. Rattmann, I., Kleff, V., Sorg, U. R., Bardenheuer, W., Brueckner, A., Hilger, R. A., Opalka, B., Seeber, S., Flasshove, M., and Moritz, T. (2006) Gene transfer of cytidine deaminase protects myelopoiesis from cytidine analogs in an in vivo murine transplant model. *Blood* **108,** 2965–2971.

31. Flasshove, M., Frings, W., Schroder, J. K., Moritz, T., Schutte, J., and Seeber, S. (1999) Transfer of the cytidine deaminase cDNA into hematopoietic cells. *Leuk. Res.* **23,** 1047–1053.

32. Magni, M., Shammah, S., Schiro, R., Mellado, W., la-Favera, R., and Gianni, A. M. (1996) Induction of cyclophosphamide-resistance by aldehyde-dehydrogenase gene transfer. *Blood* **87,** 1097–1103.

33. Bunting, K. D., Townsend, A. J. (1996) Protection by transfected rat or human class 3 aldehyde dehydrogenase against the cytotoxic effects of oxazaphosphorine alkylating agents in hamster V79 cell lines. Demonstration of aldophosphamide metabolism by the human cytosolic class 3 isozyme. *J. Biol. Chem.* **271,** 11891–11896.

17

Lentiviral Gene Transduction of Mouse and Human Stem Cells

Zhaohui Ye, Xiaobing Yu, and Linzhao Cheng

Summary

This chapter describes the methods we use to transduce mouse and human hematopoietic stem cells (HSCs) and human embryonic stem cells (hESCs). We provide detailed protocols for producing high-titer lentiviral supernatants by transient transfection and for measuring viral titers. Methods to concentrate viral supernatants to achieve a higher titer are also described. The protocols given here have been used successfully to transduce engrafting mouse and human HSCs as well as progenitor cells. These cells maintained stable transgene expression after engraftment in mice and in vivo differentiation. Human ESCs can also be transduced with a high efficiency, and transgene is expressed stably after hematopoietic differentiation.

Key Words: Lentivirus; hematopoietic stem cells; hematopoietic progenitor cells; embryonic stem cells.

1. Introduction

Hematopoietic stem cells have the unique capability of repopulating the entire hematopoietic system because of their self-renewal and pluripotent differentiation potentials, thus represent an important target for treatment of various blood and immune disorders. Stable gene transfer to these stem cells therefore has great potential to achieve both long-term and short-term therapeutic effects for the treatment of these diseases. Oncoretroviral (also called as gamma-retroviral) and lentiviral vectors represent two major choices for efficient transduction and stable integration of transgenes into hematopoietic stem and progenitor cells *(1)*. Lentiviral vectors (lentivectors or LVs) offer

From: *Methods in Molecular Biology, vol. 430: Hematopoietic Stem Cell Protocols*
Edited by: K. D. Bunting © Humana Press, Totowa, NJ

several advantages over traditional gamma-retroviral vectors. LVs efficiently transduce slowly dividing cells, including HSCs, resulting in stable gene transfer and expression *(2,3)*. Additionally, recently developed self-inactivating (SIN) LVs allow promoter-specific transgene expression *(2,4)*. The SIN safety modification of LVs, which permanently disables the viral promoter within the viral long-terminal repeat (LTR) after integration, enables transgene expression in the targeted cells to be controlled solely by internal promoters *(1–7)* without reducing viral titers *(2,5)*. High-titer lentivector supernatants ($\geq 10^7$ infectious units per milliliter) can be easily made by transient transfection of commonly used 293 cells or derivatives by $CaPO_4$ precipitation or liposome-mediated methods. In addition to the transducing (template) lentivector that serves as transgene template, two other (helper) plasmids required for viral assembly are used in transfection. One is to express the HIV-1 *gag/pol* gene and the second is to express the VSV-G envelope protein that is good for essentially any vertebrate cell types. The VSV-G pseudo-typed recombinant viruses are much more stable and can be concentrated easily. Upon incubation with target cell in culture or injected in vivo, VSV-G pseudo-typed recombinant viruses will fuse to cell membrane, un-load two copies of RNA templates, reversely transcribe into DNA form, and eventually integrate into cellular chromosomes favorably at a transcribed region. Then integrated transgene and its promoter will be regulated as part of cellular chromosomes.

Since the first human hESCs line was derived in 1998, it has attracted significant attention because of its ability to self-renew in culture and its potential to differentiate into all types of cells in the body, including blood lineages *(8–10)*. It has been shown that hESCs can be transduced with lentivirus vectors at a high efficiency and that their hematopoietic progeny maintain stable transgene expression *(11,12)*. Genetically modified human ESCs therefore offer great tools for study of molecular events in early hematopoiesis and provide potentially unlimited sources of HSCs for clinical applications.

In this chapter, we describe the procedures to produce lentiviral supernatants and to transduce both mouse and human hematopoietic stem/progenitor cells. A protocol for transducing human embryonic stem cells (hESCs) is also described.

2. Materials

1. Medium for culturing 293T cells (used to produce lentiviral vectors): D-MEM with high glucose (4.5 g/ml), 10% fetal bovine serum (FBS), 1× penicillin-streptomycin.
2. Medium for virus collecting in 293T cells: Essentially any culture medium. We commonly use D-MEM + 1% FBS or any serum-free medium (with insulin + transferrin supplements).

3. Medium for culturing primary human hematopoietic cells:
   QBSF-60 serum-free medium with L-glutamine (Quality Biological, Inc., Gaithersburg, MD) (*see* **Note 1**); gentamycin, 20 ng/ml (human) thrombopoietin; 100 ng/ml human SCF or KIT ligand, 50–100 ng/ml (human) FLT3 ligand (collectively called TSF).

4. Medium for culturing primary mouse hematopoietic stem/progenitor cells:
   QBSF-58 serum-free medium with L-glutamine (Quality Biological, Inc., Gaithersburg, MD; *see* **Note 1**), gentamycin, 100 ng/ml mouse KIT ligand as well as 20 ng/ml (human) thrombopoietin, 50–100 ng/ml (human) FLT ligand (*see* **Note 2**).

5. Medium for culturing hESCs:
   KNOCKOUT™ D-MEM or DMEM/F12, 20% knockout serum replacement, 2 mM L-glutamine, 0.1 mM non-essential amino acids (all from Invitrogen, Carlsbad, CA), 0.1 mM of β-mercaptoethanol, and 4 ng/ml of bFGF (Pepro Tech, Rocky Hill, NJ or other sources).

6. 1x phosphate-buffered saline (PBS) Ca^{+2} and Mg^{+2} free.

7. 0.05% Trypsin–ethylenediaminetetraacetic acid (EDTA).

8. Red blood cell (RBC) lysis buffer: 8.3 g NH_4Cl, 1.0 g $KHCO_3$, 1 mM EDTA in 1 l solution.

9. Lineage depletion kit for mouse cells (StemCell Technologies, Vancouver, Canada).

10. CD34-positive selection kit (Miltenyi Biotec, Bergisch Gladbach, Germany)

11. Matrigel™ matrix (Becton Dickinson, Bedford, MA): Matrix for human ESC culture.

12. Poly-D-lysine (500–550 KDa polymers, BD 35-40210): to make 293T cells more adherent.

13. Polybrene (as an attachment factor to enhance virus-cell fusion): dissolve hexadimethrine bromide (Sigma, St. Louis, MO, Cat. No. H9268) in water to make final concentration to 8 mg/ml. Sterilize solution by 0.22-μm filter.

14. Sterile polystyrene 5-ml tube (Falcon 2054).

15. Stericup™ (Millipore, Billerica, MA) or other sterilization filtration units, pore size 0.22 and 0.45 μm.

16. Amicon Ultra Centrifugal Filter Devices (Millipore, Billerica, MA). The unit with a filter of 100,000 MWCO is an easy way to concentrate retro- and lentivectors by up to 100- to 200-fold.

17. Regents for 293T transfection: basically you may use any type of liposomes you might have tried before. We commonly use lipofectamine 2000 (Invitrogen), which is insensitive to serum in media. However, the classic $CaPO_4$ method is also fine, albeit validation of the solution and practice is often required to ensure high and consistent efficiency. The following is a recipe for two solutions required for the classic $CaPO_4$ method.

18. 2 M $CaCl_2$ solution: dissolve 22.2 g of $CaCl_2$ in water to final volume of 100 ml. Sterilize solution by 0.22-μm filter. Store at 4°C.

19. Hank's balanced salt solution (2×): dissolve 16.4 g of NaCl, 11.9 g of HEPES acid, and 0.21 g of Na_2HPO_4 in 800 ml of water, adjust pH value to 7.05 with 5N NaOH solution, add water to a final volume of 1 l. Sterilize solution by 0.22-μm filter. Store at 4°C.

3. Methods

3.1. Production of Lentiviral Supernatant

The protocol described here utilizes co-transfection of 293T cells with three plasmids: the lentiviral vector coding for the viral genome that contains the transgene, CMVΔR8.91 expressing the required three lentiviral (HIV-1) proteins *(5)*, and MD.G expressing the VSV-G envelope proteins *(6)*.

3.1.1. Transfection of 293T Cells Using Lipofectamine 2000

1. Coat tissue culture plates or dishes with poly-D-lysine (50–100 μg/ml) for 60 min at room temperature, then wash twice with PBS (*see* **Note 3**).
2. (Day 0) Plate 293T cells using 293T medium, see **Table 1** for the cell number needed (*see* **Note 4**).
3. (Day 1) Change 293T medium to virus collecting medium (*see* **Note 5**). Transfect 293T cells with the three plasmids (*see* **Table 2**) using lipofectamine 2000 following manufacturer's instruction (*see* **Note 6**): (for a 6-well plate or for a 150-mm dish, underlined)

   a) 4 or 24 μg DNA + 0.15 or 1.2 ml of OPTI-MEM (Falcon 2054 polystyrene 5-ml tube).
   b) 6 or 36 μl of lipofectamine 2000 + 0.15 or 1.2 ml of OPTI-MEM, incubate for 5 min at RT (in polystyrene 5-ml tube, Falcon 2054).
   c) Mix the diluted DNA with diluted lipofectamine 2000; incubate for 20 min at room temperature.
   d) Add the DNA–lipid complexes dropwise into the media.
   e) After 6 h of incubation, change the medium to fresh virus collecting medium (*see* **Notes 7** and **8**).

Table 1
Cell Numbers Needed for Plating 293T Cells in Virus Production

Plate	Cell number per well, × 10^6	Media volume (ml)
35 mm = one well in 6-well plate	0.5	2
60 mm	1	4
100 mm	4	8
150 mm	≥6	≥12

Table 2
Amount of Plasmids DNA Needed for Transfection

Plates	35 mm	150 mm
Total plasmid DNA:	$\leq4\,\mu g$	$24\,\mu g$
1. LV-transducing vector	1	9
2. CMVΔR8.91 (expressing three required HIV proteins)	2	12
3. MD.G (expressing the VSV-G envelope proteins)	0.5	3

4. (Day 2–4) Harvest viral supernatant (*see* **Note 9**). Centrifuge the collected supernatant at 2500 x *g* for 10 min to get rid of cells/debris. If it is critical to eliminate cell contamination, pass the supernatant through 0.45 um filter units. The virus can be used immediately or stored for later use (*see* **Note 10**).

3.1.2. Transfection by CaPO₄ Method

1. Plate 293T cells as described in the lipofectamine 2000 method (*see* **Subheading 3.1.1.**).
2. When cell density reaches approximately 70–80%, make transfection cocktails according to **Table 2** (DNA amounts) and **Table 3** in 5- or 15-ml conical tube. Add water, CaCl₂, and DNA in the tube first. The quantity of water is based on the total volume of other ingredients.
3. When adding the Hank's balanced salt solution (HBSS), bubble air through the cocktail with pipette for about 10–20 times until the solution becomes slightly cloudy.
4. Let the mixed cocktail stand in room temperature for 10–20 min.
5. Drop cocktail directly onto media currently on the cells. Swirl media on plates once.
6. After 6 h of incubation, change the medium to fresh virus-collecting medium (*see* **Note 8**).
7. Collect viral supernatant the same way as described in **Subheading 3.1.1**.

Table 3
Transfection Cocktails for CaPO₄ Method

Plate size (mm)	CaCl₂ (µl)	Total DNA (µg)	2× HBS	Total final volume
35	31	4–5	250 µl	500 µl
60	62	8–10	500 µl	1 ml
100	124	15–20	1 ml	2 ml
150	372	45–60	2.5 ml	5.0 ml

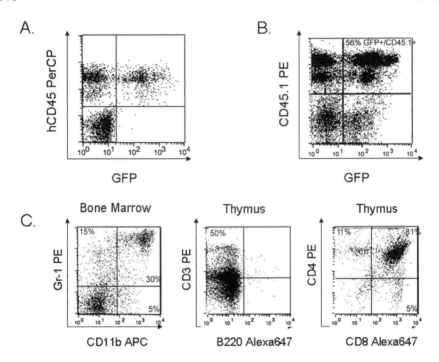

Fig. 1. (A) Human cord blood CD34+ cells were transplanted into sub-lethally irradiated NOD/SCID mice 48 h after transduction with lentivirus expressing GFP. After 8 weeks, bone marrow cells from the transplanted NOD/SCID were analyzed by using FACS for GFP transgene expression in human hematopoietic cells expressing the human (h) CD45 marker. (B) Mouse bone marrow lineage-depleted cells from CD45.1 background mice were transduced with GFP-expressing lentivirus and transplanted into lethally irradiated CD45.2 mice. Analysis of recipients' bone marrow 8 weeks after transplant showed 56% of GFP+ donor (CD45.1+) cells. (C) Further analysis of transduced donor cells (GFP+CD45.1+) cells in the bone marrow and thymus with markers of various lineages.

3.2. Concentrating VSV-G Pseudo-Typed Lentiviral Supernatants (Optional) (see Note 11)

1. Add 15 ml of viral supernatant to the top portion of Amicon filter and centrifuge at 2,500 x *g* for 15 min at 4°C (*see* **Note 12**).
2. Aspirate the flow-through from the tube and add more supernatant, centrifuge at the same condition.
3. Repeat the process until the desired fold of concentration is reached. Use pipettes or pipette tips to take out the supernatant from the upper part of the centrifugal device. Use the virus immediately or store it (*see* **Note 10**).

A. GFP   B. Oct-4

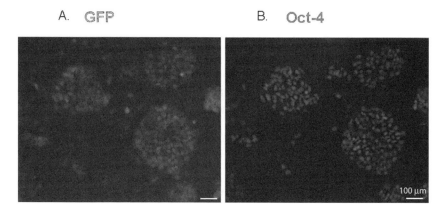

Fig. 2. Human ES cell line H1 transduced with GFP-expressing lentivirus maintained at the undifferentiated state. After selection, the transduced H1 human ES cells (G-GFP) for expanded for years in culture. To illustrate the undifferentiated state, cultured G-GFP human embryonic stem (ES) cells as undifferentiated colonies were fixed and permeabilized before staining for Oct-4 by a specific monoclonal antibody. The fluorescence of GFP signal (A) and Oct-4 staining after a secondary reagent (red, in B) were recorded from the same field.

4. Alternatively, the viral supernatant can be concentrated by ultracentrifugation 20,000 × g at 4°C for 90 min. The viral pellet can be re-suspended in PBS at 4°C overnight.

3.3. Measuring Viral Titers Based on Transgene Expression

1. (Day 0) Plate 1×10^5 of 293T cells per well in 6-well plate.
2. (Day 1) Replace the culture medium with serial diluted (1:10, 1:100, 1:1000, dilute with culture medium) viral supernatants in each well with 2 ml of medium. Add polybrene to final concentration of 8 µg/ml.
3. (Day 3 or later) Analyze the cells by using FACS based on the cell-surface marker expression or fluorescent protein (e.g., GFP). Otherwise, you have to monitor the integrated vector at DNA level, because measuring the titer by measuring RNA levels in viral supernatants or transduced cells is problematic.
4. Calculating infectious or transduction units (TU): If 10% of cells are GFP positive with 0.02 ml (1/100) of original viral supernatants, the titer is: 10% × cell number at Day 1($\sim 2 \times 10^5$)/0.02 ml = 1×10^6 TU/ml (*see* **Note 13**).

3.4. Transduction of Hematopoietic Stem Cells

The protocols described here have been successfully transduced mouse bone marrow lineage-depleted cells, or human CD34+ cells from bone marrow,

G-CSF-mobilized peripheral blood or cord blood. The method for transducing mouse and human hematopoietic stem/progenitor cells are the same except that the different culture media (*see* **Subheading 2**) were used. Two alternative methods are described below.

3.4.1. A Simple and Quick Method

1. Count cell numbers and prepare cell cultures lasting for 48 h.
2. Plate the freshly isolated or thawed cells in 12-well plates with the appropriate QBSF medium with growth factors TSF. Cells were plated at a density of 1–2 million per well with 1–2 ml of medium. Add concentrated (>100-fold) lentivirus ($\geq 10^8$ TU/ml) and co-culture the cells with virus in the presence of 8 µg/ml polybrene.
3. After 48 h of co-culture, the cells can be harvested, washed, and used for further functional analysis. The essence of this method is to transduce HSCs quickly and preserve the engraftment activities.

3.4.2. An Enhancement (Spin-Inoculation) Method

1. One day before transduction, isolate cells or thaw the cryopreserved human CD34+ cells and culture them (10^6 cells/ml) overnight in QBSF media supplemented with appropriate growth factors (*see* **Subheading 2.4**).
2. The next day, harvest the cells and re-suspend them in a sterile 5-ml tube (Falcon 2054) with lentivirus-containing media; add polybrene to final concentration of 8 µg/ml. Incubate for 15–20 min. Tighten up the cap and centrifuge at 2,000 × *g* for 3 h using a bench-top centrifuge at room temperature.
3. After centrifugation, add an equal volume of growth factor-containing media to the tubes to dilute the viruses and polybrene. Re-suspend the cells in a pellet, loosen the cap of the tube, and culture overnight in a tissue culture incubator. If necessary, repeat transduction again.
4. Spin down the cells and transfer the cells with media-containing growth factors to tissue culture plates. Culture the cells for 1–2 more days.
5. The cells are now ready for either FACS analysis or subsequent in vitro and in vivo studies (**Fig. 1**). The essence of this method is to increase transduction efficiency (even when the viral titer is relatively low or un-concentrated) and to monitor transgene expression, such as GFP, at the end of transduction (3–4 days after the start of transduction).

3.5. Transduction of hESCs

3.5.1. Making Matrigel Plates for hESC Transduction

1. Thaw frozen Matrigel vial (10 ml) in 4°C overnight (*see* **Note 14**). Injecting 10-ml medium by a sterile syringe will reduce the viscosity significantly. Aliquot extra into chilled tubes and for long-term storage at –80°C.

2. Dilute Matrigel further (1:15) with cold medium and plate it into 6-well plate with 1 ml per well. The final dilution is 1:30.
3. Leave the plate in room temperature for 1 h. The plates are now ready to use or can be stored at 4°C for 1 week. Aspirate excess Matrigel and wash once before use.

3.5.2. Culturing hESCs on Matrigel Plates

1. Split the hESCs from feeder plates using 0.05% Trypsin/EDTA or 1 mg/ml Collagenase IV solution by following standard protocols.
2. Aspirate the medium from the Matrigel plate and wash the wells once with PBS.
3. Plate the cells onto Matrigel plates with 2.5 ml of the conditioned medium (CM) with growth factors. Change CM everyday.

3.5.3. Transduction of hESCs

1. When hESCs reach 60–80% confluence, add concentrated lentiviral supernatant ($\sim 10^8$ TU) into 2 ml of CM in the presence of 4–8 µg/ml of polybrene.
2. Incubate the hESCs with viral supernatant for 4–6 h in tissue culture incubator.
3. Aspirate the virus-containing medium and add 2.5 ml of fresh CM.
4. When cells reach confluence on Matrigel plates, transfer them to normal feeder plates for expansion.
5. Cells can be analyzed 3 days after transduction for transgene expression (**Fig. 2**).

4. Notes

1. We have good experience with the media from Quality Biological, Inc.; however, you can also use other serum-free culture media you prefer.
2. Human FLT ligand and thrombopoietin are active for both human and mouse cells, and a higher concentration (100 and 20 ng/ml, respectively) is preferred. Human and mouse SCF (or Kit ligand) do not interact cross the species well (10- to 1,000-fold less efficient); thus, species-specific Kit ligand is important. Although lentivectors can transduce quiescent cells, adding growth factors during the transduction (and in prior culture before the transduction step) will significantly activate cells and enhance transduction efficiency.
3. 293T cells are loosely adherent cells. They will easily detach from the bottom of plates when confluent. Coating the plates with poly-D-lysine is recommended. When coating the plates, add enough solution to cover the bottom of the plates. If not used immediately, coated plates can be stored in cold room for weeks. Poly-D-lysine solution can be re-used up to three times.
4. The numbers given in **Table** 2 are for your reference. You can titrate cell numbers to obtain the ideal density for optimal efficiency. We found that 70% confluence is ideal start point for transfection.
5. Depending on the cell type that is going to be transduced with, different media specific for culturing that cell type can be used to collect viruses (if no virus

concentration is planed). In our experience, the viral supernatant collected in the medium described here is easy to be concentrated following the protocol provided in this chapter; the concentrated virus (>100-fold) can then be diluted in any preferred medium for transduction, which often gives better results.

6. The protocol described here is modified from lipofectamine's protocol. We found that lipofectamine 2000 (serum-insensitive) gives consistent transfection efficiency. You can, however, use your favorite transfection methods for 293T cells.

7. Serum-free medium with insulin-transferrin is fine, too. Collection media containing a higher percentage of serum (>5%) are hard to be concentrated efficiently.

8. If you start the transfection late in the day, you can skip the first medium change. Collect the supernatant the next day.

9. Collect the supernatant at day 4 only if the cells are still healthy and attached to the plate. The titer may be significantly lower.

10. If the viral supernatant is to be used within 3–5 days, it can be stored at 4°C. If longer term storage is desired, the supernatant should be aliquoted and stored at –80°C. Avoid repeated freeze and thaw process, which will significantly reduce the titer.

11. For most viral constructs, the titer of the collected supernatant before concentration is around 10^6–10^7 TU/ml, which is adequate for many experiments. If a significant higher titer is required or the titer of a specific construct is consistently low, then concentrate viral supernatants.

12. Depending on the concentration of the virus as well as the composition of the collecting medium, the centrifugation time may vary from 10 to 20 min; Start with 10 min to see how much more you should centrifuge.

13. This calculation is based on the assumption that the 293T cells would double the cell number overnight after initial plating. Alternatively, an extra well of cells can be plated on day 0, and the cells can be harvested and counted at the time of adding virus to get more accurate number.

14. It is important to keep the Matrigel solutions cold throughout the whole process until it is plated on the plate; higher temperature will solidify the gel prematurely and reduce the quality of the plates.

References

1. Hawley RG. (1996) Therapeutic potential of retroviral vectors. *Transfus Sci* **17**, 7–14.

2. Cui Y, Golob J, Kelleher E, Ye Z, Pardoll D, and Cheng L. (2002) Targeting transgene expression to antigen-presenting cells derived from lentivirus-transduced engrafting human hematopoietic stem/progenitor cells. *Blood* **99**, 399–408.

3. Yu X, Zhan X, D'Costa J, Tanavde VM, Ye Z, Peng T, Malehorn MT, Yang X, Civin CI, and Cheng L. (2003) Lentiviral vectors with two independent

internal promoters transfer high-level expression of multiple transgenes to human hematopoietic stem-progenitor cells. *Mol Ther* **7**, 827–838.

4. Lois C, Hong EJ, Pease S, Brown EJ, and Baltimore D. (2002) Germline transmission and tissue-specific expression of transgenes delivered by lentiviral vectors. *Science* **295**, 868–872.

5. Zufferey R, Dull T, Mandel RJ, Bukovsky A, Quiroz D, Naldini L, and Trono D. (1998) Self-inactivating lentivirus vector for safe and efficient in vivo gene delivery. *J Virol* **72**, 9873–9880.

6. Zufferey R, Nagy D, Mandel RJ, Naldini L, and Trono D. (1997) Multiply attenuated lentiviral vector achieves efficient gene delivery in vivo. *Nat Biotechnol*, **15**, 871–875.

7. Naldini L, Blomer U, Gallay P, Ory D, Mulligan R, Gage FH, Verma IM, and Trono D. (1996) In vivo gene delivery and stable transduction of nondividing cells by a lentiviral vector. *Science* **272**, 263–267.

8. Thomson JA, Itskovitz-Eldor J, Shapiro SS, Waknitz MA, Swiergiel JJ, Marshall VS, and Jones JM. (1998) Embryonic stem cell lines derived from human blastocysts. *Science* **282**, 1145–1147.

9. Kaufman DS and Thomson JA (2002) Human ES cells: haematopoiesis and transplantation strategies. *J Anat* **200**, 243–248.

10. Zhan X, Dravid G, Ye Z, Hammond H, Shamblott M, Gearhart J, Cheng L. (2004) Functional antigen-presenting leucocytes derived from human embryonic stem cells in vitro. *Lancet* **364**, 163–171.

11. Ma Y, Ramezani A, Lewis R, Hawley RG, and Thomson JA. (2003) High-Level sustained transgene expression in human embryonic stem cells using lentiviral vectors. *Stem Cells* **21**, 111–117.

12. Zhou BY, Ye Z, Chen G, Gao Z, Zhang YA, and Cheng L. (2006) Inducible and reversible transgene expression in human stem cells after efficient and stable gene transfer. *Stem Cells*, **25**, 779–789.

18

Retroviral Integration Site Analysis in Hematopoietic Stem Cells

Olga S. Kustikova, Christopher Baum, and Boris Fehse

Summary

Stable transgene insertion into a host genome irrevocably and unambiguously marks individual cells and all their descendants, i.e., the respective cell clone. Based thereon, retroviral gene marking has become an important tool for investigating the in vivo fate of different cell types, both in animal models and in clinical gene transfer. Moreover, identification of (vector) insertion sites in malignant clones transformed because of insertional activation of proto-oncogenes after experimental as well as therapeutic retroviral gene transfer has resulted in new insights into oncogenic transformation of hematopoietic stem cells (HSCs). However, because of the high sensitivity of the PCR-based methods for insertion site detection, researchers are often confronted with large numbers of different insertion sites/cell clones whose contribution to the given state is hard to judge. A relatively simple ligation-mediated polymerase chain reaction (LM-PCR) method allows the preferential analysis of insertion sites in those cell clones that significantly contributed to the cell pool analyzed. In murine bone marrow transplantation models, we have shown that this method is very useful to analyze the impact of retroviral insertion sites on both malignant and benign clonal dominance of individual repopulating HSC.

Key Words: PCR; insertional mutagenesis; clonality; gene transfer; retroviral vector; RVISs.

1. Introduction

Retroviral (including lentiviral) vectors are widely used for genetic modification of hematopoietic cells. Retrovirally delivered DNA integrates into a given, semi-randomly selected site of a cellular chromosome as a monomer, with a predictable configuration of the transgene's long terminal repeats (LTRs).

From: *Methods in Molecular Biology, vol. 430: Hematopoietic Stem Cell Protocols*
Edited by: K. D. Bunting © Humana Press, Totowa, NJ

One of the modules within the LTRs, the so-called U3 region, is particularly amenable to genetic modification, thus allowing the deletion and introduction of sequences of interest. Moreover, larger expression cassettes can be introduced in between the LTRs, provided that the sequences required for particle incorporation and reverse transcription are maintained *(1)*.

As retroviral vector integration (RVI) may disrupt cellular genes or their regulatory regions, the clonal fitness of engineered cells may be enhanced or impaired. Sequences included in the vector such as viral enhancer/promoter elements may further increase the impact of the RVI site (RVISs) on clonal behavior. RVISs may thus induce clonal imbalance, potentially resulting in (oligo-)clonal dominance of normal or malignant hematopoiesis *(2–6)*. Clonal imbalance appears to progress over time and as a function of environmental stress. The molecular identification of RVISs may thus provide important insights into cellular genes which impact self-renewal and transformation under various experimental conditions. This is of particular relevance, but not limited to studies addressing stem cell biology, leukemogenesis, and the development of gene therapy for hematopoietic disorders.

As mice as well as larger animals and humans may have approximately 16,000 hematopoietic stem cells *(7)*, an exhaustive analysis of all retrovirally engineered clones that contribute to hematopoiesis remains a challenge, despite major progress *(6,8,9)*. A shortcoming of the high sensitivity of PCR-based methods currently used for the analysis of RVISs is the large amount of different sequences obtained and the resulting difficulty to judge the impact of a given single insertion. For example, in animals with a long-term reconstituted hematopoiesis (after bone marrow transplantation), thousands of insertion sites may be present in different long-living immune cells (such as memory T and B lymphocytes). However, most investigators would obviously be more interested in insertions that happened in HSC. Therefore, application of a method preferentially detecting insertion in dominant clones (those that succeed in interclonal competition) would be preferable. We found that a simplified procedure of ligation-mediated PCR (LM-PCR) preferentially amplifies RVISs of abundant cell clones.

LM-PCR was originally introduced to increase the sensitivity of studies addressing promoter methylation *(10,11)* and subsequently found wide-spread application in many fields of molecular biology (>350 entries in PubMed as of Oct 2006). Schmidt et al. *(12)* have adapted this method for retrieval of various forms of RVISs by introducing a solid-phase-based enrichment step of the initial primer extension (PE) product, thus increasing the sensitivity and specificity to retrieve template DNA for subsequent linker ligation before nested PCR. Our LM-PCR protocol was derived from the one introduced by Schmidt et al. Minor procedural modifications improved the utility of this method for the amplification products that reflect dominant clones *(13)*. The resulting protocol is relatively fast, inexpensive, and sufficiently sensitive to obtain dominant RVISs thus allowing to follow the clonal composition and fate of genetically modified HSCs or (pre-)leukemic clones.

2. Materials

2.1. LM-PCR Strategy for RVISs Identification and Primer Design

1. Electronic version of the vector sequence.
2. Program for restriction site analysis of the sequence, e.g., DNA Star program (DNASTAR, Inc., Madison, WI, USA).
3. Program for primer design, e.g., http://www.premierbiosoft.com/netprimer/index.html

2.2. Genomic DNA Preparation

QIAamp DNA blood kit (QIAGEN; Hilden, Germany) or any comparable method for high-quality DNA isolation.

2.3. Preparation of Asymmetric Polylinker Cassette

1. Polylinker oligonucleotides (*see* **Note 1**):
   Linker-Oligo 1: 5´-GACCCGGGAGATCTGAATTCAGTGGCACAGCAGT TAGG-3´;
   Linker-Oligo 2: 5´-CCTAACTGCTGTGCCACTGAATTCAGATCTCCCG-3´.
2. 5× annealing buffer: 0.5 M Tris–HCl pH 7.4, 0.35 M MgCl$_2$. Store at –20 °C.
3. Water bath set at 70 °C.

2.4. Restriction Digest and Precipitation

1. Four-cutter restriction enzyme: for example, Tsp509 I (10 U/µl, New England BioLabs [NEB], Frankfurt a. Main, Germany).
2. 3 M NaAc pH 5–6.
3. Glycogen (20 µg/µl, Roche, Penzberg, Germany).
4. Absolute ethanol and 70% ethanol.

2.5. Primer Extension

1. Native Pfu DNA polymerase (2.5 U/µl, Stratagene, La Jolla, CA, USA).
2. 10× Native Pfu buffer (Stratagene).
3. dNTP mix (25 mM each, Stratagene).
4. QIA Quick PCR kit (QIAGEN).
5. Primer 5´ biotin-rvLTR I, e.g., for SF vectors:
   5´ Biotin-CTGGGGACCATCTGTTCTTGGCCTC 3´.

6. Thermocycler.
7. Rotator: sample mixer (Dynal Biotech ASA, Oslo, Norway).

2.6. Biotin–Streptavidin Interaction

1. Dynabeads M-280 Streptavidin (10 µg/µl, Dynal, Hamburg, Germany).
2. 2× BW buffer: 10 mM Tris–HCl pH 7.5, 1 mM ethylenediaminetetraacetic acid (EDTA), 2.0 M NaCl. Store aliquots at –20°C.
3. Dynal Magnetic Particle Concentrator [Magnetic tube holder] (Dynal).

2.7. Polylinker Cassette Ligation

1. T4 DNA ligase (400 U/µl, NEB).
2. 10× NEB ligation buffer (NEB).
3. 40 µM asymmetric polylinker mix. Store in single-use aliquots at –20°C.

2.8. First and Nested PCR

1. Extensor Hi-fidelity PCR Master mix (AB gene, Hamburg, Germany).
2. Primers OC1, OC2 (12):
   OC1: 5′-GACCCGGGAGATCTGAATTC-3′,
   OC2: 5′-AGTGGCACAGCAGTTAGG-3′.
3. Primers RvLTR 2, RvLTR 3 (for SF vectors):
   Primer RvLTR 2: 5′-GCCCTTGATCTGAACTTCTC-3′,
   Primer RvLTR 3: 5′-CCATGCCTTGCAAAATGGC-3′.

2.9. Separation of PCR Products in an Agarose Gel

1. Agarose.
2. TBE buffer.
3. Gel tanks.
4. Gel extraction kit (QIAGEN).

2.10. Purification of Dominant Products and Direct Sequencing

1. 3 M NaAc pH 5–6, glycogen (20 µg/µl) and absolute ethanol.
2. Big Dye Terminator v1.1 Cycle Sequencing Kit (Applied Biosystems, Foster City, CA, USA) or CEQ Dye Terminator Cycle Sequencing with Quick Start Kit (Beckman Coulter, Fullerton, CA, USA).
3. Sequencing primer for SF vectors: 5′-CTTGCAAAATGGCGTTAC-3′.

2.11. Database Analysis of RVISs

Computer with internet access.

3. Methods

LM-PCR for RVISs identification comprises the following major procedures: (**Subheading 3.1.**) retroviral vector sequence analysis, choice of restriction enzymes and primer design; (**Subheading 3.2.**) preparation of genomic DNA from peripheral blood, spleen, or bone marrow; (**Subheading 3.3.**) preparation of asymmetric polylinker cassette; (**Subheading 3.4.**) restriction digestion of genomic DNA; (**Subheading 3.5.**) PE step; (**Subheading 3.6.**) target DNA enrichment; (**Subheading 3.7.**) linker cassette ligation; (**Subheading 3.8.**) first PCR and nested PCR to amplify the target DNA (insertion site); (**Subheading 3.9.**) insertion pattern visualization by gel-electrophoresis; (**Subheading 3.10.**) purification and sequencing of dominant products; (**Subheading 3.11.**) database search to identify retroviral integration sites (RVISs); (**Subheading 3.12.**) protocol peculiarities.

3.1. Retroviral Vector Sequence Analysis, Choice of Restriction Enzymes, and Primer Design:

1. Delineate the structure of the integrated into the host genome retroviral provirus.
2. Choose whether the 5´- or 3´-genomic flanking region of the provirus will be the object of investigation. On the basis of this decision, design LTR-specific primers and select (a) four-cutter enzyme(s) for digestion of genomic DNA (Cave! The enzyme must not be sensitive to any form of methylation).
3. Check whether the size of the internal control is expected to be within the range of 130–500 bp (*see* **Fig. 1**).

3.2. Preparation of Genomic DNA from Peripheral Blood, Spleen, or Bone Marrow

1. To extract genomic DNA from at least 0.5×10^6 cells of mouse bone marrow, spleen, or liver or 100 µl of peripheral blood, use the QIAamp DNA blood kit (QIAGEN) or any alternative procedure that yields high-quality genomic DNA.
2. For long-term storage, keep the genomic DNA at –20 °C, for short term at +4 °C.
3. In these procedures, pay attention to avoid potential contamination with plasmids or PCR products to which the PCR primers used below may anneal.

3.3. Preparation of Asymmetric Polylinker Cassette

1. Mix in an Eppendorf tube:
   40 µl: H$_2$O;
   20 µl: Linker-Oligo 1 (200 pmol/µl);
   20 µl: Linker-Oligo 2 (200 pmol/µl) (*see* **Note 1**).

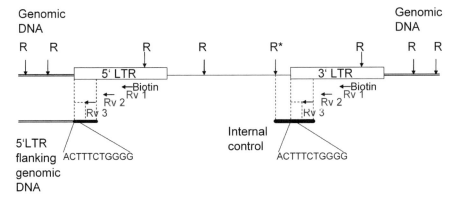

Fig. 1. Schematic representation of primer design for investigation the 5′ long terminal repeat (LTR) flanking genomic region. The optimal amplification range of ligation-mediated polymerase chain reaction (LM-PCR) is within the range of 100–800 bp. The size of the internal control should be 130–500 bp. It can be calculated by adding the distance from the 5′ end of the 3′ LTR to the first recognition site of the restriction enzyme within the vector (R*) to the length of the co-amplified linker and LTR sequences (including the specific primers). R, 4-bp-cutter restriction enzyme recognition site; ACTTTCTGGGG, reverse LTR-specific sequence; Rv 1, Rv 2, Rv 3, reverse primers (LTR specific).

2. Keep the mixture for 5 min at 70 °C in a water bath. Then add at room temperature (RT) 20 µl of 5× annealing buffer, mix, and incubate for another 5 min at 70°C. Switch off the water bath and leave the tube in the self-cooling water overnight.
3. Next morning, prepare RT aliquots of 5–10 µl each, which may be stored at –20°C. Use each aliquot only once.

3.4. Restriction Digestion of Genomic DNA

1. Restriction digest is carried out in a final volume of 30 µl in an Eppendorf tube (*see* **Note 2**):
   0.1–1.0 µg genomic DNA;
   3 µl: 10× restriction buffer;
   0.5 µl (5U): restriction enzyme; ad 30 µl H$_2$O.
2. Incubate for 2 h at 37 °C or at 65 °C (depending on the chosen restriction enzyme).
3. Precipitate overnight at –20°C with 0.3 M NaAc pH 5–6 (final concentration), 2 vol of absolute ethanol, and 20 µg of glycogen.
4. Next morning, spin down at 16000 g (Eppendorf centrifuge) for 20 min, carefully discard supernatant, add 200 µl of 70% ethanol, centrifuge again for 5 min, discard the supernatant.
5. Resuspend the air-dried pellet in 10 µl of H$_2$O (*see* **Note 3**).
6. Restricted genomic DNA is best stored at –20°C.

3.5. PE Step

1. Mix in a PCR tube (end volume 20 μl) (*see* **Note 2**):
   10 μl: DNA (from **Subheading 3.4, step 5**);
   2 μl: 10× Native Pfu buffer;
   0.16 μl: 25 mM dNTP;
   1 μl: Primer rvLTR I–biotin (0.25 pmol/μl);
   1 μl: (2.5 U) Native Pfu DNA polymerase (Stratagene);
   5.84 μl: H_2O.
2. Incubate in a thermocycler with following program configuration:
   first step: 95°C, 5 min;
   second step: 64°C, 30 min;
   third step: 72°C, 15 min.
3. After PE reaction, clean the product using QIA Quick PCR kit (QIAGEN), elute in 40 μl of H_2O in an Eppendorf tube.
   The product of PE may be stored at –20 °C. The protocol may be interrupted at this point before proceeding to the next step.

3.6. Target DNA Enrichment

1. 200 μg of Dynabeads M-280 Streptavidin (*see* **Note 4**) should be prepared for the enrichment of the PE product. Therefore, transfer 20 μl of standard bead solution (10 μg/μl) into an Eppendorf tube and wash two times with 150 μl 2× BW buffer as follows: add 2× BW buffer, mix gently by finger knocking, put onto a magnetic tube holder, remove the buffer, take out from holder, repeat the procedure.
   Resuspend in 40 μl of 2× BW buffer.
2. Add the washed Dynabeads (200 μg in 40 μl of 2× BW buffer) to 40 μl of purified PE reaction (*see* **Subheading 3.5, step 3**).
3. Incubate the solutions with gentle rotation of the tubes at RT for 2–4 h. Then wash the mixture twice with 100 μl H_2O (the same way as with 2× BW buffer, *see* **Subheading 3.6., step 1**). Resuspend by pipetting in 5 μl H_2O (*see* **Note 3**).

3.7. Linker Cassette Ligation

1. Prepare all components for linker cassette ligation on ice: thaw the aliquot of asymmetric polylinker mix (*see* **Subheading 3.3., step 3**) on ice; thaw ligation buffer at RT, vortex (to dissolve the DTT pellet), and spin down, put on ice.
2. Mix the following components to an end volume of 10 μl in a PCR tube or an Eppendorf tube (*see* **Note 2**):
   5 μl: DNA (from **Subheading 3.6., step 3**);
   1 μl: 10× NEB ligation buffer;

1 μl: asymmetric polylinker mix (40 μM);

0.2 μl (80 U): Ligase (NEB 400 U/μl);

2.8 μl: H_2O (4 °C).

Put the ice-cold ligation reaction into a thermostat (best a thermocycler) and incubate at 16 °C overnight.

3. Next morning, wash the ligation mix two times with 100 μl of H_2O using the magnetic tube holder (*see* **Subheading 3.6., step 1**), re-suspend in 10 μl of H_2O by pipetting. After washing, the ligation mix may be stored at –20 °C for several months.

3.8. First PCR and nested PCR to amplify the target DNA (Insertion Site)

1. Program configuration for first and nested PCR:

   94°C, 2 min;

   94°C, 15 sec;

   60°C, 30 sec;

   68°C, 2 min;

   30 cycles (**steps 2–4**);

   68°C, 10 min.

2. First PCR end volume: 25 μl; mix in a PCR tube (*see* **Note 2**):

   1 μl: DNA (from **Subheading 3.7., step 3**);

   12.5 μl: Extensor Hi-fidelity PCR Master mix (AB gene);

   1 μl: Primer OC 1 (25 pmol/μl);

   1 μl: Primer RvLTR 2 (25 pmol/μl);

   9.5 μl: H_2O.

3. Nested PCR: the same cycling conditions as for the first reaction. To prepare DNA template for nested PCR dilute the first PCR mix 1:500. Mix the following components in a PCR tube (*see* **Note 2**):

   1 μl: DNA;

   12.5 μl: Extensor Hi-fidelity PCR Master mix (AB gene);

   1 μl: Primer OC 2 (25 pmol/μl);

   1 μl: Primer RvLTR 3 (25 pmol/μl);

   9.5 μl: H_2O.

4. For both PCRs, the use of a ready-to-use Extensor Hi-Fidelity PCR Master mix is highly recommended. This master mix contains a mixture of polymerases (with proofreading activity), reaction buffer, dNTPs, $MgCl_2$, and gel-electrophoresis loading buffer. The Extensor Hi-Fidelity PCR Master mix provides better fidelity and yields than standard Taq DNA polymerase.

3.9. Insertion Pattern Visualization by Gel-Electrophoresis

For insertion pattern visualization, it is recommended to use 2% agarose 0.5× TBE gel electrophoresis. Pay attention to the size of internal controls of

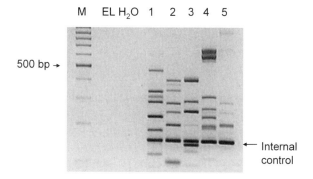

Fig. 2. An example of ligation-mediated polymerase chain reaction (LM-PCR) for retroviral vector integration site (RVIS) identification. The intensity of the internal control correlates with the number of RVIS (e.g., if there are 3 RVISs, the internal control may be about three times more intense than in the presence of a single RVIS). M, marker; EL, empty lane; H_2O, water control; 1–5, samples from a gene-marking experiment with serial Bone Marrow Transplantation (BMT) as described *(4,14)*.

your probes, which serve as an in-reaction quality measure for your LM-PCRs. An example result is shown in **Fig. 2** (*see* **Note 5**).

Cut out dominant bands and extract DNA fragments for further analysis using Gel extraction kit (QIAGEN) or any other procedure for extraction of DNA from agarose gels.

3.10. Purification and Sequencing of Dominant Products

To improve the quality of direct sequencing, it is better to precipitate the eluted fragments using 3 M NaAc pH 5–6, glycogen, and absolute ethanol. Conditions for sequencing should be established according to the used sequencing reagents and device. In addition to the LTR-specific, a linker-specific primer may be used for sequencing (*see* **Note 6–8**).

3.11. Database Search to Identify RVISs

1. Sequenced PCR products may only be viewed as RVIS if they contain the expected vector-specific sequences. Short fragments should also reveal polylinker sequences.
2. Before database search, sequences should be edited using DNA analysis software tools to remove bases that correspond to the retroviral vector and to the polylinker. Submit your edited sequences to genome blast programs such as http://www.ncbi.nlm.nih.gov/BLAST or http://www.ensembl.org.

3.12. Protocol Peculiarities

The present protocol contains a number of minor modifications as compared with the originally published one by Schmidt et al. *(12)*. In our hands, those changes led to higher yields of PCR products and better reproducibility:

1. Depending on the proportion of obtained and identifiable (**Subheading 3.11.**) insertions as related to the expected numbers (determined, for instance, based on Southern blot hybridization), it might be necessary to repeat all above steps in **Subheadings 3.1.–3.11** using alternative restriction enzymes. Based on our experience, analyses performed with two different restriction endonucleases do result in an approximately 90% overlap (absolutely identical RVISs coordinates) but also reveal some additional RVISs. Thus, repetition with an additional enzyme represents a very good quality control for your LM-PCR protocol and clonal dominance stage.

2. Alternatively, one may also repeat LM-PCR on "the other side" of the vector (for instance 3´ instead of 5´) using a different set of specific primers. In a situation of clonal dominance, the majority of insertions should confirm data obtained with the initial protocol.

3. The product of the PE step is not concentrated using a spin column but just purified with QIA Quick PCR kit (QIAGEN) (*see* **Subheading 3.5., step 3**).

4. After linker cassette ligation (*see* **Subheading 3.7.**), we proceed directly to first PCR (without additional NaOH-mediated denaturation).

5. The use of an enzyme mix (containing a proofreading polymerase) for the amplification steps (*see* **Subheading 3.8.**) is advantageous over single *Taq* polymerase. This is consistent with numerous reports showing that PCR with mixtures of *Taq* polymerase with enzymes with proofreading activity result in both higher DNA yields and PCR fidelity than PCRs with standard *Taq* DNA polymerase alone. Based on our experience, the Extensor Hi-fidelity PCR Master mix (ABgene) is a good choice.

4. Notes

1. Oligonucleotides for polylinker preparation should lack 5´ phosphates to prevent self-ligation. Given the length of the linker primers, it is important to obtain oligonucleotides of high quality.

2. Master mix preparation is recommended for all reactions before adding individual samples. Always start collecting reaction components from water; do not forget to carefully resuspend and spin all components before use.

3. Take care to collect all pellets from the walls by pipetting, especially when working with dynabeads.

4. Never vortex or centrifuge the Dynabeads M-280 Streptavidin.

5. The source of the genomic DNA may change the results: Clones that dominate the bone marrow are not necessarily over-represented in peripheral blood, spleen, or thymus. This is of special relevance in cases of pre-malignant or malignant

hematopoiesis. For cases with intact hematopoiesis, we prefer bone marrow as the source of genomic DNA.

6. Sometimes, direct sequencing of eluted PCR products may not work well. In this case, we recommend sequencing after subcloning of the purified PCR product. If you decide to subclone the entire, unpurified PCR for subsequent shotgun sequencing, it may be hard to determine which sequence corresponds to the dominant PCR product and which to minor products that may be of the same or similar size. Shotgun cloning also leads to a significant retrieval of the internal control band unless you excise this band before eluting DNA.

7. In cases without clonal dominance (e.g., because of low-marking levels, low selection pressure, short observation time, no major impact of RVIS on clonal fitness), insertion sites of murine endogenous retroviruses (M-ERV) may become detectable by LM-PCR analysis of murine cells. Therefore, inclusion of a mock control from the same mouse strain is recommended. M-ERV-derived signals may be particularly misleading if their amplification takes place in the optimal amplification range (*see* **Fig. 1**). In those cases, we recommend to change the restriction enzyme, because this can completely or partially remove signals reflecting the presence of M-ERV.

8. Pay attention to potential sampling errors *(10,11)*. We recommend repeating the LM-PCR for each sample to ensure reproducibility of the band pattern.

General Note: Formal demonstration of HSC transduction requires detection of the same RVISs in sorted hematopoietic subpopulations, preferably myeloid cells and T lymphocytes *(9)*.

Acknowledgments

This work was supported by grants from the Deutsche Forschungs-gemeinschaft (DFG SPP1230, to B.F. and C.B.), the European Union (CONSERT-LSHB-CT-2004-005242, to C.B.), and the National Cancer Institute (R01-CA107492-01A2, to C.B.). We thank Nora Zingler (Heinrich-Pette-Institute, Hamburg, Germany) for technical advice.

References

1. Baum, C., Schambach, A., Bohne, J., and Galla, M. (2006) Retrovirus vectors: toward the plentivirus? *Mol. Ther* . **13**, 1050–1063.
2. Li, Z., Dullmann, J., Schiedlmeier, B., Schmidt, M., von Kalle, C., Meyer, J., Forster, M., Stocking, C., Wahlers, A., Frank, O., Ostertag, W., Kuhlcke, K., Eckert, H. G., Fehse, B., and Baum, C. (2002) Murine leukemia induced by retroviral gene marking. *Science* **296**, 497.
3. Hacein-Bey-Abina, S., Von Kalle, C., Schmidt, M., McCormack, M. P., Wulffraat, N., Leboulch, P., Lim, A., Osborne, C. S., Pawliuk, R., Morillon, E., Sorensen, R., Forster, A., Fraser, P., Cohen, J. I., de Saint Basile, G., Alexander, I.,

   Wintergerst, U., Frebourg, T., Aurias, A., Stoppa-Lyonnet, D., Romana S, Radford-Weiss, I., Gross, F., Valensi, F., Delabesse, E., Macintyre, E., Sigaux, F., Soulier, J., Leiva, L. E., Wissler, M., Prinz, C., Rabbitts, T. H., Le, Deist, F., Fischer, A., and Cavazzana-Calvo, M. (2003) LMO2-associated clonal T cell proliferation in two patients after gene therapy for SCID-X1. *Science* **302**, 415–419.

4. Kustikova, O., Fehse, B., Modlich, U., Yang, M., Düllmann, J., Kamino, K., von Neuhoff, N., Schlegelberger, B., Li, Z., and Baum, C. (2005) Clonal dominance of hematopoietic stem cells triggered by retroviral gene marking. *Science* **308**, 1171–1174.

5. Seggewiss, R., Pittaluga, S., Adler, R. L., Guenaga, F. J., Ferguson, C., Pilz, I. H., Ryu, B., Sorrentino, B. P., Young, W. S. 3rd, Donahue, R. E., von Kalle, C., Nienhuis, A. W., and Dunbar, C. E. (2006) Acute myeloid leukemia associated with retroviral gene transfer to hematopoietic progenitor cells of a rhesus macaque. *Blood* **107**, 3865–3867.

6. Ott, M. G., Schmidt, M., Schwarzwaelder, K., Stein, S., Siler, U., Koehl, U., Glimm, H., Kühlcke, K., Schilz, A., Kunkel, H., Naundorf, S., Brinkmann, A., Deichmann, A., Fischer, M., Ball, C., Pilz, I., Dunbar, C., Du, Y., Jenkins, N. A., Copeland, N. G., Luthi, U., Hassan, M., Thrasher, A. J., Hoelzer, D,, von Kalle, C., Seger, R., and Grez, M. (2006) Correction of X-linked chronic granulomatous disease by gene therapy, augmented by insertional activation of MDS1-EVI1, PRDM16 or SETBP1. *Nat. Med.* **12**, 401–409.

7. Abkowitz, J. L., Catlin, S. N., McCallie, M. T., and Guttorp, P. (2002) Evidence that the number of hematopoietic stem cells per animal is conserved in mammals. *Blood* **100**, 2665–2667.

8. Schmidt, M., Carbonaro, D. A., Speckmann, C., Wissler, M., Bohnsack, J., Elder, M., Aronow, B. J., Nolta, J. A., Kohn, D. B., and von Kalle, C. (2003) Clonality analysis after retroviral-mediated gene transfer to CD34+ cells from the cord blood of ADA-deficient SCID neonates. *Nat. Med.* **9**, 463–468.

9. Schmidt, M., Hacein-Bey-Abina, S., Wissler, M., Carlier, F., Lim, A., Prinz, C., Glimm, H., Andre-Schmutz, I., Hue, C., Garrigue, A., Le Deist, F., Lagresle, C., Fischer, A., Cavazzana-Calvo, M., and von Kalle, C. (2005) Clonal evidence for the transduction of CD34+ cells with lymphomyeloid differentiation potential and self-renewal capacity in the SCID-X1 gene therapy trial. *Blood* **105**, 2699–2706.

10. Pfeifer, G. P., Steigerwald, S. D., Mueller, P. R., Wold, B., and Riggs, A. D. (1989) Genomic sequencing and methylation analysis by ligation mediated PCR. *Science* **246**, 810–813.

11. Steigerwald, S. D., Pfeifer, G. P., and Riggs, A. D. (1990) Ligation-mediated PCR improves the sensitivity of methylation analysis by restriction enzymes and detection of specific DNA strand breaks. *Nucleic Acids Res.* **18**, 1435–1439.

12. Schmidt, M., Hoffmann, G., Wissler, M., Lemke, N., Mussig, A., Glimm, H., Williams, D. A,, Ragg, S., Hesemann, C. U., and von Kalle C. (2001) Detection and direct genomic sequencing of multiple rare unknown flanking DNA in highly complex samples. *Hum. Gene Ther.* **12**, 743–749.

13. Kustikova, O., Geiger, H., Li, Z., Brugman, M. H., Chambers, S. M., Shaw, C. A., Pike-Overzet, K., de Ridder, D., Staal, F. J. T., von Keudell, G., Cornils, K., Nattamai, K., J., Modlich, U., Wagemaker, G., Goodell, M. A., Fehse, B., and Baum, C. (2007) Retroviral vector insertion sites associated with dominant hematopoietic clones mark "stemness" pathways. *Blood* **109**, 1897–1907.

14. Li, Z., Fehse, B., Schiedlmeier, B., Düllmann, J., Frank, O., Zander, A. R,, Ostertag, W., and Baum, C. (2002) Persisting multilineage transgene expression in the clonal progeny of a hematopoietic stem cell. *Leukemia* **16**, 1655–1663.

19

A Detailed Protocol for Bacterial Artificial Chromosome Recombineering to Study Essential Genes in Stem Cells

Andriy Tsyrulnyk and Richard Moriggl

Summary

Bacterial artificial chromosome (BAC) recombineering is a novel technique for DNA manipulation. It starts from an original chromosomal gene locus that is modified to introduce a transgene under the expression control of the original gene locus. In most cases a cell type specific promoter is chosen and the transgene is placed in a way that the exon containing the start codon is replaced. Alternatively, BACs such as the Rosa26 BAC are chosen because of their known open chromatin and ubiquitous promoter activity that allows a broad expression profile of the transgene in the whole body. Thus, transgenes can be overexpressed within their natural transcriptional regulatory circuit. BAC transgenes have a high tendency to maintain their appropriate chromatin status because the endogenous locus was expressed in different cell types. Here, we give a detailed protocol based on the original idea to choose a BAC approach until the injection of the modified BAC DNA that leads to the generation of novel transgenic mouse lines. As an example for a BAC mouse model suitable for the analysis of stem cell or hematopoietic stem cell functions, we chose modification of the locus for the transcription factor Stat3. Stat3 variants replace the wild-type Stat3 gene to study their function in particular in the earliest cell types of the body.

Key Words: BAC-recombineering protocol; haematopoietic stem cells; ES cells; Stat3 transgenic.

1. Introduction

Bacterial artificial chromosome (BAC) recombineering is also called Red/ET recombineering, and the methodology was mainly developed in the laboratory of A. Francis Stewart [1,2]. The laboratory of Neil Copeland also adapted this

From: *Methods in Molecular Biology, vol. 430: Hematopoietic Stem Cell Protocols*
Edited by: K. D. Bunting © Humana Press, Totowa, NJ

prophage system by transferring it to *Escherichia coli* DH10B cells, a BAC host strain. In addition, arabinose inducible *cre* and *flp* genes were introduced into these cells to facilitate BAC modification using loxP and FRT sites *(3)*.

Recombination occurs through homology arms (HA), which are stretches of DNA shared by the two molecules that recombine. Homologous recombination allows the exchange of genetic information between two DNA molecules in a precise, specific, and faithful manner and in a quality that is optimal for DNA engineering regardless of the size. Cloning of plasmids that carry large inserts, such as bacterial artificial chromosomes (BACs) *(4)*, offers great potential: When a BAC is appropriately manipulated and handled, it is very stable and the transgene insert can be relatively big. All known regulatory regions essential for desired transgene expression or additional reporter genes for labeling of the transgene expression can be included. In general, the modified BAC is used for the generation of transgenic mice that allows a more physiological expression of a transgene. The copy number of a BAC transgenic mouse is limited leading to more physiologic expression patterns, but this depends also on the chosen transcriptional regulatory region to integrate the transgene into the BAC. Normally, not more than 1–10 copies of a BAC are found to be integrated into the genome, whereas with classical transgenes several hundred copies can be randomly integrated.

We want to briefly introduce the two most commonly used BACs in the field: the Rosa26 BAC *(5)* and the collagen 1a1 BAC *(6)*, both of which have open chromatin, strong transcriptional regulatory regions, and wide expression patterns in almost all tissues. It is generally accepted that the Rosa26 BAC allows expression of transgenes during embryo development, whereas the collagen 1a1 BAC allows expression only in adult tissues. Differences in the expression pattern, strength of the promoter, or an open chromatin matter for the biological questions. Published successful BAC transgenic mice describe many stories to individual success *(7–10)*. Choosing an unknown BAC contains a certain risk, but in the case of going for a new BAC with a new transgene engineered through BAC recombineering, new scientific avenues can be paved.

We point out that a further step of complexity was reached through the use of site-specific recombinase systems such as the CRE-loxP system *(11)*. A STOP cassette flanked by loxP sites can be placed in front of a transgene, therefore, allowing transcription only after deletion of the STOP cassette through CRE recombinase action. Alternatively, gene inversion techniques catalyzed again by site-specific recombinase systems can be used to switch a transgene from an inactive to an active configuration. Overall, these techniques are sophisticated and they certainly parallel modern BAC-recombineering strategies to generate new transgenic mouse models that are for example inducible. Because of the higher complexity in the description of such techniques, we avoid them here and focus on the basic BAC approach. Next, we give an example for a gene essential in stem cell functions.

Embryonic stem (ES) cell self-renewal depends upon extrinsic signals from leukemia inhibitory factor (LIF) and bone morphogenetic proteins (BMPs) *(12,13)*. Many cytokines such as LIF, oncostatin M (OSM), interleukin-6 (IL-6), IL-11, IL-31, or ciliary neurotrophic factor (CNTF) and growth factors such as epidermal growth factor (EGF) or platelet-derived growth factor (PDGF) activate Stat3. The transcription factor Stat3 has been the focus of increasing attention in the field of cancer research, because many transforming tyrosine kinases such as V-SRC, NEU, caKITR($D_{816} \rightarrow$ V), BCR-ABL, FLT3 ITD, or NPM-ALK activate Stat3 through tyrosine phosphorylation, but also oncoproteins such as RAS or Polyoma middle T antigen can lead to activation of Stat3 proteins *(14–16)*. BMPs regulate the expression of Id transcription factor genes and together with Stat3 they are essential for ES cell self-renewal *(17)*. Stat3-deficient mice are embryonic lethal, but conditional deletion of Stat3 or specific knock in mice for the Stat3alpha or Stat3beta isoforms have revealed unique functions for Stat3 proteins in different cell types *(18,19)*. In myeloid cells, conditional deletion of Stat3 resulted in negative regulation of granulopoiesis *(20,21)*, whereas hyperactivation of gp130 signaling is associated with hyperactivation of Stat3 proteins and haematopoietic abnormalities such as thrombocytosis *(22)*. A more recent study displayed that early regenerative activity of transplanted hematopoietic stem cells (HSCs) could be either up- or down-regulated by genetically modulating the levels of activated Stat3 *(23)*. The effect was intrinsic to HSC functions, and it was dependent on the HSC self-renewal under haematopoietic recovery conditions of enforced cytokine and growth factor activation. In this regard, it is also noteworthy that most hematopoietic transplant systems of mouse origin that use bone marrow or fetal liver cells that were retrovirally transduced with gene variants use IL-6 in the transplant protocol *(24,25)*.

The Stat3 transcription factor is a good example for an essential gene in ES or HSC cells. Here, it serves as our example for the generation of a BAC transgenic mouse model, where the original exon with the start codon of wild-type Stat3 will be replaced with a Stat3 mutant transgene that is either hyper- or hypo-activated (*see* **Fig. 1**; *26,27*). We describe the strategy to replace the endogenous Stat3 locus with a gain or a loss of function mutant. This allows the study of the Stat3 transcription factor in a structure–function relationship for the whole organism or the hematopoietic system in particular. The Stat3 gain- or loss-of-function approach is only an example for two interesting mutants of Stat3 that can be recombined into the Stat3 BAC. We provide the mutants with reporter gene activity to mark cells expressing them. First, the BAC is selected through a bioinformatics approach, followed by manipulations that fall

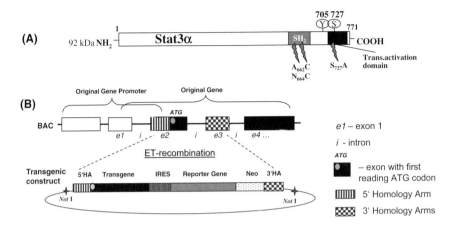

Fig. 1. Schematic overview for bacterial artificial chromosome (BAC) recombineering. (A) To study embryonic stem (ES) cell or hematopoietic stem cell (HSC)
function, two Stat3alpha variants are given as an example for the transgene-cDNA. A
gain of function mutant of Stat3 with a double mutation (A662→C and N664→C),
which is persistently active because of disulfide bridge formation *(26)* or a loss of
function mutant of Stat3 with diminished transactivation potential because of a mutation
in a critical Phospho-Ser residue at amino acid position 727 *(27)*. (B) The exon
containing the first ATG and the close fragment of the original BAC gene is replaced
by the artificial transgenic construct. The recombination occurs specifically through 5´
and 3´ Homology Arms. The modified BAC containing the transgene is under control
of the original promoter of the gene.

into routine molecular biology and genetics. Integrity checks of the BAC DNA
are important before injection can be carried out to generate a new transgenic
mouse model.

2. Materials
2.1. Cloning
1. DNA-software programs: e.g., NTI-vector, DNA-star, Clone manager.
2. Plasmid backbone with suitable restriction sites (RS) for the cloning of transgenic
   construct. It should contain the multi-cloning site (MCS), resistance cassette,
   and two NotI RS from both ends of the MCS (needed for cutting of transgenic
   construct from the plasmid backbone).
3. Restriction endonucleases with buffers, polymerase chain reaction (PCR) reagents.

2.2. Recombineering
1. Plasmids: BAC, pR6K.αβγBAD, 705Flp (*see* **Subheading 3.4**).
2. Techniques for pulse field electrophoresis, electroporation, and chromatography.

2.3. Linearizing and Purification of the BAC

1. Sea-Plug Low-melting point agarose.
2. Agarase buffer: Tris–HCl pH 6.5, 10 mM, EDTA pH 8.0, 1 mM, NaCl, 100 mM, spermine 30 mM, spermidine 70 mM.
3. CL4b sepharose.
4. Microinjection buffer: Tris–HCl pH 7.5, 10 mM, EDTA pH 8.0, 0.1 mM, NaCl 100 mM, spermine 30 mM, spermidine 70 mM.

3. Methods

3.1. BAC-DNA Library

A BAC is a large fragment of chromosomal DNA *(1,4,28)*. The size of the BAC is approximately 200 kb. There are many well-known online BAC-DNA libraries. Commonly used are NCBI and ENSEMBL.

We demonstrate how to design the BAC recombineering using as an example the replacement of the mouse *Stat3* gene by a mutated version (*see* **Fig. 1**).

As a start, the information and chromosomal localization of the gene needs to be found. For example, use the database ENSEMBL:

1. Choose ENSEMBL (www.ensemble.org); in the middle top, you will find the windows for gene searching. In the left window, choose the species (for example, *Mus musculus*). In the right window, you need to write the name of the gene you want to modify and replace (for example, *Stat3*). Then push button GO
2. The next Web page shows the links with gene-info.
3. Go into one of these links (usually the first-one). There you will find:

   a. *Description* (*see* **Table 1**, arrow 1): The full name of the gene.
   b. *Genomic Location* (*see* **Table 1**, arrow 2): Depicts where the gene localization is on the chromosome (Start and end of the gene in the chromosome).

Table 1
Ensemble Online DNA Library: Screening for Stat3 Gene Information

Gene	Stat3 (MGI Symbol) To view all Ensembl genes linked to the name click here
Vega Gene ID	OTTMUSG00000002128 [View Gene OTTMUSG00000002128 in Vega]
Author	This locus was annotated by Havana
Gene Type	Known Protein coding
Genomic Location ←2	This gene can be found on Chromosome 11 at location 100,701,188-100,755,830 ←6 This corresponds to 100,899,536-100,753,978 in VEGA coordinates The start of this gene is located in Contig AL591466.8.
Version & Date	Version 1 Gene last modified on 01/07/2004 (Created on 01/07/2004)
Description ←1	signal transducer and activator of transcription 3
Curation Method	Finished genomic sequence is analysed on a clone by clone basis using a combination of similarity searches against DNA and protein databases as we a series of ab initio gene predictions (GENSCAN, Fgenes). In addition, comparative analysis using vertebrate datasets is used to aid novel gene discover The data gathered in these steps is then used to manually annotate the clone adding gene structures, descriptions and poly-A features. The annotation is based on supporting evidence only.
☐ Transcripts	OTTMUST00000004321 OTTMUSP00000002052 Stat3-001 [Transcript info] [Exon info] [Peptide info] OTTMUST00000004322 OTTMUSP00000002053 Stat3-002 [Transcript info] [Exon info] [Peptide info] OTTMUST00000004323 no translation Stat3-003 [Transcript info] [Exon info]

4. From this page you can go to:

    a. *Transcript info* (*see* **Table 1**, arrow 3): It contains the complete transcript sequence.

    b. *Exon info* (*see* **Table 1**, arrow 4): You are presented with detailed gene information: full sequences, size of exons and introns, first reading fame ATG, stop codon, direction of transcription (+/–), and other valuable information.

    c. *Peptide info* (*see* **Table 1**, arrow 5): Displays the amino acid sequence of the protein for the selected gene (*see* **Note 1**).

5. To check for the suitable BAC: in the *Genomic Location* section, push the interactive button (*see* **Table 1** arrow 6):

    a. Next page: in the section *Detailed view* in the window *Decoration* set *BAC map* and *Gene legend* only (*see* **Table 2**, arrow 1). Close the menu.

    b. Page will be automatically reloaded, and at the bottom of the *Detailed view* section, you will be provided with all suitable BACs that contain the gene (*RP23-366E18*, etc.).

    c. We recommend checking the BAC using another DNA library database (commonly used is NCBI):

6. As an example, check for the BAC RP23-366E18:

    a. Go to the home page of *NCBI* (www.ncbi.nlm.nih.gov)

    b. In window *Search* set *Nucleotide* and in the window *For* write the name of the BAC (as example *RP23-366E18*) and then push button *GO*.

    c. The next page shows you the BAC (look again if the name of the BAC is correct, **Table 3**, arrow 1). Here, you can find the sequence of the BAC (*see* **Note 2**).

    d. To check whether the BAC contains the selected gene (Stat3), go with the PC-mouse to the *links* and push the left button on the PC-mouse. Then, choose from menu *Map Viewer* (*see* **Table 3**, arrow 2).

    e. In the next page, you need to set up the optimal demonstrative conditions: push the button *Maps&Options* and set in the window *Maps Displayed (left to right)*: *Gene, BES Clone, GenBank DNA* using *ADD/REMOVE* buttons.

    f. The reloaded page will present graphically the design of the BAC map (*see* **Table 4**, arrow 1): its access number (*see* **Table 4**, arrow 2), which fragment of the chromosome it covers (*see* **Table 4**, arrow 3), location of the gene (*see*

Table 2
Ensemble Online DNA Library: Screening for the Stat3 BAC

⊟ **Detailed view**

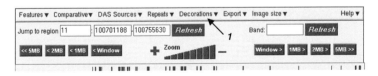

Table 3
NCBI Online DNA Library: Check Stat3 BAC and Stat3 Gene Localization (BAC and Gene Maps)

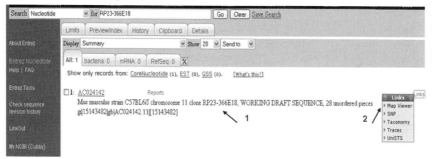

Table 4, arrow 4). At the bottom of the page, in *Summary of Maps*, you can find all information in table format (Table View) for both: gene and BAC.

3.2. Cloning Strategy for the Transgene to be Introduced Into the BAC

A BAC is the exact copy of a chromosomal fragment. Thus, the gene or the part of it you want to modify will be present in the BAC as a genomic variant containing exons and introns. The transgenic construct that should be recombined into the BAC contains the modified gene as a transcriptional variant—transgene (without introns, starts with first reading ATG and ends with stop codon). In our case, it is either a mutated gain or a loss-of-function mutant of Stat3 (*see* **Fig. 1**; *26,27*). BAC transgenic mice expressing the Stat3 variant can later be used to study Stat3 function in either a wild-type context or Stat3-deficient animals *(18,19)*. Additionally, HA, a reporter gene, and a resistance cassette have to be cloned and assembled into a BAC (*see* **Note 3**).

When the transgenic construct is cloned, it will be recombined into the BAC by replacement of the original gene fragment in the BAC. You do not need to replace the entire exon–intron original gene locus in the BAC with the transgenic construct. It is enough to replace the first two exons (*see* **Fig. 1**). Usually, the second or third exon contains the first reading frame ATG of the gene, and it is important to remove the exon with the START (ATG) codon. Only a linearized DNA molecule can be recombined into the BAC. Thus, the transgenic construct should contain the RS on both ends (usually used is Not I).

3.2.1. Design of the Transgene

Basically, every gene can be modified and recombined into the BAC. For high expression of the transgene, we recommend to clone the 5´ UTR sequence

Table 4
NCBI Online DNA Library: Detailed Maps of Genes Close to the Stat3 Locus and Graphic View of
the BAC Localization on the Relevant Region on Mouse Chromosome No. 11

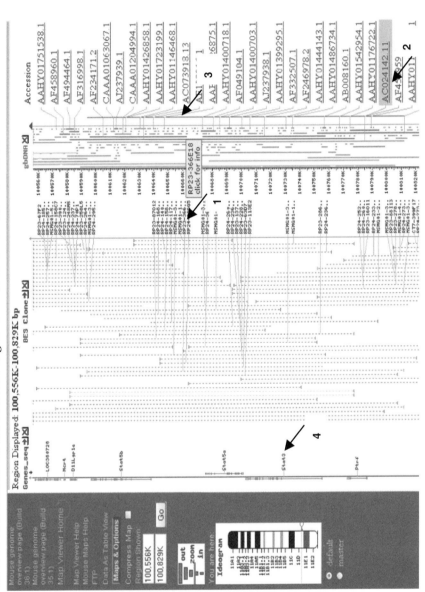

GACTCACAACCCCAGAAACACCACCATG...

Fig. 2. Recommended start sequence of the transgene. For better expression of the transgene, we recommend to clone in front of the first reading **ATG** (in bold) a 5′ UTR of the beta-globin sequence (cursive) and an optimal *KOZAK* sequence for appropriate ribosome binding and translation (underlined).

of the β-globin gene and an optimal KOZAK sequence before the ATG START-codon of the transgene to allow for efficient ribosome binding and translation (*see* **Fig. 2**). It is important to check whether a suitable BAC is available (*see* **Subheading 3.1.**).

3.2.2. Selection of the Homology Arms (HAs)

HAs are short fragments of DNA (usually 200–400 bp in length are chosen) *(1,2)*. HAs gate the fragment of DNA that should be replaced/recombined. Through HAs, the recombinase mediates site-specific recombination between two molecules of DNA. The 5′ HA is localized exactly in front of the first reading ATG of the gene, and thus, it is defined. It contains the original KOZAK sequence and part of the promoter (*see* **Fig. 3**). A fragment on the BAC used for 3′ HA can be chosen, e.g., in our case, we choose a part of exon 4 of the Stat3 gene (*see* **Fig. 3**). The distance between the 5′ and 3′ HA in the BAC ranges usually from 100 to 1,000 bp. It is important to remove at least the first reading frame exon with the original START codon. We recommend always checking the sequence of the next exon for an alternative ATG START codon(s). In such a case of an additional START codon in the second exon, it is better to remove not only the first reading frame exon but also the second exon with the alternative ATG. Removal of START codons in the first and second exon is essential to avoid unwanted translational initiation.

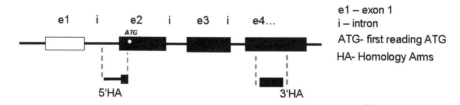

Fig. 3. Design of Homology Arms for recombineering. In case of Stat3, the 5′ HA will contain the sequence of the first intron and part of the second exon until the ATG. The 3′ HA can be chosen for example on exon 4 or close to it.

3.2.3. Choosing a Reporter Gene

Reporter (marker) genes are only a wise tool for detection of the transgenic construct expression in the cells or tissues. They all have different advantages or limitations (described elsewhere). The following reporter genes can be used:

a) For fluorescent microscopy: Green fluorescent protein (GFP), variants with non-overlapping emission spectra (Cyan-FP, Yellow-FP) and red fluorescent protein from *Discosoma* sp. dsRED.
b) For Fluorescence-activated cell sorting (FACS): human truncated cell-surface molecules, the cluster designation (CD) antigens—htCD2, htCD4, htCD5, and htCD8 are widely used.
c) A TAG-sequence can be used for direct detection of the transgene product (Western blot, immuno-cyto/histochemistry). The short DNA sequence is fused to the transgene; in most cases, it is safe to fuse the TAG to the C-terminus (in case of Stat3 or Stat5 proteins, N-terminal protein tags alter cytokine inducibility and DNA-binding affinity, but a TAG at the C-terminus behaves like wild-type Stat3 or Stat5 molecules), and the amino acids encoded by the TAG are recognized by specific antibody. The most commonly used ones are the *FLAG, HA, VSV*, or the *MYC* TAG.

3.2.4. Assembly of the Resistance Cassette

The resistance cassette in the transgenic construct is used for antibiotic selection of the modified BAC in both bacteria and ES cells. Thus, it should be expressed in eukaryotic cells as well (Neo). After selection, this cassette can be excised by the FLP-recombinase through *frt/frt* sites (*see* **Subheading 3.6.**).

The Basic BAC modification includes the following steps:

1. Design of the strategy for transgenic construct cloning: sequences of the transgene, HA, reporter gene, RS, etc. It can be done using DNA-software programs (*see* **Subheading 2.1.**).
2. Cloning of the transgenic construct.
3. In vitro testing of the functional activity of the transgenic construct.
4. Preparing competent bacteria containing BAC and recombineering machinery (*rec+*).
5. Digestion of the transgenic construct from a plasmid backbone and electroporation into the BAC *rec+* bacteria.
6. Screening for the transgene recombined BAC (tg-BAC).
7. ES-cell targeting or oocyte microinjection of purified tg-BAC.
8. Screening for the offspring.

3.2.5. Cloning: PCR Amplification, Restriction Digests, and Ligation

Every cloning step should be checked by restriction digest and sequencing!

1. PCR amplification
   Cloning is restricted to available single cutter restriction enzyme sites on the DNA. In this case, required RS can be added by PCR amplification with primers

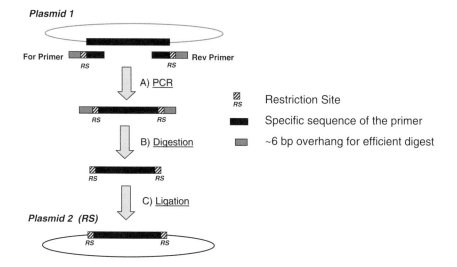

Fig. 4. Introduction of new restriction sites (RS). The DNA fragment that should be cloned needs suitable RS. They can be added by PCR using primers containing these RS.

containing these unique sites and PCR fragments can then be digested, purified, and cloned (*see* **Fig. 4**). The amplification should be done using high-fidelity PCR kits with combined polymerases for large fragment amplification and high proofreading activity.

PCR cycle conditions:

a.	Initial Denaturation	95 °C	3 min
b.	Denaturation	94 °C	0.40 min
c.	Primer annealing (*see* **Note 4**)	45–65 °C	0.50–1.10 min
d.	Extension	68–72 °C	1 min/1 kb product
e.	Go to step 2	30–35 times	
f.	Final Extension	68–72 °C	5 min

Components of the PCR Reaction Mixture:

a.	Sterile deionized water	
b.	1× PCR Buffer	
c.	Forward Primer	0.1–1 µM
d.	Reverse Primer	0.1–1 µM
e.	2 mM dNTP mix	0.2 mM of each
f.	Taq polymerase (*see* **Note 5**)	1.25 u/50 µl
g.	25 mM MgCL$_2$	1–4 mM
h.	DNA	50–100 ng

2. Restriction digests

The DNA molecule can be digested by endonucleases at its specific RS. The enzymes can be ordered from many companies (*see* **Note 6**). Digestion Mix (20 µl):

   a. 2 µl 10× Buffer
   b. 0.1–2 µg DNA
   c. 1 µl enzyme (*see* **Note 7**)
   d. Add H$_2$O to 20 µl
   e. Incubate for 2–4 h at 37 °C. Meanwhile, prepare the agarose gel.
   f. Purify the DNA band from the gel and process to ligation.

3.3. Mini/Maxiscale BAC-DNA Preparation

The BAC DNA for Restriction Digests or PCR amplification of the BAC fragments can be obtained by miniscale DNA preparation. Always do a glycerol stock of the bacteria you used (*see* **Note 3**).

Using reagents from "DNA-Plasmid purification Mini/Maxi kit" (*see* **Note 8**):

1. 5 ml liquid LB-culture grown overnight with antibiotic.
2. Spin down 4,000–5,000 × g/10 min at 4 °C.
3. Resuspend cell pellet in 250 µl of resuspension buffer (see manufacturer's protocol).
4. Add 250 µl lysis buffer, incubate 5 min at RT (mix gently by inverting the tube).
5. Add 400 µl Neutralization Buffer, incubate 5 min on ice (mix gently by inverting the tube).
6. Centrifuge maximum speed, 10 min at 4 °C.
7. Transfer 900 µl of supernatant into new tube, add 600 µl isopropanol (mix gently by inverting the tube), incubate 5 min on ice.
8. Centrifuge maximum speed 30 min at 4 °C.
9. Wash pellet with 300 µl ice-cold 70% ethanol and centrifuge with maximum speed, 10 min at 4 °C in a table centrifuge.
10. Discard Ethanol and air-dry DNA pellet at RT.
11. Dissolve DNA: add 40–50 µl TE Buffer pH = 8.0 or water, incubate 1 h at RT (do not pipet up and down and do not vortex! This could damage the DNA).
12. Gently pipet up and down 3–5 times.
13. Keep at 4 °C (can be stored up to 1 month; *see* **Note 9**).

For Pulse-Field Gel Electrophoresis (PFGE) and Southern blotting, BAC DNA can be obtained by Maxiscale DNA preparation using Plasmid DNA purification kits (*see* **Note 8**). We recommend pre-warming the elution buffer up to 60 °C. The dissolving of the DNA should be done as described above.

3.4. Recombination of the Transgene into the BAC

In Red/ET recombineering, target DNA molecules are precisely altered by homologous recombination in strains of *E. coli*, which express phage-derived protein pairs, either RecE/RecT from the Rac prophage or Redα/Redβ from the λ phage. These protein pairs are functionally and operationally equivalent. RecE and Redα are 5′→3′ exonucleases, and RecT and Redβ are DNA-annealing proteins. A functional interaction between RecE and RecT or between Redα and Redβ is also required in order to catalyze the homologous recombination reaction. Thus, bacteria should contain not only the BAC but also plasmid (pR6K.αβγBAD), which expresses recombineering enzymes. It is important to use correct bacteria strains that host the BAC. We strongly recommend reading of the manuscript published by Muyrers and coworkers *(2)*.

3.4.1. Preparation of BAC Containing Competent Cells

1. Pick colonies from DH10B bacteria bearing the desired BAC clone (e.g., from a streaked out glycerol stock prepared after initial identification of a BAC) and grow in 5 ml of LB medium with antibiotic for the BAC overnight while shaking.
2. Transfer 1–2 ml into 50 ml LB medium (e.g., in 250 ml flask) with antibiotic and grow 2–4 h at 37 °C and vigorous shaking.
3. Prepare 10% glycerol with dH_2O and cool down on ice for at least 2 h before using. Place Falcon and Eppendorf tubes on ice.
4. When the cells reach log phase ($OD_{600} = 0.4$), pour them into the 50-ml falcon tubes and cool down on ice for 10 min.
5. Spin the bacteria at $4,000–5,000 \times g$, 10 min at 4 °C.
6. Discard supernatant and put on ice, resuspend bacteria in 50 ml ice cold 10% glycerol.
7. Repeat washing 2 times.
8. Discard supernatant and resuspend cells in the remaining liquid. The final volume of the bacteria suspension should be 0.6–0.8 ml.
9. Transfer the aliquots (50–70 µl) into each pre-cooled Eppendorf tube and freeze in liquid N_2 or use directly for electroporation.
10. All steps should be done on ice!

3.4.2. Co-Transformation of BAC Characterized Clones with pR6K.αβγBAD Plasmid

1. Pre-cool cuvettes on ice (5–10 min).
2. Thaw competent bacteria on ice and add 0.1–0.5 µg of pR6K.αβγBAD.
3. Electroporate bacteria at 2.3 kV (Bio-Rad Gene Pulser (Bio-Rad, California, USA), 25 µF with Pulse controller set to 200 Ω).
4. Add 1 ml of LB medium without antibiotic and transfer back into the Eppendorf tube.

5. Incubate at 37 °C for 1.5 h (shaking in thermomixer, maximum 350 × g).
6. Spin the bacteria at 4,000–5,000 × g for 7 min and remove supernatant, but leave about 50–100 μl inside.
7. Resuspend in the remaining supernatant and plate on plates with two antibiotics (first for BAC, second for pR6K.αβγBAD plasmids).
8. Incubate plates overnight at 37 °C.

3.4.3. Preparation of BAC and pR6K.αβγBAD Containing Competent Cells

1. Pick single colony from DH10B Bacteria harboring both the pR6K.αβγBAD and desired BAC clone (from transformation described above) and grow bacteria in 5 ml LB medium with two antibiotics overnight. For this step, we recommend picking of the bacteria colony directly after transformation with pR6K.αβγBAD, but not using glycerol-stock or old (2–3 days) plates with BAC and pR6K.αβγBAD bacteria. We have observed that in competent bacteria made from glycerol-stock or old plates, pR6K.αβγBAD expressed less and the recombineering efficiency decreased. We always pick bacteria from the plate directly after transformation on the next day.
2. Transfer 2 ml into 50 ml of LB medium with two antibiotics and grow them at 37 °C for 2–4 h.
3. Prepare 10 % glycerol with dH2O and cool down on ice for at least 2 h before using. Place Falcon and Eppendorf tubes on ice.
4. When the bacteria reach OD_{600} = 0.3, add Arabinose (final concentration 0.2%, best from a 20 % stock) and grow them at 37 °C for 1 h.
5. Pour into 50-ml falcon tubes and cool down on ice for 10 min.
6. Spin the bacteria at 4,000–5,000 × g and place them for 10 min on ice.
7. Discard supernatant and put on ice, resuspend bacteria in 50 ml ice-cold 10% glycerol.
8. Repeat washing two times.
9. Discard supernatant and resuspend bacteria in the remaining liquid. The final volume of the bacteria suspension should be 0.6–0.8 ml.
10. Transfer the aliquots (50–70 μl) into each pre-cooled Eppendorf tube and freeze in liquid N_2 or use directly for electroporation.
11. All steps should be done on ice!

3.4.4. Electroporation of the Transgene Vector into pR6K.αβγBAD and BAC-Containing Bacteria

Because the ET recombination occurs through free ends of both HAs of the insert, the transgenic construct should be digested from the plasmid at the sites close to the HA from both ends. The cloning strategy should foresee the RS that can be used, generally Not I is used. Excision and purification of the transgenic construct from the plasmid backbone can be done by digest and then by agarose gel electrophoresis. For cutting of the gel slice with the DNA fragment on UV-light table, we recommend to do it as fast as possible because of UV-damage of DNA.

1. Pre-cool cuvettes on ice (5–10 min).
2. Thaw competent bacteria on ice and add 1–2 µg of linearized transgenic construct.
3. Electroporate bacteria at 2.3 kV (Bio-Rad Gene Pulser, 25 µF with Pulse controller set to 200 Ω).
4. Add 1 ml of LB media with Arabinose (final concentration 0.2%) without antibiotic and transfer back into the Eppendorf tube.
5. Incubate at 37 °C for 1.5 h (shake in table-thermomixer, maximum 350 × g).
6. Spin bacteria at 4,000–5,000 × g for 7 min, remove supernatant, and leave about 50–70 µl inside.
7. Resuspend in the remaining supernatant and plate on plates with three antibiotics (first for BAC, second for pR6K.αβγBAD plasmids, and third for the Transgene-Neo resistance cassette).
8. Incubate plates overnight at 37 °C (*see* **Note 10**).

3.5. Screen of Recombinant BAC/PAC Colonies

3.5.1. PCR Analysis

The faster way to screen the colonies is "colony PCR." The DNA of the colony can be obtained directly by touching the labeled colonies on the plate and transferring the picked bacteria to the PCR mix tube labeled in the same way. During the first step of the PCR at 95 °C, the cells will be destroyed and DNA released into the PCR mix. The positive colony defined by PCR can be found on the plate later according to the labeling and then expanded. You can use the cheapest PCR components, because this is a screening procedure.

3.5.2. Restriction Digests

The transgenic construct recombined into the BAC changes the size of the BAC and brings new RS (*see* **Fig. 5**). This allows identification of the recombined tg-BAC from the wt-BAC by restriction digest and agarose electrophoresis. The optimal enzyme can be found using DNA-software programs (*see* **Subheading 3.2.**). In case the quality of the BAC DNA is good, one can see the DNA-bands that are different in size between tg-BAC and wt-BAC on the gel and as predicted by the DNA-software program (*see* **Fig. 6**).

1. Digest 10–30 µg of each: wt-BAC and tg-BAC (pre-selected by PCR) 4–5 h.
2. Load on the gel DNA marker, digest wt-BAC, and then digest tg-BACs from several colonies resuspended in loading buffer.
3. Run in 1× TAE buffer, 40–50 V, overnight.

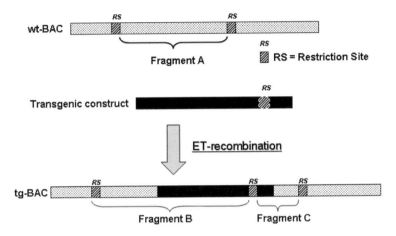

Fig. 5. Changes of the bacterial artificial chromosome (BAC) size and restriction fragment pattern after recombination. The recombined transgenic construct in the BAC brings new restriction sites into the BAC and changes the size of the BAC. The size difference between digested wt- and tg-BAC fragments (fragments A, B, and C) can be used for screening to fish the correctly modified tg-BAC by simple Mini Prep and restriction digest.

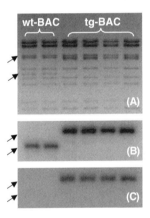

Fig. 6. Identification of transgene-containing bacterial artificial chromosomes (tg-BACs) from the wild-type BAC (wt-BACs) by differences in digested DNA fragment sizes. (**A**) Agarose gel electrophoresis of digested wt-BACs (first two lanes) and four colonies of modified tg-BACs. The difference in digested DNA fragments is obvious (black arrows). (**B**) The Southern blotting with an external probe confirms that the transgenic construct recombined site specifically. (**C**) The Southern blotting with an internal probe proves that non-specific recombination did not happen (there should be only one band on the tg-BAC lane; if multiple bands appear, they are additional non-specific integrations of the transgene into the BAC, which happens in rare cases).

4. Check on UV light and save/print a picture.
5. The positively recombined-BAC can be found according to the expected size changes of DNA-bands.

3.5.3. Southern Blot

Southern blotting is used to check whether the transgenic construct was recombined specifically into the BAC (external probe) and whether no additional unspecific recombinations happened (internal probe).

3.5.3.1. BLOTTING

1. Digest the BAC and separate the bands using gel-electrophoresis (*see* **Subheading 3.6.2.**).
2. Depurinate: 20–30 min in 0.25 M HCl.
3. 20 min in 0.4 M NaOH.
4. Blotting sandwich: chamber with 0.4 M NaOH – ground plate, Whatman paper soaked with 0.4 M NaOH-gel with face down – Whatman areas that are not covered with gel have to be covered with plastic (transparent film) – blotting membrane (without air bubbles), 2× Whatmann paper – thick layer of absorbent tissue, ~0.5 kg weight.
5. Blot overnight at room temperature (RT).
6. Wash twice in 2× Saline Sodium Citrate buffer (SSC).
7. Cross-linking for better conserving of DNA on the membrane (not obligatory), but we recommend 2× Auto cross-linking with a STRATAGENE cross linker.
8. Keep membrane covered with plastic.

The probe labeling and hybridization is described elsewhere. For Random Priming, we recommend Rediprime™ II DNA Labelling System, GE Healthcare; for radioactive probe purification: Micro Bio-spin Chromatography Columns, Bio-Rad; and for hybridizations: Rapid hybridisation buffer, Amersham Biosciences. All details are described best in the manufacturer's protocols.

The Southern blot membrane should display the same expected specific bands like visualized on the agarose gel (*see* **Fig. 6**). The external probe will visualize the bands in both: wt-BAC and tg-BAC, which will have a different size (*see* **Fig. 6**). The internal probe (specific for transgenic construct only) will detect the band on the tg-BAC lane only.

3.6. Flp-Mediated Excision of the Resistance Cassette

705 Flp plasmids are based on the pSC101 temperature-sensitive origin. This origin maintains a low copy number and replicates at 30 °C. These plasmids will be lost from cells when they are incubated at temperatures above 37 °C.

Additionally, Flp are expressed from the lambdaPR promoter, which expresses weakly at 30 °C and strongly at 37 °C. Flp will excise the frt/frt-gated resistance cassette, which is subsequently lost. Thus, 705-Flp can be used to give a transient burst of expression after which they will be eliminated so the recombined product can be isolated uncontaminated by other plasmids.

3.6.1. Co-transformation of Recombinant tg-BAC Clones with 705-Flp Plasmid

1. Electroporate the bacteria (containing the tg-BAC) with 705-Flp (competent cells can be obtained as described above; *see* **Subheading 3.8.**).
2. Incubate at 30 °C for 1.5 h with shaking.
3. Plate on LB plates with two antibiotics: first for BAC, second for 705-Flp.
4. Incubate the plates at 30 °C for 2 days.
5. Incubate plates for 2 h at 37 °C.
6. Pick single colonies and grow in 5 ml of LB media with one antibiotic (for BAC only) overnight at 37 °C.
7. Streak onto LB plate with one antibiotic (for BAC) and culture at 37 °C overnight (*see* **Note 10**).
8. Pick single colonies and grow in two tubes of 5 ml of LB media: first tube media with one antibiotic for BAC and second tube media with two antibiotics: one for BAC and second for transgene resistance cassette that has to be removed.
9. The colony that did not grow in second tube (two antibiotics) can be used for further analysis. The further analysis of the tg-BAC can be done by Southern blotting (*see* **Subheading 3.6.** and **Note 11**).

3.7. Linearization and Purification of tg-BAC DNA for Pronuclear Injection

For the final step of purification, we recommend to use freshly prepared BAC-DNA only.

3.7.1. PFGE using the CHEF-DR® III Pulsed Field Electrophoresis Systems from BioRad Using Low-Melting Point Agarose Gels

The big fragments of DNA (100–250 kb) can be purified using PFGE (*see* **Fig. 7**). As the DNA separation will take up to 16 h, the running buffer and water for washing of the EF-chamber should be freshly prepared and autoclaved. Because the BAC-DNA purified by PFGE will be injected into oocytes, never touch or contaminate the PFGE-chamber, running buffer or gel with ethidium bromide (EB), nor with any other substance known to interfere negatively with the quality of DNA preparations. Never use the reagents or equipment used for routine (EB-containing) electrophoresis, instead provide a unique set of reagents.

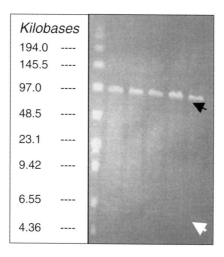

Fig. 7. The Pulse-Field Electrophoresis for direct purification of the linearized BAC DNA. The BAC (black arrow) digested with Not I was separated from the plasmid backbone (white arrow) and a marker labels the fragment sizes.

1. Set up analytical digestion: 50–100 µg. Incubate approximately 4–6 h. Gently pipette every 2 h 2–3 times up and down to allow efficient DNA digestion.
2. Clean and prepare the casting tray for the gel.
3. Cast Gel: 1 % Sea-Plug Low-melting point Agarose in 0.5× TAE. Combine for example eight slots for the sample well.
4. Rinse PFGE machine with 2–3 l autoclaved water for 10 min and drain completely.
5. Add marker (λ concatamers in the agarose plug) in respective pockets of the gel.
6. Insert gel into PFGE machine; fill carefully with 2 l with running buffer (0.5× TAE).
7. Switch on pump and after this the cooling device.
8. Wait for the cooling device to be stably at 14 °C, then load samples in the loading buffer.
9. Start PFGE. Proposed parameters: initial t (switch) = 0.5 s, final t (switch) =20 s, including angle α = 120 °C, IEI = 6 V/cm, I = 140 mA, total time T = 12–14 h.
10. Cut the portion of the gel-containing marker lanes and edges of sample lane.
11. Stain the removed portion of the gel with EB. Do not expose central portion containing the bulk of sample lane to either EB or UV.
12. Under UV light, identify the edges of product band and mark them with a blade to identify position.
13. Wrap the stained portions, align them on a glass plate to central portion and cut around the position of the product band according to marks.

14. Preserve the cut-out slice; counter stain the remainder of the gel to make sure you cut the correct band.
15. Determine the weight of the gel slice, aliquot in pieces of maximum 0.5 g in 1.5-ml Eppendorf tubes.
16. Add 1 ml of agarose buffer to each tube, let it stand at RT for 30 min spin down agarose plug gently, discard buffer. Repeat this equilibration step twice.
17. Spin down agarose plugs, place tubes at 68 °C for 10 min to allow melting of the agarose. Put tubes at 40 °C and wait for them to cool to 40 °C.
18. Add 1 μl β-agarase for every 100 mg agarose. Leave enzyme aliquot in pipette tip for a moment to warm up. Mix once with a 1-ml pipette with a cut-off tip.
19. Incubate at 40 °C for 2 h. After 1 h, gently flick tubes once.
20. Place tubes on ice, 10 min.
21. Centrifuge in tabletop centrifuge at RT for 15 min, full speed.
22. Pre-rinse ultrafiltration cartridge with microinjection buffer.
23. Transfer supernatant on ultrafiltration cartridge.
24. Place cartridge in collection tube, centrifuge at $6,000 \times g$ for 4 min in tabletop centrifuge. Check for volume in the cartridge, repeat shorter (1–2 min) centrifugation steps until some 100 μl remain in the cartridge.
25. Incubate for 1 h at RT.
26. Prepare dialysis membrane by putting it on the surface of 30-ml microinjection buffer in a 10-ml Petri dish. Let equilibrate for 2–3 h.
27. Place filters on fresh microinjection buffer, transfer concentrated DNA sample from cartridge on filter surface.
28. Dialyze for 2–3 h.
29. Collect DNA sample.
30. Estimate concentration of a 2-μl aliquot on normal 0.8% agarose gel by using a series of dilutions of a reference DNA (e.g., a large plasmid), from 1 to 10 ng/lane.
31. Dilute DNA sample to 1 ng/μl with microinjection buffer (*see* **Note 11**). Distribute in 10-μl aliquots in new Eppendorf tubes. Close them, seal them with parafilm, and store them at 4 °C until microinjection.
32. Check integrity of DNA sample in analytical PFGE.

3.7.2. Sepharose 4B-CL Chromatography

This protocol was originally described by Yang and coworkers, which we recommend to read *(29)*.

1. To linearize the tg-BAC DNA, digest 30–50 μg of cesium-banded BAC DNA overnight with Not I enzyme (or with other appropriate enzymes) in 0.5 ml total volume. Be sure to supplement the standard restriction buffer with 2.5 mM spermidine (final concentration) to help digest the BAC DNA. For circular BAC DNA, proceed directly to **step 2**.
2. Preparation of the CL4b Column (done at RT): Take a 5-ml plastic pipette, use pressured air to blow the cotton to the tip, and clamp the pipette on a stand. Shake

the CL4b sepharose (Pharmacia, Uppsala, Sweden) well, and gradually add the sepharose into the plastic pipette. Add until the packed sepharose is almost at the top (with about 1 ml space to spare). Never let the column dry.

3. Once the column is ready, use a 10-ml syringe to set a reservoir on top of the column (buffer is added to the reservoir). Then equilibrate the column with 30 ml of the injection buffer. This takes about 2–3 h.

4. Now add 5 μl standard DNA dye into the 0.5-ml digested BAC DNA. Take the reservoir out and gently add the DNA (+dye) onto the top of the column with a Pasteur pipette. Wait until the DNA plus the dye just go into the column; gently add 0.5 ml of injection buffer on top of the column. Once the injection buffer almost goes in, the reservoir is put back with 10 ml of injection buffer added into it. Now, start collecting 0.5-ml fractions with a 24-well plate. Collect about 12 fractions (until the blue dye almost goes to the bottom). The linear BAC DNA tends to come out relatively late, in fractions 6–9, and the circular BAC DNA tends to come out relatively early, in fractions 4–7.

5. Run 50 μl of each fraction (or every other fraction) on a pulse field gel to identify the appropriate fractions with intact BAC DNA and with the least degradations and with minor vector bands. The bands should be visible after EB staining. Southern blotting (optional) can be done to choose the fractions with the highest yield and least degradations. We recommend checking through Southern blotting.

6. DNA concentration is determined by serial dilution of the appropriate BAC elution fraction in 0.2 ng/ml EtBr (dilution factors are 1:2, 1:5, 1:10, 1:20, and 1:50), and compared with the fluorescent intensity of these dilutions to that of a standard serial dilution of a DNA sample of known concentration (i.e., DNA marker; final concentrations are 5, 2, 1, 0.5, and 0.1 μg/ml).

7. The appropriate purified fraction should contain 5–20 μg/ml of BAC DNA, and it is then diluted to the final concentration of 0.6–1 μg/ml with the injection buffer (*see* **step 3**) before microinjections. Purified DNA should be stored at 4 °C (should not be frozen). It is stable for weeks (no degradation was detected after 3 weeks).The tg-BAC is ready for oocyte microinjection.

If you want to electroporate the tg-BAC into ES cells, and go through ES colony screening, the linearized and purified tg-BAC should be dissolved in the sterile PBS. Protocols for ES-cell targeting and microinjection are well described elsewhere.

After oocyte injection or preparation of ES-cell, you can choose for copy number, random, or site-specific integration of the tg-BAC into genome. Now, the mice take over the important job.

4. Notes

1. Some genes have several transcriptional variants. It is important to check which variant you want to modify and introduce into the BAC.

2. The sequence is not published for all BACs. Because it is known exactly which part of a chromosome is covered by a particular BAC, you can extract the sequence of this DNA fragment from any DNA-library.

3. Cloning of each fragment (transgene, HAs, reporter gene, etc). can be done step by step only. It is very convenient to keep the bacteria containing the plasmid from each step of cloning. Sometimes, it happens that you need to go back to some cloning step and replace one fragment by another (reporter gene is not well expressed and you need to re-clone with another one, etc). For long-time storage: keep DNA at −20 °C and glycerol-stock of bacteria containing the transgenic construct (400 μl of 87% glycerol or 300 μl of 99% glycerol and 300 μl bacteria suspension from the positive mini-prep, keep at −70 °C).

4. Optimal annealing temperature is 5–7 °C lower than the melting temperature of primer–template DNA duplex (annealing time: 30–50 seconds). The approximate melting temperature (T_m) is calculated using the following formula: $T_m = 4(G + C) + 2(A + T)$. For primers that contain long unspecific fragments (loxP or Frt sites, etc.), we recommend to increase the annealing time up to 70 s.

5. Normal Taq polymerase gives mutations during the amplification. Thus, PCR of the DNA fragment, which will be cloned, should be amplified using prove-reading Taq-polymerase only. We recommend: AccuPrime (tm) Pfx DNA Polymerase (Invitrogen, California, USA) or High Fidelity PCR Enzyme Mix (Fermentas, Maryland, USA).

6. There are many well-known companies for restriction enzymes and they offer basic protocols to cloning and other helpful information online (www.neb.com; www.fermentas.com; www.roche.com)

7. Do not use more enzyme than 10% of the final reaction volume. 1 μl of enzyme is plenty for 20 μl of digestion mix. This is because the enzyme storage buffer contains antifreeze (glycerol) to allow it to survive at −20 °C. The glycerol will inhibit the digestion if present in sufficient quantities. You may be digesting your DNA with two (or more) enzymes. This is fine but you have to make sure to use the buffer that will be most compatible with all the enzymes.

8. DNA purification kits can be obtained from different companies (www.sigmaaldrich.com; www.qiagen.com; www.invitrogen.com; www.roche.com)

9. Never vortex or freeze/thaw the BAC DNA. The BAC is a large size DNA molecule and can be easily damaged mechanically. Pipetting should be done gently. We found that after 1 month at 4 °C, BAC DNA starts to degrade and it coagulates.

10. During this incubation step, the recombineering enzymes mediate site-specific homologous recombination. The transgenic construct will replace the original BAC DNA fragment at the sites of the HA.

11. Optionally, the transgenic construct recombined into the BAC (final tg-BAC) can be sequenced. The big size of the BAC DNA makes a problem. Thus,

we recommend amplifying the fragment of tg-BAC with proofreading Taq-polymerase and then sequencing this PCR product.

Acknowledgment

This work was supported by funding from the Ludwing Boltzmann Gasellschaft to AT and RM and by FWF grant SFB-F28 to RM.

References

1. Zhang Y, Buchholz F, Muyrers J.P.P., and Stewart A.F. (1998) A new logic for DNA engineering using recombination in Escherichia coli. *Nat. Genet.* **20**, 123–128.
2. Muyrers, J.P.P., Zhang, Y., Testa, G., and Stewart, A.F. (1999) Rapid modification of bacterial artificial chromosomes by ET-recombination. *Nucleic Acids Res.* **27**, 1555–1557.
3. Lee, E.C., Yu, D., Martinez de Velasco, J., Tessarollo, L., Swing, D.A., Court, D.L., Jenkins, N.A., and Copeland, N.G. (2001) A highly efficient Escherichia coli-based chromosome engineering system adapted for recombinogenic targeting and subcloning of BAC DNA. *Genomics* **73**, 56–65.
4. Shizuya, H., Birren, B., Kim, U.J., Mancino, V., Slepak, T., Tachiiri, Y., and Simon, M. (1992). Cloning and stable maintenance of 300-kilobase-pair fragments of human DNA in Escherichia coli using an F-factor-based vector. *Proc. Natl. Acad. Sci. USA* **89**, 8794–8797.
5. Giel-Moloney, M., Krause, D.S., Chen, G., Van Etten, R.A., and Leiter, A.B. (2007) Ubiquitous and uniform in vivo fluorescence in ROSA26-EGFP BAC transgenic mice. *Genesis* **45**, 83–89.
6. Hochedlinger, K., Yamada, Y., Beard, C., and Jaenisch, R. (2005) Ectopic expression of Oct-4 blocks progenitor-cell differentiation and causes dysplasia in epithelial tissues. *Cell* **121**, 465–477.
7. Wallace, H.A., Marques-Kranc, F., Richardson, M., Luna-Crespo, F., Sharpe, J.A., Hughes, J., Wood, W.G., Higgs, D.R., Smith, A.J. (2007) Manipulating the mouse genome to engineer precise functional syntenic replacements with human sequence. *Cell.* **128**, 197–209.
8. Gong, S., Zheng, C., Doughty, M.L., Losos, K., Didkovsky, N., Schambra, U.B., Nowak, N.J., Joyner, A., Leblanc, G., Hatten, M.E., and Heintz, N. (2003) A gene expression atlas of the central nervous system based on bacterial artificial chromosomes. *Nature* **425**, 917–925.
9. Wilson, C.B. and Schoenborn, J. (2006) BACing up the interferon-gamma locus. *Immunity* **25**, 691–693.
10. Ango, F., di Cristo, G., Higashiyama, H., Bennett, V., Wu, P., and Huang, Z.J. (2004) Ankyrin-based subcellular gradient of neurofascin, an immunoglobulin family protein, directs GABAergic innervation at purkinje axon initial segment. *Cell* **119**, 257–272.

11. Sauer, B. (1993) Manipulation of transgenes by site-specific recombination: use of Cre recombinase. *Methods Enzymol.* **225**, 890–900.

12. Brook, F.A. and Gardner, R.L. (1997). The origin and efficient derivation of embryonic stem cells in the mouse. *Proc. Natl. Acad. Sci. U.S.A.* **94**, 5709–5712.

13. Evans, M.J. and Kaufman, M.H. (1981) Establishment in culture of pluripotential cells from mouse embryos. *Nature* **292**, 154–156.

14. Coppo, P., Flamant, S., De Mas, V., Jarrier, P., Guillier, M., Bonnet, M.L., Lacout, C., Guilhot, F., Vainchenker, W., and Turhan, A.G. (2006) BCR-ABL activates STAT3 via JAK and MEK pathways in human cells. *Br. J. Haematol.* **134**, 171–179.

15. Yu, H. and Jove, R. (2004) The STATs of cancer–new molecular targets come of age. *Nat. Rev. Cancer* **4**, 97–105.

16. Kortylewski, M., Jove, R., and Yu, H. (2005) Targeting STAT3 affects melanoma on multiple fronts. *Cancer Metastasis Rev.* **24**, 315–327.

17. Ying, Q.L., Nichols, J., Chambers, I., and Smith, A. (2003) BMP induction of Id proteins suppresses differentiation and sustains embryonic stem cell self-renewal in collaboration with STAT3. *Cell* **115**, 281–292.

18. Maritano, D., Sugrue, M.L., Tininini, S., Dewilde, S., Strobl, B., Fu, X., Murray-Tait, V., Chiarle, R., and Poli, V. (2004) The STAT3 isoforms alpha and beta have unique and specific functions. *Nat. Immunol.* **4**, 401–409.

19. Takeda, K., Noguchi, K., Shi, W., Tanaka, T., Matsumoto, M., Yoshida, N., Kishimoto, T., and Akira, S. (1997) Targeted disruption of the mouse Stat3 gene leads to early embryonic lethality. *Proc. Natl. Acad. Sci. U.S.A.* **94**, 3801–3804.

20. Panopoulos, A.D., Zhang, L., Snow, J.W., Jones, D.M., Smith, A.M., El Kasmi, K.C., Liu, F., Goldsmith, M.A., Link, D.C., Murray, P.J., and Watowich, S.S. (2006) STAT3 governs distinct pathways in emergency granulopoiesis and mature neutrophils. *Blood* **108**, 3682–90.

21. Lee, C.K., Raz, R., Gimeno, R., Gertner, R., Wistinghausen, B., Takeshita, K., DePinho, R.A., and Levy, D.E. (2002) STAT3 is a negative regulator of granulopoiesis but is not required for G-CSF-dependent differentiation. *Immunity* **17**, 63–72.

22. Jenkins, B.J., Roberts, A.W., Najdovska, M., Grail, D., and Ernst, M. (2005) The threshold of gp130-dependent STAT3 signaling is critical for normal regulation of hematopoiesis. *Blood* **105**, 3512–20.

23. Chung, Y. J., Park, B. B., Kang, Y. J., Kim, T. M., Eaves, C. J., and Oh, I. H. (2006) Unique effects of Stat 3 on the early phase of hematopoietic stem cell regeneration. *Blood* **108**, 1208–1215.

24. Moriggl, R., Sexl, V., Kenner, L., Duntsch, C., Stangl, K., Gingras, S., Hoffmeyer, A., Bauer, A., Piekorz, R., Wang, D., Bunting, K.D., Wagner, E.F., Sonneck, K., Valent, P., Ihle, J.N., and Beug, H. (2005) Stat5 tetramer formation is associated with leukemogenesis. *Cancer Cell* **7**, 87–99.

25. Kovacic, B., Stoiber, D., Moriggl, R., Weisz, E., Ott, R.G., Kreibich, R., Levy, D.E., Beug, H., Freissmuth, M., and Sexl, V. (2006) STAT1 acts as a tumor promoter for leukemia development. *Cancer Cell* **10**, 77–87.

26. Bromberg, J.F., Wrzeszczynska, M.H., Devgan, G., Zhao, Y., Pestell, R.G., Albanese, C., and Darnell, J.E., Jr. (1999) Stat3 as an oncogene. *Cell* **98**, 295–303.

27. Decker, T. and Kovarik, P. (2000) Serine phosphorylation of STATs. *Oncogene* **19**, 2628–37.

28. Giraldo, P. and Montoliu, L. (2001) Size matters: use of YACs, BACs and PACs in transgenic animals. *Transgenic Res.* **10**, 83–103.

29. Yang, X.W., Model, P. and Heintz, N. (1997) Homologous recombination based modification in Escherichia coli and germline transmission in transgenic mice of a bacterial artificial chromosome. *Nat. Biotechnol.* **15**, 859–65.

20

Bioluminescence Imaging of Hematopoietic Stem Cell Repopulation in Murine Models

Yuan Lin, Joe Molter, Zhenghong Lee, and Stanton L. Gerson

Summary

Hematopoietic stem cells (HSCs) have been studied for decades in order to understand their stem cell biology and their potential as treatments in gene therapy, and those studies have resulted in tremendous advancement of understanding HSCs. However, most of the studies required the sacrifice of cohorts of the animals in order to obtain data for analysis, resulting in the use of large animal numbers along with difficult long-term studies. The dynamic engraftment and expansion of HSC are not fully observed and analyzed. Until recently, with the development of optical imaging, HSC repopulation can be continuously monitored in the same animal over a long period of time, reducing animal numbers and opening a new dimension for investigation. In this chapter, bioluminescence imaging of murine HSC is described for observing the dynamic repopulation process after transplantation. Photons emitted from transplanted murine HSCs expressing firefly luciferase within the mice can be visualized in light-sealed chamber with a highly sensitive digital camera after injection of substrate D-luciferin. Xenogen IVIS200 imaging system is used to record the process, and other similar imaging systems can also be used for this process.

Key Words: Bioluminescence imaging, in vivo imaging, cooled CCD camera, optical imaging, hematopoietic stem cell, luciferase, transplantation, repopulation, engraftment, D-luciferin.

1. Introduction

Hematopoietic stem cells (HSCs) have been an attractive target for gene therapy because of their self-renewal and pluripotency *(1,2)*. After transplantation into myeloablated recipients, HSC can give rise to all lineages of blood

From: *Methods in Molecular Biology, vol. 430: Hematopoietic Stem Cell Protocols*
Edited by: K. D. Bunting © Humana Press, Totowa, NJ

cells. True long-term HSC currently can only be identified through functional transplantation and repopulation assay *(3,4)*. About 5 years ago, most data for interpreting the engraftment and expansion of transplanted HSCs in small animal model were obtained and interpreted from post-mortem collection and analysis of the hematopoietic organs. However, the dynamic range of movement and the engraftment pattern of HSCs were not available until recently. With the development of imaging equipments and techniques, molecules and cells can be detected in vivo without killing the subjects *(5–7)*. HSCs can be tagged with optical imaging reporters and transplanted into recipients. With the expansion of HSCs, photons emitted from the progeny of those HSCs could provide spatial and temporal information that no other techniques could provide. For the first time, the extent of hematopoiesis in murine transplantation model can be visualized.

1.1. Benefits and Advantages of Bioluminescence Imaging

Obvious benefit and advantage of in vivo imaging techniques including the optical imaging are non-invasiveness, less study animals, complete picture of biological process, and long-term monitoring. Optical imaging includes fluorescent imaging and bioluminescent imaging. Fluorescent imaging captures photons emitted from the fluorescent probes after being excited, such as green fluorescence protein, inside of the targeting cells to track cell migration and monitor gene expression. Fluorescent imaging, especially Green Fluorescent Protein (GFP) signal, is limited by its relatively low tissue-penetrating capability and high background because of auto-fluorescence from the animals studied. However, for superficial imaging, such as the tumor xenograft model, fluorescent imaging can provide fast, reasonably sensitive, and less-expensive imaging services. To study HSC repopulation, in which cells circulate and reside deep inside the hematopoietic organs and bone marrows, bioluminescent imaging (BLI) provides higher sensitivity and almost no background signals for overall better performance *(8,9)*. BLI utilizes an enzymatic reaction to give off visible light with the ability to penetrate a couple centimeters. BLI requires the injection of a substrate to initiate the reaction, and no excitation light source is needed, which dramatically reduces imaging background from excitation or animal autofluorescence, and increases the signal to noise ratio. Besides optical imaging, there are other imaging techniques, such as MRI/MRSI and nuclear medicine imaging (SPECT, PET), that can offer similar services. However, those are clearly beyond the scope of the current text and will not be discussed here.

1.2. Bioluminescence Proteins and Their Applications

There are many different bioluminescent enzymes, which have been isolated and cloned from vast and diverse organisms. The most studied and widely used luminescent enzyme is from North American firefly (*Photinus pyralis*) *(10)*. Firefly luciferase is a 61-kDa monomeric protein, and it interacts with its substrate D-luciferin in the presence of oxygen, Mg^{2+}, and ATP to release green light with a peak wavelength at 562 nm. At body temperature of 37 °C, this peak shifts toward 600 nm. The light generated from this reaction has a broad spectrum, and only the upper 30% of the spectrum (>600 nm) is able to travel through tissues with low level of scattering and absorption *(11)*. Other useful luciferases from bacteria, jellyfish (*Aequorea*), sea pansy (*Renilla*), and click beetle (*Pyrophorus plagiophthalamus*) were also cloned and are used in various applications *(12)*. However, the light spectrums generated from those luciferases do not have a longer wavelength component as that of firefly luciferase; thus, they are less frequently used for deep tissue imaging. Nevertheless, most of luciferases have been used to study tumor growth and metastasis, viral infections, progress of infections, cellular protein activity, and protein–protein interaction *(13–16)*.

1.3. Bioluminescence Imaging for HSC

Current knowledge of HSC transplantation and repopulation has come from decades of peripheral blood and post-mortem studies. Now, the dynamic engraftment and expansion of HSC in vivo is finally able to present itself with the help of BLI. Firefly luciferase has been successfully incorporated into highly purified KTLS ($Lin^-Sca-1^+c-Kit^+Thy-1^{lo}$) murine HSC and human HSC, visualized, and the process of repopulation can be monitored over long period of time *(17,18)*. Even though transplanted HSC can home to bone marrow within 24 h of infusion *(19)* BLI signals can be seen 3–5 days after transplantation. Low number of luciferase-positive cells during transplantation limited the detection of early homing process, and onset of foci from repopulation of HSC and progenitor cells happens in a cell dose-dependent manner (Yuan Lin & Stanton L. Gerson unpublished data). Limitations of current BLI of HSC have been due mainly to massive diffusion and scattering/absorption of the light photons in living animal and two dimensional imaging, which make the quantitative measurement and precise localization of those transplanted HSCs and their progeny difficult. However, other imaging modalities, such as MRI, CT, and PET, and certain post-mortem studies can provide spatial reference and data confirmation for BLI results.

2. Materials

2.1. Mice and HSCs

1. Albino mice are the best normal recipients for BLI imaging because they have low absorption of light emitted from their bodies. A good strain would be Balb/c

background mice. Black C57 background or brown color mice can also be used for imaging in a special study model; however, fur is usually shaved from the imaging area to reduce photon absorption. Black and brown mice tend to have very dark skin as well, which can absorb the signal (emitted light photons) more than albino.

2. HSCs (KTLS in murine or $CD34^+CD38^-$ in human) can be isolated with direct or indirect isolation methods (see previous chapters for HSC isolation).

2.2. Reagents

1. Diluted rodent anesthesia mixture.
   Diluted mice anesthesia solution:

Drugs	Amount used (stock concentration)	Volume (ml)
Ketamine HCl	15 mg (100 mg/ml)	0.15
Xylazine HCl	3 mg (20 mg/ml)	0.15
Acepromazine	0.5 mg (10 mg/ml)	0.05
Sterile H_2O or saline		1.4
Total volume		1.75

IP injection Dose: 0.1-0.2 ml/25-gram-mouse (recipe and working dosage obtained from Animal Resource Center at Case Western Reserve University). Alternatively, Avertin can be used instead of rodent anesthesia mixture. Avertin (2-2-2 tribromoethanol) can be dissolved in *tert*-amyl alcohol to make stock solution and stored at room temperature in the dark. Working solution is 20 mg/ml from the dilution of the stock solution with PBS. Sterilize with Nalgene 0.22 μm filter bottle and store the working solution in 4 °C in the dark for up to several months. IP injection dose: 0.3–0.6 mg/gram in mice.

2. D-Luciferin solution: Synthetic firefly D-luciferin potassium salt (Biosynth, Naperville, IL, USA or Xenogen, Hopkinton, MA, USA) is dissolved in Dulbecco's phosphate-buffered saline (Cellgro, Herndon, VA, USA) to make injection solution at 12.5 mg/ml final concentration. Dissolved D-luciferin solution is filtered through 0.22 μm Steriflip (Millipore, Billerica, MA, USA) to ensure sterility. Aliquot into 2-ml sterile O-ring tubes (Fisher Scientific, Waltham, MA, USA, Cat. No. 0566957). Working solution should be kept in the dark and stored in a –20 °C freezer for up to 3 months. The working solution is best kept at –80 °C. This will allow you to store it up to 1 year. Only thaw out the quantity you will use for that imaging session. Avoid multiple freezing and thawing cycles.

3. Bac/Neo Antibiotics: For 50 ml of volume, dissolve 5.0 g of neomycin sulfate (Fisher Scientific) and 250,000 units of bacitracin zinc salt (Sigma-Aldrich, St. Louis, MO, USA) in sterile water. Antibiotic solution is stored in –20 °C freezer. When used, thaw and add 2 ml of solution to each drinking water bottle in the cage.

2.3. Supplies and Equipments

1. Common laboratory supplies: weight scale, alcohol swipes, 1-ml syringes with 26- to 28-G needles (1-cc Insulin syringes with built-in needles work very well), and Contura HS-40 shaver (Weller, Basingstake Hampshire, UK) if needed.
2. Black paper and black photographer's tape: mice can be imaged on top of the black paper. Photographer's tape (3M No. 235 Photographic Blocking tape). You can use tape to affix the animals to the paper when they are imaged in the supine position. This holds down the legs and allows unobstructed views of the chest and abdomen.
3. Imaging system: Charge-coupled device (CCD)-camera, light-sealed imaging chamber, and computer with image analysis software, such as ImageJ or Matlab. In this protocol, we use Xenogen IVIS200 systems with XGI8 gas anesthesia system and Living Image 2.5 software (*see* **Note 1**).

3. Methods

3.1. Isolation of HSCs Expressing Firefly Luciferase Reporter Gene

There are three different ways to obtain murine HSC expressing luciferase gene. First, luciferase plasmid can be transiently transfected into HSC ex vivo. The method provides fast and safe introduction of luciferase gene into HSC; however, the expression of luciferase protein may decrease with the repopulation after transplantation because the plasmid does not replicate with cell division. Second, luciferase gene can be stably integrated into targeted cell genome ex vivo by retroviral or lentiviral transduction (see Chapter 19 for viral transduction of HSC). This method could achieve long-term luciferase gene expression. Third, luciferase transgenic mice could also be generated to obtain murine HSC for long-term luciferase expression. The third method could be the best to generate "natural" HSC without any ex vivo manipulation, which may potentially affect HSC repopulation. Human HSC can also be isolated and introduced luciferase transgene through retroviral or lentiviral transduction.

3.2. Bone Marrow Transplantation

Irradiation of recipients can facilitate the engraftment of transplanted HSCs. Balb/c mice (Charles River, Boston, MA, USA) were irradiated 1–24 hours before transplantation at 750–800 rad with a Cs-137 radiation source. Viral transduced HSCs (murine or human) or murine transgenic HSCs with bioluminescence gene are infused into irradiated syngeneic recipients through tail-vein injection with 28-G 1-ml syringe. Antibiotics, bacitracin/neomycin sulfate, are needed in the drinking water after irradiation and transplantation to prevent bacterial infection.

3.3. Anesthesia of Experimental Animals

Two methods of anesthesia can be performed on the experimental animals: injection of diluted ketamine/xylazine/acepromazine (Henry Schein, Inc., Melville, NY, USA) mixture or using isofluorane vapor and oxygen supplied by Xenogen XGI-8 gas anesthesia system. Based on the size of the mice and the concentration of the rodent anesthesia mixture, about 100–150 µl of mixture is intraperitoneally injected into each mouse 3–5 minutes before the injection of D-luciferin. The anesthesia effect lasts about 30–40 minutes with minimum movement, and mice will completely wake up a couple of hours after imaging (*see* **Note 2**).

XGI-8 gas anesthesia system offers better imaging flexibility. It can easily anesthetizes studied subject, and the animal can wake up quickly after being removed from the anesthetic gas. Steady flow of anesthesia gas allows repeated imaging within a very short time, and longer imaging time could be performed with this gas anesthesia system. The animal will first be placed in the induction chamber with 2.0–3.0% isoflurane with 100% oxygen with a 1–1.5 l/min flow rate. Once the animal is anesthetized, it should be quickly transferred from the induction chamber to the sample stage inside the imaging chamber and its nose be placed inside a nose-cone attached to the XGI-8 system, and adjust isoflurane level to 1.0–2.0% with 100% oxygen (*see* **Note 3**).

3.4. Acquisition of Bioluminescence Images

The imaging system should be turned on hours before imaging because of the cooling and stabilizing of the CCD camera to working temperature (*see* **Notes 4** and **5**).

1. Start up the Living Image 2.5 software program by clicking the program icon, and enter user identification (user initials). The IVIS system control panel will appear.
2. Before any other information can be entered, you must click the Initialize IVIS system button in the control panel. When initialization starts, the machine will reset all the motor-controlled components, such as the camera and stages, and all the software variables of the IVIS system. After initialization, make sure the temperature box of system status in the control panel is green to indicate the temperature of the camera is ready for imaging.
3. Select the appropriate imaging parameters, such as exposure time, binning, and field of view (FOV), from the control panel. First, ensure the imaging mode was set at luminescence. Select exposure time from a few seconds to 10 minutes, depending on the intensity of the signal. Exposure time is the time period during which photons captured by the CCD camera will be added together to increase sensitivity of bioluminescent signal. For HSC imaging, close to 5 min of exposure time should be sufficient enough. Too long of exposure time will

result in overly saturated signals, which is harder to quantify (*see* **Note 6**). Binning increases pixel size of the camera, which delivers higher sensitivity at the expense of spatial resolution (Xenogen manual). Medium binning (4 × 4) for HSC imaging can provide decent resolution and good sensitivity. FOV indicates the distance between the camera and the sample stage, and it is preset to five settings (A–E). A represents the smallest whereas E represents the largest imaging area. For example, for imaging four mice, position D (18.4 cm) will be a good size. f/stop, filters, and subject height use their default value at 1, open, and 1.5 cm, respectively. Click on "Select sequential mode" to expand the control panel and click "set" twice to setup obtaining two consecutive images together (*see* **Note 7**).

4. Make sure the laser-generated alignment grid box is enabled because it can show the exact area of the image the CCD camera sees. The size of the alignment grid changes with the change of FOV.

5. Injection (i.p.) of rodent anesthesia mixture or turn on XGI-8 gas anesthesia system to anesthetize the studied animals (*see* **Subheading 3.3.**).

6. After animal is anesthetized, inject appropriate amount of D-luciferin into each mouse based on their body weight 7 minutes before imaging to allow D-luciferin to circulate inside the body (*see* **Note 8**). Average amount of D-luciferin injected is 10 μl/g of body weight. Because the stock is 12.5 mg/ml, the final concentration of D-luciferin per animal is 125 mg/kg (*see* **Note 9**).

7. Place studied animals on their back on black paper, spread out, and tape down their limbs with black photographer's tape (*see* **Note 10**). Place the black paper on the sample stage and make sure the studied animals are in the center of the alignment grid.

8. 7 minutes after injection of D-luciferin, click "Acquire Sequence" button to start collecting images.

9. When the image taken process is over, the computer will automatically display luminescent pseudo-color image overlaid on grayscale photographic image. On the right side of the image, a rainbow scale bar shows the relative intensity of bioluminescent signal from the subject. Purple and blue indicate weakest whereas yellow and red indicate strongest signals. The unit is in photon counts.

10. A window will pop up asking for the information regarding this image upon the completion of taking each image, then save the image data by choosing "Save Living Image Data" or by closing each image window, not the software window, and follow the instruction to save images.

11. Printing of the images after collection of data may require restart of the program. Enter the user ID, but initialization of the system is not required for printing and analysis. Select "Load LI Data" button from Living Image menu, and select correct images from their folders, and choose print command from the pull down menu.

12. Remove the sample animals from the imaging chamber, turn off the XGI-8 gas anesthesia system, and exit the program. Leave the imaging system on.

3.5. Analysis of Imaging Data

Imaging data obtained by IVIS200 system are automatically pseudo-colored based on bioluminescence intensity, and only the second images from the sequential images were compared (*see* **Note 7**).

1. Startup the Living Image 2.5 software and enter user ID. Do not need to initialize the system for analysis.
2. Choose "Browse for LI Data" or "Load LI Data" from "Living Image" pull-down menu and select image folders needed to be analyzed from directory, and double-click the image file name to open the analysis window.
3. To measure the bioluminescence intensity, regions of interest (ROIs) are created by choosing the number and shape from the analysis window and clicking "Create." To compare HSC repopulation, ROI can be defined as the whole body of the studied animal or the approximate locations of internal organs.
4. Move ROI over the area of interest, resize if necessary, and click "Measure" (*see* **Note 11**). The data are displayed in photons mode, and the unit is photons/s/cm^2/sr.
5. Save ROIs and measurements of BLI intensity by choosing pull down menus from "Tools" and "Living image."
6. For visual presentation, Images with the same BLI intensity scale from the same animal on different days can be selected, cropped, and compared side by side (*see* **Note 12**) (*see* **Fig. 1**).
7. Offline processing can be utilized when software on the system is not adequate for certain tasks. We have done image fusion with BLI and *x*-ray for better signal source or internal organ localization. Make sure that you have registration marks that will help you determine scale and alignment between the

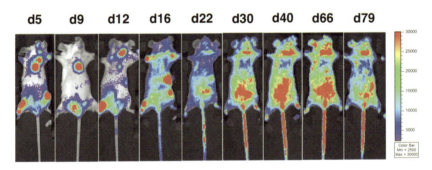

Fig. 1. Lentiviral transduced murine hematopoietic stem cells (HSC) repopulation. Balb/c bone marrow cells transduced with lentiviral vector containing luciferase gene under the control of MND promoter. Images were taken from day 5 to 79 after transplantation. Initial foci indicated engraftment and expansion of early progenitor cells. Strong signals throughout the body beyond 8 weeks contributed to the engraftment and hematopoiesis from long-term repopulating HSCs.

different imaging modalities. Additional software such as ImageJ [free from NIH (http://rsb.info.nih.gov/ij/)] or Adobe Photoshop can be used to do image fusion.

4. Notes

1. Bioluminescence imaging can be obtained by placing studied animals in a light-sealed box with the highly sensitive digital camera on top. There is no preference for particular system; however, for convenience in this protocol, we describe the procedures with Xenogen IVIS200 Living Imaging system we are currently using. Different systems are available for BLI, such as Roper, Kodak, Pixis, and many others. Follow recommendations from the manufacture for set-up and imaging times. Important considerations are light tight chamber for imaging, CCD cooling, and a proper lens that will allow the shortest lens to target distance. This increases the amount of light to the camera. Most of the protocol can be applied directly to the IVIS100 systems. Consult manuals for other manufacturer systems, both camera operations and software interface. Assure adequate cooling for the CCD camera. Routinely check for light leaks in the dark box.

2. Particular attention must be paid to keep the animals warm. Ketamine-based anesthetics reduce the mouse's ability to regulate its body temperature. Animals will get cold very quickly and can die. Keep animals warm throughout the entire time under anesthesia. Heat lamps or warming blankets can be utilized if the imaging system does not provide heating device. IVIS200 imaging system maintains the imaging chamber at a steady 37 °C, which can keep the study animals comfortable through out the imaging process.

3. Stand-alone Isoflurane systems can be easily adapted to other camera systems. Many other manufacturers of camera systems will have light trap ports on their dark boxes to allow the introduction of anesthesia, oxygen, and waste gas lines. To reduce the death of study subjects, use the minimal amount of gas to keep the animal under.

4. Test any materials for auto-luminescence before using them with the mice. Many plastics and some papers will give off a "glow" when imaged.

5. The camera needs to be cooled to reduce the background signal (dark current) that is caused by thermal energy arising from the CCD over long exposure times. Many cameras utilize either a thermoelectric Peltier-type device or a liquid refrigerant system to maintain a low temperature. The best dark current suppression happens at −100 °C or lower. Some older systems utilize liquid nitrogen. Please use caution when handling cryogens. For Xenogen IVIS200 system, the power should be left on all the time because the system needs to perform background measurements at night. The cooling system on the IVIS200 is a closed loop and is not user serviceable.

6. Selection of exposure time needs to be consistent during imaging of the same experimental group on different days. Selection of time longer than 5 minutes may require manual measurement of background (please contact manufacturer for instructions). Exposure time can be changed if the BLI signal is saturated

during the initial imaging process. However, keep in mind, during later days of imaging, saturation may happen, but changing exposure time at that point makes comparison with the data from previous imaging day impossible.

7. Consecutive imaging allows the user to observe the change of BLI signal during imaging. It is very useful to detect the peak signal throughout one imaging session. Usually, only the images taken at the same time are used for quantification and comparison. Not all images from consecutive imaging process are necessary for quantification.

8. From published data, bioluminescence signal reaches peak at 15–20 minutes after i.p. injection of D-luciferin, and half-life of D-luciferin in vivo can be as long as 3 hours *(20)* As two consecutive images are taken in our studies, 7-min waiting time is chosen. The starting imaging time can be flexible based on the experiment and the user.

9. D-Luciferin concentration can be varied from 115 to 150 mg/kg. Continuous release of D-luciferin has been studied by using a micro-osmotic pumps *(21)*.

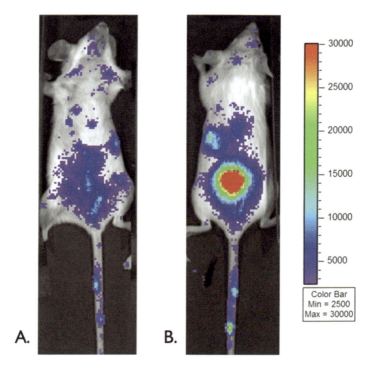

Fig. 2. Ventral and dorsal BLI images. In this Lin⁻ bone marrow transplantation study, at day 38 after transplantation, both dorsal and ventral sides of the mouse were imaged, showing the different BLI intensities. (A) Second image of consecutive ventral side imaging. (B) Dorsal side image was taken immediately after the consecutive ventral side imaging with the same exposure time.

10. Spreading the extremities of the animal can help view the body and its bone marrows more clearly. Ventral imaging provides clear picture of their limbs, chest, and abdomen. However, a dorsal image is highly recommended to capture the foci or signal blocked by the body of the animal in two-dimensional imaging (*see* **Fig. 2**).

11. In the study of HSCs engraftment and repopulation, bioluminescence signals from HSCs are too weak to be observed until they are expanding and undergoing hematopoiesis. As a method of final confirmation, internal organs of the mice can be taken out several minutes after injection of D-luciferin, each organ can be imaged, selected, and analyzed as its own ROI.

12. To arrange images side by side, make sure the maximum and minimum values on the analysis window for each image are the same before cropping.

Acknowledgment

This work was supported by DOE (DE-FG02-03ER63597 Z Lee), NIH R21 (EB001847 Z Lee), NIH R01 (CA073062 SL Gerson).

References

1. Domen, J., and Weissman, I.L. (1999) Self-renewal, differentiation or death: regulation and manipulation of hematopoietic stem cell fate. *Mol Med today* **5**, 201–208.

2. Ferguson, C., Larochelle, A., and Dunbar, C.E. (2005) Hematopoietic stem cell gene therapy: dead or alive? *Trends in Biotechnology* **23**, 589–597.

3. Jordan, C.T., and Lemischka, I.R. (1990) Clonal and systemic analysis of long-term hematopoiesis in the mouse. *Genes Development* **4**, 220–232.

4. Szilvassy, S.J., Humphries, R.K., Lansdorp, P.M., Eaves, A.C., and Eaves, C.J. (1990) Quantitative assay for totipotent reconstituting hematopoietic stem cells by a competitive repopulation strategy. *Proceedings of the National Academy of Sciences of the United States of America* **87**, 8736–8740.

5. Balaban, R.S., and Hampshire, V.A. (2001) Challenges in small animal noninvasive imaging. *ILAR journal/National Research Council, Institute of Laboratory Animal Resources* **42**, 248–262.

6. Lewis, J.S., Achilefu, S., Garbow, J.R., Laforest, R., and Welch, M.J. (2002) Small animal imaging. current technology and perspectives for oncological imaging. *European Journal of Cancer* **38**, 2173–2188.

7. Piwnica-Worms, D., Schuster, D.P., and Garbow, J.R. (2004) Molecular imaging of host-pathogen interactions in intact small animals. *Cellular Microbiology* **6**, 319–331.

8. Choy, G., Choyke, P., and Libutti, S.K. (2003) Current advances in molecular imaging: noninvasive in vivo bioluminescent and fluorescent optical imaging in cancer research. *Molecular Imaging* **2**, 303–312.

9. Ottobrini, L., Lucignani, G., Clerici, M., and Rescigno, M. (2005) Assessing cell trafficking by noninvasive imaging techniques: applications in experimental tumor

immunology. *The Quarterly Journal of Nuclear Medicine and Molecular Imaging* **49**, 361–366.

10. Sadikot, R.T. and Blackwell, T.S. (2005) Bioluminescence imaging. *Proceedings of the American Thoracic Society* **2**, 537–540, 511–532.

11. Rice, B.W., Cable, M.D., and Nelson, M.B. (2001) In vivo imaging of light-emitting probes. *Journal of Biomedical Optics* **6**, 432–440.

12. Contag, C.H., and Bachmann, M.H. (2002) Advances in in vivo bioluminescence imaging of gene expression. *Annual Review of Biomedical Engineering* **4**, 235–260.

13. Hardy, J., Francis, K.P., DeBoer, M., Chu, P., Gibbs, K., and Contag, C.H. (2004) Extracellular replication of Listeria monocytogenes in the murine gall bladder. *Science* **303**, 851–853.

14. Negrin, R.S., Edinger, M., Verneris, M., Cao, Y.A., Bachmann, M., and Contag Ch, H. (2002) Visualization of tumor growth and response to NK-T cell based immunotherapy using bioluminescence. *Annals of Hematology* **81 Suppl 2**, S44–S45.

15. Paulmurugan, R., Umezawa, Y., and Gambhir, S.S. (2002) Noninvasive imaging of protein-protein interactions in living subjects by using reporter protein complementation and reconstitution strategies. *Proceedings of the National Academy of Sciences of the United States of America* **99**, 15608–15613.

16. Zhang, G.J., Safran, M., Wei, W., Sorensen, E., Lassota, P., Zhelev, N., Neuberg, D.S., Shapiro, G., and Kaelin, W.G., Jr. (2004) Bioluminescent imaging of Cdk2 inhibition in vivo. *Nature Medicine* **10**, 643–648.

17. Cao, Y.A., Wagers, A.J., Beilhack, A., Dusich, J., Bachmann, M.H., Negrin, R.S., Weissman, I.L., and Contag, C.H. (2004) Shifting foci of hematopoiesis during reconstitution from single stem cells. *Proceedings of the National Academy of Sciences of the United States of America* **101**, 221–226.

18. Wang, X., Rosol, M., Ge, S., Peterson, D., McNamara, G., Pollack, H., Kohn, D.B., Nelson, M.D., and Crooks, G.M. (2003) Dynamic tracking of human hematopoietic stem cell engraftment using in vivo bioluminescence imaging. *Blood* **102**, 3478–3482.

19. Plett, P.A., Frankovitz, S.M., and Orschell, C.M. (2003) Distribution of marrow repopulating cells between bone marrow and spleen early after transplantation. *Blood* **102**, 2285–2291.

20. Lipshutz, G.S., Gruber, C.A., Cao, Y., Hardy, J., Contag, C.H., and Gaensler, K.M. (2001) In utero delivery of adeno-associated viral vectors: intraperitoneal gene transfer produces long-term expression. *Molecular Therapy* **3**, 284–292.

21. Gross, S., Abraham, U., Prior, J. L., Herzog, E. D., and Piwnica-Worms, D. (2007) Continuous delivery of D-Luciferin by implanted micro-osmotic pumps enables true real-time bioluminescence imaging of luciferase activity in vivo. *Molecular Imaging* **6**, 121–130.

Index

Printed in the United States of America